Soil Fertility Management

Soil Fertility Management

Edited by **Lester Bane**

SYRAWOOD
PUBLISHING HOUSE
New York

Published by Syrawood Publishing House,
750 Third Avenue, 9th Floor,
New York, NY 10017, USA
www.syrawoodpublishinghouse.com

Soil Fertility Management
Edited by Lester Bane

International Standard Book Number: 978-1-68286-205-6 (Hardback)

Printed in China.

Contents

	Preface	IX
Chapter 1	**Rangeland characteristics of a supercritically degraded landscape in the semi-arid area of South Africa** Clifford Tafangenyasha, Amos T. Mthembu, Hector Chikoore, Nothile Ndimande, Sifiso Xulu and Nonkululeko Gwcensa	1
Chapter 2	**The importance of Turkish hazelnut trace and heavy metal contents for human nutrition** Faruk Özkutlu, Yusuf Ziya Doğru, Nedim Özenç, Gizem Yazici, Metin Turan and Fatih Akçay	9
Chapter 3	**Different land use types in the semi-arid rangelands of Kenya influence soil properties** Kevin Z. Mganga, Nashon K. R. Musimba, Dickson M. Nyariki, Moses M. Nyangito, Wellington N. Ekaya, William M. Muiru and Agnes, W. Mwang'ombe	18
Chapter 4	**Status and distribution of extractable micronutrients in Haplustults in Yamaltu-Deba Local Government Area, Gombe state, Nigeria** S. Mustapha, H. K. Mamman and N. A. Abdulhamid	23
Chapter 5	**Effects of crop rotation on soil macronutrient content and pH in potato producing areas in Kenya: Case study of KARI Tigoni station** Jane Muthoni and Jackson N. Kabira	28
Chapter 6	**Effects of different sources of nitrogen on potato at Tigoni, Kenya** Jane Muthoni and Jackson N. Kabira	35
Chapter 7	**Studies on sub-soil characteristics of sand deposits in some parts of Bayelsa State, Eastern Niger Delta, Nigeria** P. O. Youdeowei and H. O. Nwankwoala	43
Chapter 8	**Effects of tillage and manure application on soil physicochemical properties and yield of maize grown on a degraded intensively tilled alfisol in southwestern Nigeria** Adeyemo A. J. and Agele S. O.	48
Chapter 9	**Spatial variability mapping of some soil properties in El-Multagha agricultural project (Sudan) using geographic information systems (GIS) techniques** M. Y. Mohamed and B. M. Abdo	60
Chapter 10	**The integration of variable rate technologies for a soil-applied herbicide in leafy green production** Kurt D. Nolte	68

Chapter 11 **Morphogenetic characterization of soils formed from basement complex rock in the humid tropical rainforest of Nigeria** **76**
O. T. Ande

Chapter 12 **Yield and nitrogen accumulation in five cassava varieties and their subsequent effects on soil chemical properties in the forest/savanna transitional agro-ecological zone of Ghana** **81**
S. Adjei-Nsiah

Chapter 13 **Effects of farm yard manure on cadmium and lead accumulation in Amaranth (*Amaranthus oleracea* L.)** **87**
M. Alamgir, M. G. Kibria and M. Islam

Chapter 14 **Comparative influence of benomyl on rhizosphere and non-rhizosphere bacteria of cowpea and their ability to solubilise phosphate** **91**
Ekundayo F. O.

Chapter 15 **Determinants of adoption of conservation tillage practices in maize-cowpea cropping systems: The case of Makueni District, Kenya** **100**
Jane Mutune, John Mburu1, Rose Nyikal and Geoffrey Kironchi

Chapter 16 **Extractable micronutrients status in relation to other soil properties in Billiri Local Government Area** **108**
Ibrahim, A. K., Usman, A., Abubakar, B., and Aminu, U. H.

Chapter 17 **Soil management strategies for rubber cultivation in an undulating topography of Northern Cross River State** **112**
J. R. Orimoloye, I. K. Ugwa and S. O. Idoko

Chapter 18 ***Balanites aegyptiaca*, a potential tree for parkland agroforestry systems with sorghum in Northern Ethiopia** **118**
Hailemariam Kassa, Kindeya Gebrehiwet and Charles Yamoah

Chapter 19 **The physical and chemical characteristics of soils of Northern Kenya Aridlands: Opportunity for sustainable agricultural production** **126**
E. M. Muya, S. Obanyi, M. Ngutu, I. V. Sijali, M. Okoti, P. M. Maingi and H. Bulle

Chapter 20 **Manure application setback effect on phosphorus and sediment in runoff** **134**
Ahmed Al-wadaey, S. Charles Wortmann, A. Charles Shapiro, G. Thomas Franti and E. Dean Eisenhauer

Chapter 21 **Soil classification and micromorphology: A case study of Cholistan Desert** **141**
Farooq Ahmad

Chapter 22 **The potential of four non traditional legumes in suppressing the population of nematodes in two Ghanaian soils** **149**
K. Osei, J. O. Fening, S. R. Gowen and A. Jama

Chapter 23 **The effects of soil density on the vegetation of the Umfolozi catchment in South Africa** **155**
Clifford Tafangenyasha, Amos T. Mthembu, Hector Chikoore, Nothile Ndimande, Sifiso Xulu and Nonkululeko Gcwensa

Chapter 24 **Characterizing sites for the design and restoration of Kalacha irrigation scheme, Chalbi district, Kenya: Socio-economic and biophysical considerations** 166
E. M. Muya, K. Gitau, S. Obanyi, M. Ngutu, S. Kuria, H. Sharfi, D. G. Dub, H. Bulle, D. M. Miano, J. Muriethi, G. Keya, I. V. Sijali, S. Mamo, P. M. Maingi, M. Okoti and Eng. M. J. Omedi and A. Esilaba

Chapter 25 **Reducing compaction effort and incorporating air permeability in Proctor testing for design of urban green spaces on cohesive soils** 174
A. M. Ibrahim, N. Persaud, R. W. Zobel and A. Hass

Chapter 26 **Resolving the gully erosion problem in Southeastern Nigeria: Innovation through public awareness and community-based approaches** 183
Obidimma C. Ezezika and Olorunfemi Adetona

Chapter 27 **The utilization of wood ash as manure to reduce the use of mineral fertilizer for improved performance of maize (*Zea mays* L.) as measured in the chlorophyll content and grain yield** 189
F. O. Adekayode and M. R. Olojugba

Chapter 28 **Effect of quantity and placement distances of inorganic 15-15-15 fertilizer in improving soil fertility status and the performance and yield of maize in a tropical rain forest zone of Nigeria** 195
F. O. Adekayode and M. O. Ogunkoya

Chapter 29 **Comparative study of different organic manures and NPK fertilizer for improvement of soil chemical properties and dry matter yield of maize in two different soils** 204
Adeniyan O. N., Ojo A. O., Akinbode, O. A. and Adediran J. A.

Chapter 30 **Changing weather and introduction of pigeon pea intercrop in formerly cold pyrethrum zone** 209
Mwangi, J. A., Waweru, S. N., Oeba, V. O. and Wepukhulu, S. B.

Chapter 31 **Sources and distribution of polycyclic aromatic hydrocarbons in post flooded soil near Afam power station, South East Niger Delta, Nigeria** 212
Orok E. Oyo-Ita, Inyang O. Oyo-Ita and Samuel U. Ugim

Chapter 32 **Integrating indigenous knowledge and soil science approaches to detailed soil survey in Kaduna State, Nigeria** 224
Raji, B. A., Malgwi, W. B., Berding, F. R. and Chude, V. O

Permissions

List of Contributors

Preface

Over the recent decade, advancements and applications have progressed exponentially. This has led to the increased interest in this field and projects are being conducted to enhance knowledge. The main objective of this book is to present some of the critical challenges and provide insights into possible solutions. This book will answer the varied questions that arise in the field and also provide an increased scope for furthering studies.

Soil fertility management has gained a global significance for increasing agricultural productivity and ensuring food security. This book provides comprehensive insights into the field of soil fertility management and its importance in maintaining agricultural sustainability. Most of the topics introduced in this book cover new methods to evaluate soil fertility and focus on developing remediation techniques to maintain soil health. It strives to provide a fair idea about this discipline and to help develop a better understanding of the latest advances within the fields of soil nutrients and minerals, climate adaptation measures, etc. Scientists and students actively engaged in this field will find this book full of crucial and unexplored concepts.

I hope that this book, with its visionary approach, will be a valuable addition and will promote interest among readers. Each of the authors has provided their extraordinary competence in their specific fields by providing different perspectives as they come from diverse nations and regions. I thank them for their contributions.

Editor

Rangeland characteristics of a supercritically degraded landscape in the semi-arid area of South Africa

Clifford Tafangenyasha*, Amos T. Mthembu, Hector Chikoore, Nothile Ndimande, Sifiso Xulu and Nonkululeko Gwcensa

Department of Geography and Environmental Science, University of Zululand, P, Bag X10003, KwaDlangezwa 3886, South Africa.

The objective of this study was to investigate rangeland condition in a semi-arid area with doubling erosion rates. Work was conducted using aerial photographic analysis and field investigations were conducted using 10 paired 20 x 20 m plots and 2 1 x 1 m quadrats randomly located with each plot. Degraded sites showed significant differences in soil compaction but not in soil water infiltration time. Degraded sites were characterized by absence of woody plant cover. Grass cover and litter cover were lower on degraded plots than on undegraded plots. Bare soil cover was more extensive on degraded plots than on undegraded plots. The sheet eroded plots had very low woody cover, grass cover and litter cover, low organic matter, low porosity and high soil density. Soil density on degraded sites was significantly different from undegraded sites (t test; $t = 3.482$; $df = 18$; $p < 0.05$). Assessing rangeland characteristics and erosion risk in the study is part of a global management of agricultural soils that should aim at increasing productivity.

Key words: Umfolozi catchment, soil erosion, semi-arid landscapes, range-condition.

INTRODUCTION

According to Snel and Bot (1994) dry-land ecosystems are under threat from a combination of socio-economic and biophysical changes that are culminating in a downward spiral of land productivity. Land degradation has been identified as one of the key global problems today that is at the root cause of poverty, food insecurity and malnutrition (FAO, 1979; Tengberg, 2002), but is ultimately a symptom of human overpopulation. Water erosion is South Africa's most widespread soil degradation problem and affects 70% of the land (Department of Environment and Tourism, 2008).

In landscapes prone to land degradation, gullying, sedimentation, landsliding and bare soil cover, soil pedestals, crusts, exposed roots, and organ pipe erosion are often the major symptoms. The Umfolozi catchment is very vulnerable to disturbance due to slow ecological recovery, growing human populations, episodic droughts, fire and climate change. The hardpans, organ pipe erosion, sheet erosion and gullies in the Umfolozi catchment offer many impediments to vegetation establishment by contributing a harsh substrate for the establishment and germination of seedlings. The erosion in the Umfolozi catchment may be described as supercritical being characterized by abrupt changes in productivity on the scarred and scalloped terrain. Top soil is transported to streams and this dramatic loss results in less than half the return in crop yields of normal. A total of 172.4 ha (6.3%) of land were found to be severely eroded in the study area. The purpose of this investigation is to describe the range-condition of the study area. The research hypothesis was that rangeland condition may be contributing to the increasing soil erosion in the study area.

Study area

The study area (Figure 1) lies between 28°00'00" S and

*Corresponding author. E-mail: ctafangenyasha@yahoo.com.

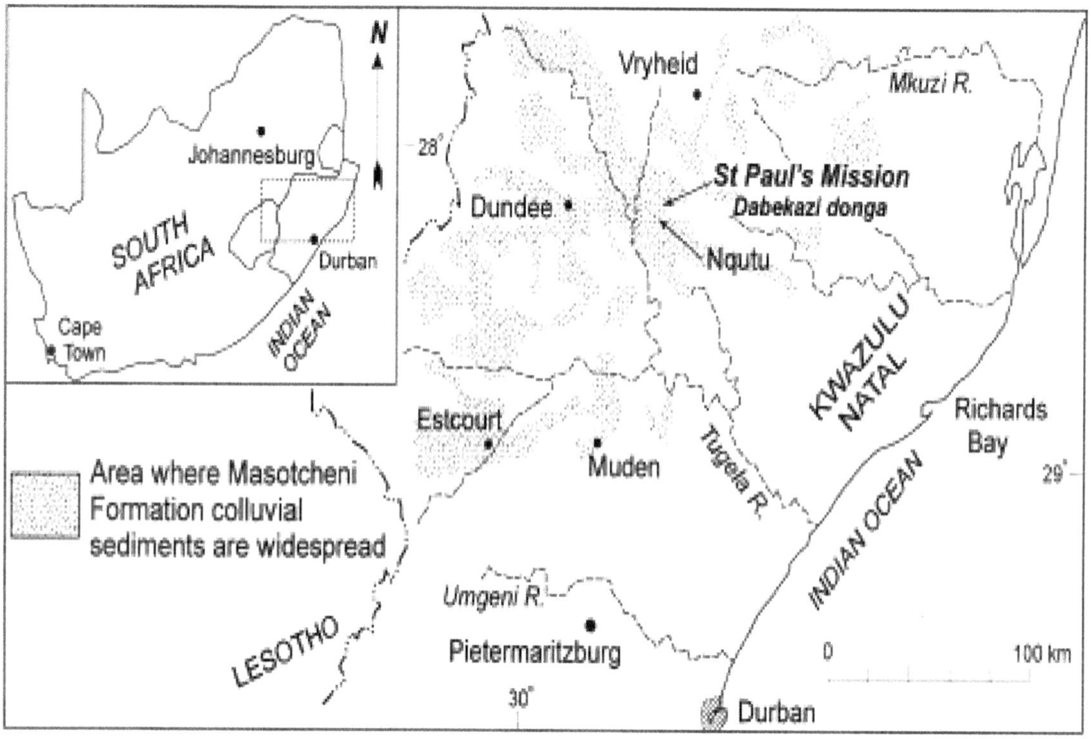

Figure 1. Present study area situated about 10 km northeast of St Paul Mission near Nqutu in central KwaZulu-Natal (after Botha and Fedoroff, 1995).

28°10'00" S; 30° 37'00" E and 30°55'00"E), 28 km southwest of Vryheid town in central KwaZulu-Natal province. The study area and the adjacent areas have been described by Rienks et al. (2000). The surface rock is dominated by the complex Masotcheni Formation colluvial sediments and associated palaeosols. The surface lithology is defined by a stepped sandstone bedrock floor sloping at 4 - 12° with the flanking spurs and concave rocky hillslope. The colluvium within the Permian Vryheid Formation is underlain by sandstone that is exposed by sheet erosion. Other rock formations underlying the colluvium include mica, siltstone and carbonaceous shale (Rienks et al., 2000). The area is drained by Mvunyana River that feeds into the White Umfolozi River and the ultimate discharge is easterly into the Indian Ocean (Figure 1).

Land use in the Umfolozi catchment is predominantly dryland agriculture that combines rain fed agriculture and communal livestock grazing. Mondlo Communal Land in which the study area gets an average rainfall of 400 - 500 mm a year. Most of the soils are sandy. Due to a lack of ground cover, water runoff is a problem and many gullies have formed due to the soft erodible nature of the soils. Surface instability due to hardpans is a key problem that contributes to more sheet erosion. Detachment and transportation of soil particles downslope in suspension as run-off flows concentrate in channels and estuary. Rills and gullies form and expand as soil loss progresses. Deep gullies, locally termed 'dongas', are a characteristic geomorphic feature of the hilly topography of central KwaZulu-Natal province in eastern South Africa, and pose severe constraints on the agricultural use of land by rendering much of the farmland inaccessible (Rienks et al., 2000). The landscape is scarred into badlands by hardpans, organ pipe erosion and sheet erosion. The study area encompassed both the badlands and areas in good range-condition. The Global Positioning sites (GPS sites) are given in Table 1.

The badlands are characterized by a stepped relief that may be construed as a sequence of the erosional phases. Large expanses of the land are being lowered by pediplanation and lowered and then dissected. The long duration of erosion has allowed for the incision of landscapes by deep weathering and transportation of sediment. The erosion is widespread and increasing, threatening the remaining productive landscapes, infrastructure and wiping away of livelihood options necessary to eke a living.

MATERIALS AND METHODS

Measurements of the extent of erosion on panchromatic aerial photography 2830 BB1 Nondweni 1982 taken at a scale of 1:10 000 by the Air Survey of Company of Africa Limited were done with the aid of a 4.0 x 5.0 cm plastic template grid with 0.5 x 0.5 mm divisions with each small grid covering 2500 m^2 on the ground. A

Table 1. Occurrence of the eroded land in the major land classes of the study area.

Land class	Total land area (ha)	Percent of land area (%)	Eroded land (ha)
Settlement	650	25	241.3
Cultivation	680	25	5.8
Grazing land	1360	50	142.3
Total	2720	100	172.4

total of 74 points on aerial photos were examined for presence/ absence of soil erosion with the aid of a grided plastic. The grided transparent template was employed to determine the frequency and distribution of sheet erosion in 0.5 x 0.5 mm grid cells. The delimited sheet eroded areas were transferred from the aerial photo to a Surveyor General's 1: 50 000 topographical map using physical features for mapping purposes. The 74 randomly selected points on the aerial photo covered variations in the extent of land degradation. The sheet erosion was measured in the major land classes that include settlement, grazing and cultivated areas on the aerial photo. Observations on 74 0.5 x 0.5 cm grid covered 6% of the study area.

Range condition study

The data were divided into two groups that include the control and test sites to test the effects of anthropogenic stressors. The control comprised areas identified as having negligible soil erosion as seen on aerial photo and that on the ground these sites were characterized by an absence of rills and gullies and those sites had well pronounced plant cover with negligible bare soil cover. A site comprised paired plots that include test (degraded) plots and control (undegraded) plots. The test sites exhibited poor plant cover, pronounced bare soil cover, rills and gullies on aerial photo and ground. The number of sampling locations was limited to 10 sites. All the 10 sites were randomly selected using the random number tables. Each plot measured 20 by 20 m in which 2 1 m x 1 m quadrats were randomly located to study groundcovers. The variables assessed in each quadrat include woody cover, grass cover, grass tuft density, woody plant density, woody plant girth, woody plant height, bare soil cover, litter cover, exposed root density, woody plant fire damage, woody plant drought damage, woody plant insect damage and woody plant animal damage. In the 20 x 20 m plot the following were assessed: woody plant density, woody plant girth, woody plant height, exposed roots, woody plant fire damage, woody plant drought damage, woody plant insect damage and woody plant animal damage. In addition, soil factors included soil compaction and soil water infiltration. Range condition evaluations were done in the dry season between 10th October 2008 and 1st November 2008.

A single infiltration ring measuring 10.5 cm in diameter, length 19.8 cm and a volume of 65.94 cm^3 was used to measure time taken by a falling head of water of 700 ml water to infiltrate the soil at 10 sites, on each of the 20 plots during the summer of 2008. Mean infiltration rate was calculated on each occasion. The water infiltration time in seconds was used as an index of water percolation time. The infiltration test was used to quantify the ability of water to move into and through a soil. Because of the great number of factors which can affect the flow of water through soils, it was used on a relative basis. The results from that particular set of tests are then compared to each other. Native plants influence infiltration changes through root spread. To test water infiltration time in the selected plots a level location was sought to test water infiltration rate into the soil. A site with gravel on stony ground was difficult or impossible to test water infiltration rate because of difficulties in sinking the cylinder into the soil. A heavy lawn sod (e.g. *Cynodon dactylon* and *Hyparrhenia* sp) created similar difficulties because of the dense mat of roots. Work was conducted around living plants taking care to expose bare soil by removing any leaf litter. The soils were little disturbed wherever possible during the water infiltration tests. The cylinder was sunk into the soil approximately five to seven centimeters to create a tight seal between the bottom of the cylinder and the soil. A light iron hammer was used on a placed wooden board on top of the cylinder when hammering to keep from denting its top. The cylinder was hammered in circles around the top to keep the cylinder perpendicular with the soil surface. During the test, if water leaked out the bottoms and sides of the cylinder, tests would be repeated. The time water took to infiltrate the soils was recorded in minutes. If the soil surface was uneven inside the ring, the time in minutes was observed until half of the surface was exposed and just glistening. Timing was stopped when the surface was just glistening.

Estimates of soil compaction were obtained in the randomly selected points for sampling quadrats using a penetrometer. A steel rod freely housed in a cylindrical pipe would slide from a permanently marked position above its housing to penetrate the ground. The depth in centimeters that a steel rod penetrated the ground defines the compaction of the soil, which is taken to be the stability of soil aggregates as a result of clay particles combining with organic matter. The soil compaction measured in centimeters was used as an index of resistance to soil penetration. Soil samples were collected outside each quadrat but within the 20 x 20 m plots.

Soil water infiltration was tested using a single infiltration ring that provided an index of infiltration. A level location was sought to give the best results by allowing the water to infiltrate evenly into the soil. The soil surface was disturbed as little as possible before setting infiltration ring. The time taken for all of the water to move into the soil with complete elimination of all puddles was recorded in seconds using a stopwatch.

A T-test was used to test the hypothesis that the means of soil compaction and soil water infiltration measured between the degraded and undegraded sites are significantly different. Stepwise multiple regression was carried out using SPSS (Nie et al., 1975). Stepwise multiple regressions identified the order and degree of contribution of the independent variables in terms of prediction of eroded land transformed percentage.

RESULTS

Human factors in the erosion of the soils in the Umfolozi catchment

The accelerated soil erosion survey on aerial photo indicated that there were some 172.4 ha of eroded land in the 2720 ha study area (Table 1). This constitutes 6.3% of the area. Of the total eroded land, 142.3 ha were in the grazing lands, with 241.3 ha being in the settlement areas where people live and 5.8 ha in the cultivated land

Table 2. Location of sample plots.

Sites	Longitude	Latitude	Elevation (m)	Sites	Longitude	Latitude	Elevation (m)
1	30°46'35"	S27°59'59"	1211	12	30°42'37"	28°02'58"	1132
2	30°46'35"	27°59'59"	1214	13	30°42'37"	28°0258"	1136
3	30°46'35"	27°59'52"	1214	14	30°42'37"	28°02'58"	1136
4	30°46'35"	28°08'00"	1216	15	30°42'08"	28°04'22"	1206
5	30°46'39"	27°59'52"	1210	16	30°42'07"	28°04'22"	1203
6	30°46'40"	27°52'53"	1210	17	30°42'06"	28°04'22"	1206
7	30°46'40"	27°59'52"	1209	18	30°41'54"	28'°04'14"	1241
8	30°42'43"	28°02'54"	1121	19	30°41'54"	28°04'13"	1247
9	30°42'42"	28°02'54"	1123	20	30°41'55"	28°04'30"	1250
10	30°42'53"	28°02'55"	1131	21	30°41'53'	28°04'13"	1250
11	30°42'42"	28°02'56"	1129				

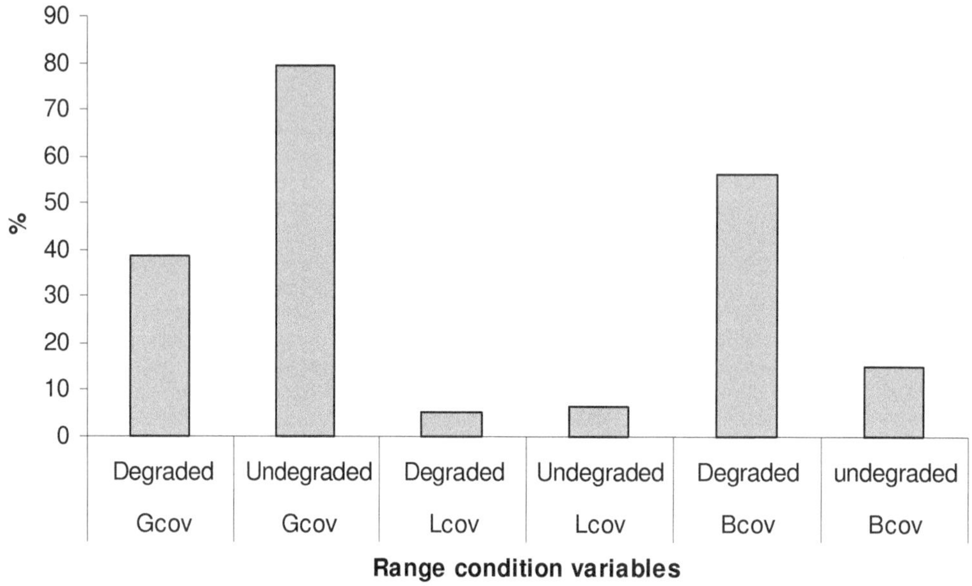

Figure 2. Mean grass cover (%), litter cover (%) and bare soil cover (%) estimated in the randomly selected plots using 1 m^2 metal frame quadrats on degraded and undegraded plots in the Umfolozi catchment (Gcov = grass cover, Lcov = litter cover, Bcov = bare soil cover).

(Table 1). About 25% of the settlement area and 25% of the cultivated areas were eroded, whereas 50% of the grazing lands were eroded (Table 1). There is much erosion in the settlement areas (Table 2).

Range condition assessment of the study area

Mean percentage grass cover was higher on degraded sites than on undegraded sites (Figure 2). Mean percentage litter cover was similar between degraded and undegraded sites (Figure 2). The mean percentage value of bare soil cover was higher on degraded sites than on undegraded sites (Figure 2). Grass cover, litter cover and bare soil cover percentage values were log transformed prior to t tests. Bare soil cover was significantly different ($t = 4.293$, df =14, $p<0.05$) between degraded and undegraded sites. Grass cover was also significantly different ($t =3.400$, df =10, $p<0.05$) between degraded and undegraded sites. Water infiltration time (seconds) and soil compaction (cm) tests were done on the randomly selected sites in the study area. Results show that soil compaction values were significantly different (Student T test, $t=3.482$, df=18, $p<0.05$) between degraded and undegraded plots (Figure 3, Table 3). Soil compaction values on degraded soils were 16% more than on undegraded plots (Figure 3). Undegraded areas show soil compaction values ranging from 3.4 to 6.15 cm

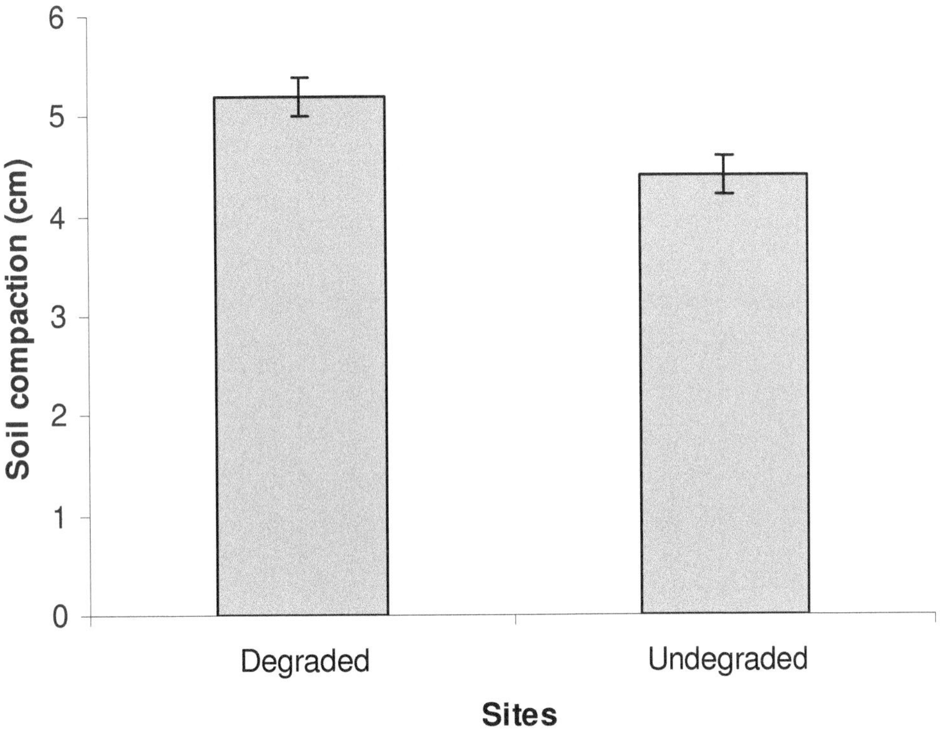

Figure 3. Soil compaction (cm) measured in degraded and undegraded plots in the Umfolozi catchment.

Table 3. T-test results of soil water infiltration time (seconds) and soil compaction (cm) measured in the degraded plots and undegraded plots in the soils of the Umfolozi catchment.

Variable	Degraded plots	Undegraded plots	t-value
Water infiltration (seconds)	175	238.9	ns
Soil compaction (cm)	5.23	4.4	*

with a mean value of 4.4 cm (Figure 3). Degraded plots show infiltration rates ranging from 3.5 to 7.0 cm with a mean value of 5.23 cm (Figure 4).

Water infiltration showed no significant differences ($p>0.05$) between degraded and undegraded sites (Table 3). Infiltration rates may vary in response to changes in soil moisture content, porosity, bulk density, surface strength, soil mineralogy (e.g. presence of montmorillonite) and soil moisture. Values of simple r (r values for each variable separately) in the regression analysis (Table 4) are given for all the regressor variables, allowing for the fact that these factors do not act in isolation. Given the large data set (n=75), strong and significant correlations were evident for Ivy's erosion class (spatial extent of the soil erosion) and slope angle, that is the two physical factors (Table 4). Distance to water, distance to hill and distance to road appeared to play no role in the erosion process in stepwise multiple regression (Table 4). Consequently, with increases in rangeland usage, so the erosion becomes more widespread. Stepwise multiple regression (Table 4) identified the order and degree of contribution of the five independent variables in terms of prediction of eroded land percentage. Ivy's erosion class proved to be the most important factor influencing erosion, with a large and highly significant F-value and an r square of 0.142 (Table 4). Slope angle was the second variable to enter the equation, increasing the r square to 0.216 (with a very large F-value associated with the r2 change). Stepwise multiple regressions (Table 5) identified the order and degree of contribution of the five independent variables in terms of prediction of eroded land percentage. Soil compaction proved to be the most important factor influencing erosion, with a large and highly significant F-value and an r square of 0.606 (Table 5). Bare soil cover was the second variable to enter the equation, increasing the r square to 0.517 (with a very large F-value associated with the r2 change).

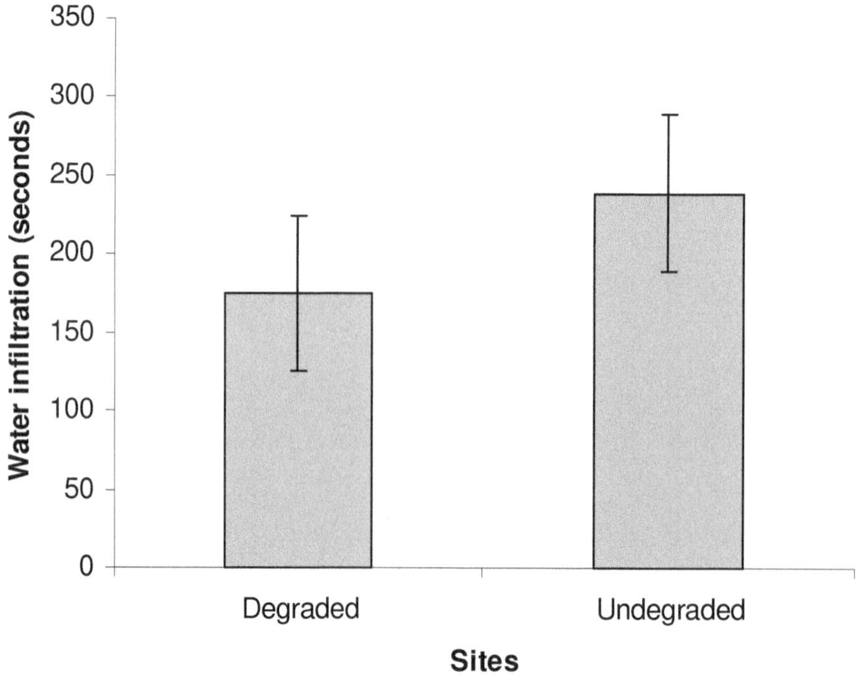

Figure 4. Water infiltration time (seconds) measured on degraded and undegraded plots in the Umfolozi catchment.

Table 4. Results of multiple regression with stepwise variable selection using plot level percent of erosion as the dependent variable and Ivy's erosion class, distance to water, distance to hill, distance to road and slope angle as predictor variables (*=P<0.05; **=P<0.005).

Predictor variable	Simple r	r square	F-value	Sig.
Ivy's erosion class	0.376	0.142	13.53	**
Distance to water			0.348	ns
Distance to hill			0.009	ns
Distance to road			0.068	ns
Slope angle	0.465	0.216	6.817	*

Table 5. Results of multiple regression with stepwise variable selection using plot level percent of erosion as the dependent variable and soil infiltration, soil compaction, grass cover, bare soil cover and litter cover as predictor variables.

Predictor variable	Simple r	r square	F-value	Sig.
Soil infiltration			1.371	ns
Soil compaction	0.779	0.606	4.105	*
Grass cover			0.072	ns
Bare soil cover	0.719	0.517	20.850	**
Litter cover			0.058	ns

DISCUSSION

The highest level of erosion in a land class (Table 1) was recorded in the human settled areas and this was followed by the grazing areas and lastly cultivated areas.

A total of 172.0 ha (6.3%) of land were found to be severely eroded in the study area. Field observations showed some gullies extending into the cropped areas. The soils of the Umfolozi catchment exhibit severe forms sheet erosion that is characterized by extensive areas of bare soil cover and gullies. The sheet eroded plots have very low woody cover, grass cover and litter cover. Woody cover, grass cover and litter cover are important environmental protective forces that act against the energy of raindrops. Many previous researchers (Botha, 1994, 1995; Watson, 1993, 1996a, 1996b) agree that the causes of land degradation are mainly anthropogenic and agriculture related. The major causes include: land clearance, deforestation, and agricultural depletion of soil nutrients, burning and absence of soil conservation measures.

The major problems which the households experience in their agricultural activities are erosion and a prolonged dry season which reduce the level of crop production (Watson, 1993). The livestock owned include the cattle, goats and sheep. Grazing areas are not fenced. The high rates of soil erosion have led to the intervention of the Department of Agriculture with contingency measures. The assistance provided by the Department of Agriculture include the testing of soil in the crop fields at a low cost (R40-00), recommendation on fertilizers to be used and training of community members on the preparation of crop fields and erosion control. The erosion is increasing and even doubling up and as part of the erosion risk assessment an understating of the present range-condition is necessary. Values of simple r (r values for each variable separately) in the regression analysis are given for all the regressor variables, allowing for the fact that these factors do not act in isolation. Strong and significant correlations were evident for plot erosion and soil compaction and bare soil cover. Soil water infiltration, grass cover and litter cover appeared to play no role in erosion process in stepwise multiple regression (Table 5). Consequently, with increases in rangeland usage, so the erosion becomes more widespread.

Distance to water, distance to hill and distance to road appeared to play no role in the erosion process in stepwise multiple regression. Consequently, with increases in rangeland usage, so the erosion becomes more widespread. Clearly, the human factors and physical factors stand out as being a greater influence on erosion in this analysis. Stepwise multiple regression identified the order and degree of contribution of the independent variables in terms of prediction of eroded land percentage and seedling density. When erosion is used as the dependent variable, erosion class, slope angle, soil compaction and bare soil cover proved to be the most important factors influencing erosion. Clearly, the human factors and physical factors stand out as being a greater influence on erosion in this analysis.

The human factors in the Umfolozi catchment have been blamed on the loss of protective factors in the study area. Degraded sites are characterised by low cover of the sward, high bare soil cover, low litter cover, sparse plant cover, pronounced soil density, low soil organic matter and low porosity. Our results contrast with Dougill and Cox (1995a) who suggest that vegetation changes are poor indicators of damage to resilience. In Botswana, vegetation changes reflect loss of resilience (Dougill and Cox, 1995b). Dougill and Cox (1995) further suggest that there is no evidence to suggest that grazing leads to soil degradation, and as such links between cattle and 'land degradation' should treated with care. A new perspective on interactions among climate, plants, and herbivore suggests that rangelands influenced by highly stochastic weather and grazing disturbance are degraded, not only by continuous grazing. Lavorel and Garnier (2002) point to the importance of climate in understanding rules relating to site variables to favoured plant traits. The functional relationships between these plant functional attributes and recognizable plant functional types in vegetation and the assembly and disassembly rules for the coexistence of these different plant functional types in major plant communities. Wilson et al (1984) discuss how grazing, but rather by the long-term absence of grazing. Arid rangeland plants that are grazed continuously may have lower residual biomass and ground cover, but they may also have greater production and better survival than ungrazed plants. Consequently, grazing, rather than being destructive, is necessary for proper management of arid zone pastures. non-equilibrium grazing models describe opportunistic land-use strategies in response to unpredictable disturbance (e.g., climate, fire, and grazing) and resulting changes in forage production. One really has show a change in the shape of the stocking rate and productivity relationships to demonstrate degradation in, for example, a pastoral production context but the criteria may be very different for other land use systems.

Meadows and Hoffman (2003) concluded that the most severely degraded areas in South Africa, including large areas of the former 'homeland' states, are likely to become even more susceptible under predicted climate change scenarios. In contrast, non-equilibrium rangeland systems are thought to be driven primarily by stochastic abiotic factors, notably variable rainfall, which results in highly variable and unpredictable primary production. Livestock populations are thought to have negligible feedback on the vegetation as their numbers rarely reach equilibrium with their fluctuating resource base. Recent studies suggest that most arid and semi-arid rangeland systems encompass elements of both equilibrium and non-equilibrium at different scales, and that management in the Umfolozi catchment needs to take into account temporal variability and spatial heterogeneity. Arid rangeland plants that are grazed continuously may have lower residual biomass and ground cover, but they may also have greater production and better survival than ungrazed plants. Consequently, grazing, rather than being destructive, is necessary for proper management of arid zone pastures (Oba et al., 2000). Opportunistic grazing where stocking rates are aimed at taking advantage of the growing vegetation in the wet season with a lowering of the stocking rates in the dry season could reduce pressure on the degraded landscapes in the study area. Ellis and Swift (1988) and Briske et al. (2003) outlined the need to develop alternate paradigms on the stability of African pastoral ecosystems but such an approach requires detailed knowledge systems of rangeland condition. The range condition characteristics provide an opportunity of identifying critical rangeland constraints in the recovery of the degraded landscape.

ACKNOWLEDGEMENTS

The research staffs are grateful for the University of Zululand Research Grant S936/07 made possible by the Research Office that made the first two expeditions successful. Chief Mdlalose is particularly thanked for encouraging ideas and permission to do the research in the area under his authority.

REFERENCES

Botha GA, Wintle AG, Vogel JC (1994). Episodic late Quartenary palaeogully erosion in northern KwaZulu-Natal, South Africa. Catena, 23: 327-340.

Botha GA, Fedoroff N (1995). Palaeosols in Late Quaternary colluvium, northern KwaZulu-Natal, South Africa. J. Afr. Earth Sci., 21: 291–311.

Briske DD, Fuhlendorf SD, Smeins FE (2003). Vegetation dynamics on rangelands; a critique of the current paradigms. J. Appl. Ecol., 40: 601-614.

Department of Environment and Tourism (2008). State of the environment: Land degradation. Government of South Africa, Pretoria: 5.

Dougill AJ, Cox J (1995a). Land degradation and grazing in the Kalahari: New analysis and alternative perspectives. ODI Pastoral Development Network Paper 38c, ODI, London.

Dougill AJ, Cox J (1995b). When is ecosystem change land degradation? A reply Dougill, A. and Cox, J. ODI Pastoral Develop Netw Pap 38c.

Ellis J, Swift D (1988). Vegetation dynamics on rangelands: a critique of the current paradigms.

FAO (1979). A provisional methodology for soil degradation assessment. FAO. Rome. Hudson, N. Soil Conservation 1995 (3 Edition). B.T. Batsford, London.

Lavorel S, Garnier E (2002). Predicting changes in community composition and ecosystem functioning from plant traits: revisiting the Holy Grail. Functional Ecol., 16: 545-556.

Meadows ME, Hoffman TM (2003). Land degradation and climate change in South Africa. Geogr. J., 169: 168-177.

Nie NH, Hull CH, Jenkins JG, Steinbrenner K, Bent DH (1975). Statistical Package for the Social Sciences. McGraw-Hill New York.

Oba G, Stenseth N, Lusigi W (2000). New Perspectives on Sustainable Grazing Management in Arid Zones of Sub-Saharan Africa. BioScience, 50: 35-51.

Rienks SM, Botha GA, Hughes JC (2000). Some physical and chemical properties of sediments exposed in a gully (donga) in northern KwaZulu Natal, South Africa and their relationship to the erodibility of the colluvial layers. Catena, 39: 11-31.

Snel M, Bot A (1994). Some suggested indicators for land degradation assessment of drylands. Report on Agenda 21 to Commission on sustainable development, U.N. New York, p. 25.

Tengberg A (2002). Land degradation assessment in drylands-Lada Project. FAO. Rome.

Watson HK (1993). An assessment of the validity of the general perception of soil erosion in the Mfolozi catchment, South Africa, Geookodynamik, 14: 75-84.

Watson HK (1996a). Short and long term influence on soil erosion of settlement by peasant farmers in KwaZulu Natal. S. Afr.Geogr. J., 78, 1 6.

Watson HK (1996b). Geology as an indicator of land capability in the Mfolozi area, KwaZulu-Natal, S. Afr. J. Sci., p. 92.

Wilson AD, Harrington GN, Beale IF (1984). Grazing management. In Management of Australia's Rangelands In: Harrington G N, Wilson A D, Young M D Eds), Melbourne Management of Australia's Rangelands, CSIRO.

2

The importance of Turkish hazelnut trace and heavy metal contents for human nutrition

Faruk Özkutlu[1], Yusuf Ziya Doğru[2], Nedim Özenç[3], Gizem Yazici[1], Metin Turan[5]* and Fatih Akçay[6]

[1]Universiy of Ordu, Faculty of Agriculture, Department of Soil Science, Ordu-Turkey.
[2]Universiy of Atatürk, Narman Vocational School, Erzurum-Turkey.
[3]Giresun Hazelnut Research Institute, Giresun-Turkey.
[4]University of Sabancı, Faculty of Engineering and Natural Science, Istanbul-Turkey
[5]Universiy of Atatürk Faculty of Agriculture, Department of Soil Science, Erzurum-Turkey.
[6]Universiy of Atatürk, Faculty of Medicine, Department of Biochemistry, Erzurum-Turkey.

The trace and heavy metal concentration of eighteen different varieties of hazelnut were studied for variety and geographical region discussed for human nutrition and health. These results showed that average Al, As, B, Cr, Co, Cd, Cu, Fe, Mn, Mo, Ni, Se, Pb, and Zn concentrations in the varieties varied in the following ranges 0.91-9.74, 0.02-0.27, 14.60-29.90, 0.02-0.04, 0.07-0.61, 0.01-0.03, 13.76-21.71, 35.12-49.40, 41.11-116.81, 0.09-0.31, 0.81-2.97, 0.02-0.05, 0.03-0.07 and 23.52-35.52 mg kg^{-1}. According to the daily microelement requirements, 100 g Uzunmusa cv, Kargalak cv., Cakildak cv. hazelnut provided about 50% for Fe and Cd, 41% for Mo, 32% for Zn, 21% for Se, 21% for Cr, 5% for B, 1% Ni of the recommended daily amount. On the other hand 100 g hazelnut can supply easily, a daily consumption of Mn, Cu, As, Pb, Al and Co levels of human beings.

Key words: Human nutrition, heavy metal, recommended daily amount, trace element.

INTRODUCTION

Hazelnut (*Corylus avellana* L.) belongs to the Betulaceae family and is a popular tree nut worldwide, mainly distributed along the coasts of the Black Sea region of Turkey, Southern Europe (Italy, Spain, Portugal, and France), and in some areas of the US (Oregon and Washington) (Turkish Hazelnut Exporter's Union, 2008).

Turkey is by far the leading producer of hazelnuts, with 75% of world production (Anonymous, 2010). Italy holds the second place with 15%, followed by the USA (2.9%), Spain (2.4%) and Greece (0.3%). In the last two decades, Turkey has increased its share, mainly at the expense of Italy and Spain, and to a lesser extends the USA. Although production is concentrated mainly in the Black Sea region, it has shifted from the eastern part of the region to the western area, thereby increasing production share from 36 to 45% (FAO, 2004). Turkey dominates world hazelnut production and is the centre of origin of the hazelnut. Hazelnuts make a significant contribution to Turkey's exports, as well as being a source of income for a large number of family farms in the Black Sea region.

Hazelnuts are widely used in the food industry such as confectionery, baking, ice-cream and dairy, candy and chocolate products. Hazelnuts can be added to a wide array of dishes from cereals and breads to yogurts, soups, salads and main dishes to confections, ice creams and pastries. They are famed for their use in pies and their affinity with chocolate. Hazelnuts also are popular with vegetarians for their inclusion in nut loaves, nut bread and rissoles; they also make delicious nut butter (Mehlenbacher, 1990; Simsek and Aslantas, 1999; Simsek and Aykut, 2007). Moreover, hazelnut is widely considered as a potential rich source of energy (620-650 kcal), minerals (2-3.5%) and vitamins (vitamin B1, vitamin B2, vitamin B6, niacin and vitamin E) for human consump-

*Corresponding author. E-mail: m_turan25@hotmail.com.

Table 1. Sampling coordinates, location, and altitudes for selected study sites.

Sampling location	Coordinate	Growth Altitude, m
ACI	400° 54´ 32″ N ; 0380° 21´ 01″ E / 400° 54´ 30″ N ; 0380° 21´ 12″ E	1/0
ALLAHVERDI	400° 54´31″ N ; 0380° 21´ 01″ E /400° 52´34″ N ; 0340° 21´ 01″ E	1- 31
CAVCAVA	400°54´ 32″ N ; 0380° 21´ 01″ E /400°54´ 33″ N ; 0380° 21´ 21″ E	0/0
CAKILDAK	400° 54´ 32″ N ; 0380° 21´ 01″ E/400° 54´ 29″ N ; 0380° 21´ 03″ E	0/20
FOSA	400°54´ 32″ N ; 0380° 21´ 01″ E/400°54´ 32″ N ; 0380° 21´ 12″ E	0/0
INCEKARA	400 °54´ 32″ N ; 0380° 21´ 01″ E/400 °54´ 28″ N ; 0380° 21´ 02″ E	0/21
K – 24	400° 54´ 34″ N ; 0380° 21´ 06″ E/400° 54´ 33″ N ; 0380° 21´ 12″ E	0/0
KALINKARA	400° 54´ 32″ N ; 0380° 21´ 01″ E/400° 54´ 29″ N ; 0380° 21´ 02″ E	0/22
KAN	400°54´ 32″ N ; 0380°21´ 01″ E/400°54´ 27″ N ; 0380°20 59″ E	0/30
KARGALAK	400 °54´ 27″ N ; 0380° 20´ 59″ E/400 °54´ 27″ N ; 0380° 20´ 59″ E	0/32
KUS	400° 54! 32″ N ; 0380° 21´ 01″ E/400 °54´ 27″ N ; 0380° 20´ 59″E	1 /35
MINCANE	400° 54´ 33″ N ; 0380° 21´ 01″ E/400° 54´ 30″ N ; 0380° 21´ 14″ E	1 /3
PALAZ	400 °54´ 33″ N ; 0380° 21´ 01″ E/400 °54´ 27″ N ; 0380° 21´ 01″ E	1 /33
SIVRI	400°54´ 32″ N ; 0380° 21´ 01″ E/400°54´ 28″ N ; 0380° 21´ 04″ E	1 /32
TOMBUL	400° 54´ 33 ″N ; 0380° 21´ 01″ E/400° 54´ 28 ″N ; 0380° 21´ 01″ E	0 /26
UZUNMUSA	400°54´ 32″ N ; 0380° 21´ 01″ E/400°54´ 29″ N ; 0380° 21´ 03″ E	0 /23
YASSI BADEM	400°54´ 31″ N ; 0380° 21´ 01″ E/400°54´ 29″ N ; 0380° 21´ 03″ E	1 /15
YUVARLAK BADEM	400° 54´ 31″ N ; 0380° 21´ 01″ E/400° 54´ 28″ N ; 0380° 21´ 02″ E	1 /20

sumption (Ackurt et al., 1999; Souci et al., 2000; Koksal et al., 2006). In spite of the known major mineral composition potassium (K), phosphorus (P), manganese (Mn), calcium (Ca), magnesium (Mg), etc.) of hazelnuts, there is scarce literature on its trace and heavy element contents, like As, Al, Co, Cr, Se, B, Pb and Ni.

In particular, hazelnut is known to be a concentrated food for major elements and it may become an interesting food for its trace elements for the human diet. Many conscious people prefer supplementary foods with rich contents. Therefore, in recent years, there has been an increasing interest in the use of trace elements as micronutrient supplements in medical treatment to pre-vent various diseases such as cancer, cardiovascular diseases, AIDS, Alzheimer's disease, Kashin-Beck disease, Keshan disease, osteoporosis, osteoarthritis, asthma, goiter, cataracts, stroke, arthritis, ageing, ane-miaete, and to complete mineral deficiency (Hunt, 1996; Reilly, 1998; Rayman, 2000). It is important to evaluate the plant type and growth condition in view of the human healthy standard and so, many papers have been published in recent years on the biological functions of microelements in humans.

Objective of this study was to investigate 18 Turkish hazelnut cultivars in view of the trace and heavy metal concentration and to compare the interspecies differences of the levels of these elements for human nutrition and health.

MATERIALS AND METHODS

This study was conducted in Northern part of Turkey which has the highest production area between 2007 and 2009, to asses the heavy metal concentrations in soil and hazelnut for human nutrition. Soil type of this area was classified as alfisol (Soil Survey Staff, 1992). Mean annual temperature, precipitation (rainfall) and relative humidity were 13.8 °C, 1183 mm and 74.7%, respectively. Wind direction was S-WSW14.

Eighteen sampling locations were selected for detailed analyses (Table 1). Soil and plant samples were taken from Giresun Hazelnut Research gardens which were located at different parts of the Black Sea Region of Turkey, at harvest of August 2007 and 2009, 10 plant samples (One-hundred-gram fruit per tree) for each cultivar (Aci, Allahverdi, Cavcava, Cakildak, Fosa, Incekara, Kalinkara, Kan, K-24, Kargalak, Kus, Mincane, Palaz, Sivri, Tombul, Uzunmusa, Yassibadem, and Yuvarlakbadem) were collected randomly from portions of about 3.0 kg within a 2000 m^2 area. Plant samples were packed into polyethylene bags and care were taken to avoid contamination.

Hazelnut fruit samples were washed carefully with deionized water to remove adsorbed dust, then oven dried for 48 h at 68°C and powdered. Micro-elements (As, B, Se, Fe, Mn, Zn, Cu, Ni, Pb, Cr, Mo, Co, Al and Cd) were determined after wet digestion of dried and ground sub-samples using a HNO_3-H_2O_2 acid mixture (2:3 v/v) with three step (first step; 145°C, 75%RF, 5 min; second step; 180°C, 90%RF, 10 min and third step; 100°C, 40%RF, 10 min) in microwave (Bergof Speedwave Microwave Digestion Equipment MWS-2) (Mertens, 2005a). Tissue macro and micro elements were determined using an Inductively Couple Plasma spectrophotometer (Perkin-Elmer, Optima 2100 DV, ICP/OES, Shelton, CT 06484-4794, USA) (Mertens, 2005b).

RESULTS AND DISCUSSION

Hazelnut is known to be a concentrated food for major elements and it may become an interesting food for its trace elements for the human diet. Trace elements are essential for human nutrition at low doses but may also be toxic for humans, animals and plants at higher doses (Tannenbaum et al., 1985; Ladipo, 2000). The toxic value for human of hazelnut is mainly depending on plant types and plant growth condition. So, it is important to evaluate the plant type and growth condition in view of the human health standards.

Hazelnut is the best source of essential element, amino acids, vitamin B and E among tree nuts (Holland et al., 1995) and serves as a good source of natural anti-oxidants (Ackurt et al., 1999; Bonvehi and Coll, 1997; USDA, 2008; Yurttas et al., 2000). Vitamin E or R-tocopherol is a lipid soluble phenolic antioxidant. The antioxidant activity of phenolics is based on their ability to donate a hydrogen atom to free radicals. Because these compounds are able to scavenge free radicals, they are believed to have potential in the prevention of cancer, atherosclerosis, and diabetes (Thompson, 1994). Hazelnut has also been reported to serve as a good source of essential elements, minerals, amino acids, and the B complex vitamins.

Boron deficiency has been reported in studies in rats, chickens, and humans. Boron appears to affect calcium and magnesium metabolism and may be needed for membrane function. Boron deficiency signs may be related to the level of vitamin D and possibly other nutrients in the diet. Boron has long been known to be essential for the growth of most plants.

The highest As and B element concentrations of hazelnut were found in the Uzunmusa cv, (0.39 and 29.9 mg kg^{-1}), the lowest were Aci and Palaz (0.07 and 14.6 mg kg^{-1}) in 18 cultivars, respectively. The B intake from hazelnut may promote bone and joint health, particularly in women (Hunt, 1996). Raisins, leafy vegetables, fruits such as grape, sour cherry, quince, peach, legumes and nuts such as almond, pistachio, and walnut are recommended as good sources of B by various researchers (Souci et al., 2000; Simsek et al., 2003a). Among the 18 cultivars, the B contents of hazelnuts are similar to those reported by Simsek et al. (2003b). Palaz varieties may be served as an excellent source of B. Consuming the recommended daily amount (RDA) of 100 g hazelnut supplies 5% of B (Tables 2 and 3).

Ni concentrations in the samples were found between 0.81 and 2.97 mg kg^{-1}. The Ni content of the Kargalak variety was found to be about two times higher than reported values (Souci et al., 2000). Consuming RDA of 100 g hazelnut supply 1% of Ni. In a recent study, nuts, cocoa and chocolate, soy beans, oatmeal, fresh and dried legumes were reported as foods that have very high Ni contents. On the other hand, the Ni content of foods is very important for dermatologists because larger amounts of Ni (600 mg/person/day) increase hand eczema in Ni-sensitive women and men (Flyvholm et al., 1984; Dahiya et al., 2005). Consequently, excessive hazelnut consumption (more than 100 g day^{-1}) may cause Ni allergy and so should be avoided.

The only known nutritional, but very vital, function of cobalt is as an integral part of vitamin B12 and is also essential for the metabolism of glucose, insulin and fatty acids. Therefore it is proposed to increase muscle mass and decrease body fat in sportsman (Berger et al., 2002). Because vitamin B, is derived from bacterial synthesis, inorganic cobalt can be considered essential for animal species that depend totally on their bacterial flora for their vitamin B. This is the case for ruminant animal species in which cobalt deficiency is well known; it might also have some relevance for strict vegetarians whose intake of the preformed vitamin is severely limited. However, there is no evidence that the intake of cobalt is ever limiting in the human diet, and no RDA is necessary. While K-24 has the highest Co levels, the lowest Co levels were found in Kargalak varieties. The Co levels of hazelnuts changed between 0.07 and 0.61 mg kg^{-1}. K-24 varieties may be served as source of recommended daily intake of 100 g hazelnut supplies 100% of Co.

Depressed growth, impaired reproductive performance, and other changes have been reported in laboratory animals which fed diets extremely low in Cd and Pb and kept in an environment allowing the strictest control of contamination. Nutritional requirements, if they exist, are very low and easily met by the levels naturally occurring in foods, water, and air. The evidence for requirements and essentiality for human is weak. The highest Cd, Al, and Pb, element concentrations of hazelnut were found in the Uzunmusa cv, and Aci (0.03, 0.03 and 0.07 mg kg^{-1}), the lowest was K-24 (3.01, 0.01 and 0.02 mg kg^{-1}) in 18 cultivars, respectively. Uzunmusa cv, and Aci varieties may be served as source of RDA intake of 100 g hazelnut supplies 50% of Cd, 100 of Al, and Pb.

Trivalent chromium is required for maintaining normal glucose metabolism in laboratory animals; it acts as a cofactor for insulin. Experimental chromium deficiency has been induced in several animal species, resulting in impaired glucose tolerance in the presence of normal concentrations of circulating insulin and, in severe cases, in a diabetes-like syndrome (Schroeder, 1966). Three cases of pronounced chromium deficiency have been reported in patients on long-term total parenteral alimentation (Jeejeebhoy et al., 1977; Brown et al., 1986); all three had in common a relative insulin resistance and peripheral or central neuropathy. Chromium-responsive impairment of glucose tolerance has been reported in malnourished children, in some but not all studies of mild diabetics, and in middle-aged subjects with impaired glucose tolerance. (IPCS, 1988.)

Chromium concentrations in human tissues decline with age, except for the lungs in which chromium accumulates. Parity, juvenile diabetes, and coronary artery disease are associated with low-chromium concentrations in hair or serum (IPCS, 1988).

The intestinal absorption of dietary chromium at daily intakes of 40 µg and more is approximately 0.5% of the total amount present; intakes less than 40 µg day^{-1} are absorbed with an increasing efficiency, up to about 2% of the total (Anderson and Kozlovsky, 1985). Absorbed chromium is excreted almost completely through the urine.

Table 2. Trace and heavy metal content of hazelnut varieties growth in Turkey (mg kg^{-1}).

	Al	As	B	Cd	Co	Cr	Cu
ACI	1.64	0.07	20.51	0.02	0.24	0.03	16.39
ALLAHVERDI	5.65	0.16	18.45	0.01	0.14	0.02	16.03
CAVCAVA	1.06	0.27	17.45	0.01	0.24	0.02	18.63
CAKILDAK	1.78	0.02	18.35	0.02	0.50	0.04	21.48
FOSA	1.61	0.17	17.03	0.02	0.19	0.02	21.38
INCEKARA	1.08	0.17	16.12	0.01	0.54	0.02	16.61
K – 24	0.91	0.13	19.12	0.01	0.07	0.03	12.80
KALINKARA	1.22	0.08	18.24	0.02	0.47	0.03	16.16
KAN	1.43	0.20	18.35	0.02	0.59	0.04	19.56
KARGALAK	9.74	0.07	15.91	0.01	0.61	0.05	17.89
KUS	1.50	0.08	18.57	0.01	0.35	0.04	15.39
MINCANE	1.28	0.14	21.35	0.01	0.18	0.04	21.71
PALAZ	1.18	0.10	14.60	0.01	0.61	0.04	19.90
SIVRI	1.27	0.06	16.10	0.01	0.60	0.04	13.76
TOMBUL	2.23	0.06	17.26	0.01	0.39	0.03	16.95
UZUNMUSA	18.11	0.19	29.90	0.03	0.37	0.02	15.36
YASSI BADEM	1.13	0.17	21.19	0.02	0.34	0.03	15.58
YUVARLAK BADEM	1.30	0.17	25.33	0.02	0.24	0.02	17.38
	Fe	**Mn**	**Mo**	**Pb**	**Zn**	**Ni**	**Se**
ACI	40.21	59.42	0.11	0.07	33.13	0.81	0.04
ALLAHVERDI	36.86	61.40	0.19	0.03	30.67	0.88	0.04
CAVCAVA	38.13	115.93	0.17	0.05	29.18	1.77	0.02
CAKILDAK	49.40	116.81	0.10	0.03	35.52	1.51	0.02
FOSA	39.84	75.46	0.11	0.07	31.92	2.46	0.03
INCEKARA	35.74	44.56	0.09	0.04	25.41	2.72	0.03
K – 24	38.97	48.83	0.29	0.02	33.39	1.36	0.04
KALINKARA	37.35	58.35	0.17	0.03	29.74	1.95	0.03
KAN	38.57	46.32	0.10	0.05	29.70	2.61	0.04
KARGALAK	36.73	46.05	0.09	0.05	27.88	2.97	0.04
KUS	36.51	52.70	0.10	0.04	26.39	2.20	0.04
MINCANE	39.94	41.11	0.21	0.04	25.35	0.83	0.05
PALAZ	35.12	63.83	0.11	0.03	26.04	1.89	0.04
SIVRI	35.25	70.08	0.09	0.04	23.52	1.56	0.04
TOMBUL	49.06	66.45	0.11	0.06	29.96	1.29	0.04
UZUNMUSA	37.62	67.75	0.31	0.04	29.84	1.54	0.05
YASSI BADEM	39.92	73.00	0.09	0.04	27.64	2.35	0.05
YUVARLAK BADEM	39.32	71.55	0.09	0.04	29.50	2.97	0.05

Chromium intake from typical western diets varies widely between 25 µg day^{-1} in elderly persons in England to approximately 200 µg in Belgian and Swedish diets, but in the most recent international studies (IPCS, 1988), intakes below 100 µg day^{-1} were reported. The concentrations of Cr ranged from 0.02 to 0.05 mg kg^{-1} for the analyzed hazelnut varieties. Cr values of some cultivars were found to be seven times and four times higher than those reported by Souci et al. (2000), respectively. According to in present study results, an average 10% of dietary Cr intake can be easily obtained from consumption of hazelnut at 100 g day^{-1}.

Some authors suggested that the Cr intake for athletes may have an effect with a high-energy diet because low-energy diets cause both deficiencies in Cr and lower body weights. Hazelnut has especially high carbohydrate (10- 22%) and lipid concentration (55-65%), obtaining the full benefits from Cr (Simsek and Aslantas 1999).

Molybdenum plays a biochemical role as a constituent of several mammalian enzymes, such as aldehyde oxidase, xanthine oxidase, and sulfite oxidase (Rajagopalan, 1988); however, induction of characteristic

Table 3. The some mineral nutrition to feel the need for human beings (Alasalvar et al., 2009).

Nutrient	RDA	RNI	UL	SUL
Fe (mg day^{-1})	8-15	11.4	54	17
Zn(mg day^{-1})	11	9.5	40	25
Mn(mg day^{-1})	2-5	>1.4	11	4
Cu(mg day^{-1})	0.9-2.0	1.2	10	10
Se (µg day^{-1})	35-55	75	400	450
Mo (µg day^{-1})	45-75	50-400	2000	NS
Cr (µg day^{-1})	25-35	>25	NS	NS
B (mg day^{-1})	<20	NS	20	9.6
Ni (mg day^{-1})	35	NS	1000	260
As (µg day^{-1})	NS	NS	NS	NS
Pb (µg day^{-1})	20	NS	NS	NS
Al (µg day^{-1})	NS	NS	NS	NS
Co (µg day^{-1})	NS	NS	NS	NS
Cd (µg day^{-1})	60	NS	NS	NS

pathological lesions in animals due to nutritional molybdenum deficiency has been difficult (Mills and Bremner, 1980).

Two last decades some of investigations have suggested a role for molybdenum in human nutrition. The foods that contributed the most to the molybdenum intake were milk, beans, breads, and cereals. Pennington and Jones (1987) found lower molybdenum content in the 1984 collection of the Food and Drug Administration's Total Diet Study ranging from 76 to 109 µg day^{-1} for adult (females and males). In this study, all of the hazelnut varieties have got high Mo potential to supplies daily human intake.

The highest Mo element concentrations of hazelnut were found in the K-24 (0.29 mg kg^{-1}), the lowest was Kargalak (0.09 mg kg^{-1}) in 18 cultivars, respectively. All of the hazelnut supplies would be expected to contribute between 90-310 µg of molybdenum per day which would constitute 41% or less of the amount of the provisional recommended intake.

Iron is a constituent of hemoglobin, myoglobin, and a number of enzymes and, therefore, is an essential nutrient for humans (Bothwell et al., 1979). In addition to these functional forms, as much as 30% of the body iron is found in storage forms such as ferritin and hemosiderin (mainly in the spleen, liver, and bone marrow), and a small amount is associated with the blood transport protein transferring.

The average loss of iron in the healthy adult man is estimated to be approximately 1 mg day^{-1} (Green et al. 1968). In adult women, there is an additional loss of about 0.5 mg day^{-1}, the amount of iron in the average menstrual blood flow averaged over 1 month (Hallberg et al., 1966). In approximately 5% of normal women, however, menstrual losses of more than 1.4 mg day^{-1} have been observed. As menstrual losses deplete iron stores, absorption of dietary iron increases. Concordant figures are found using radio-labeled iron loss from circulating erythrocytes. This method gives reliable turnover information for adults and indicates that average requirements to replace daily losses for adults ages 20 to 50 are approximately 14 µg of iron per kilogram of body weight for males (1.10 mg 79 kg^{-1}) and 22 µg for premenopausal females (1.38 mg 63 kg^{-1}) (Bothwell and Finch, 1968).

In people without genetic defects that increase iron absorption, there are no reports of iron toxicity from foods other than long-term ingestion of home brews made in iron vessels (Walker and Arvidsson, 1953). Deleterious effects of daily intakes between 25 and 75 mg are unlikely in healthy persons (Finch and Monsen, 1972). On the other hand, there are approximately 2,000 cases of iron poisoning each year in the United States, mainly among young children who ingest the medicinal iron supplements formulated for adults. The amount of Fe is found to be between 35.7 and 49.4 mg kg^{-1} in hazelnut cultivars. The highest Fe concentration was found in Cakildak (49.4 mg kg^{-1}), but the amounts of Fe in Fosa, Yassi Badem, Yuvarlak Badem and Mincane were found to be quite similar to one another. Results show that hazelnut can be served as an excellent source of Fe. Consuming the RDA of 100 g hazelnut supplies 50% of Fe (Tables 2 and 3).

Zinc, a constituent of enzymes involved in most major metabolic pathways, is an essential element for plants, animals, and humans (Hambidge et al., 1986). Relatively large amounts of zinc are deposited in bone and muscle, but these stores are not in rapid equilibrium with the rest of the organism. The body pool of readily available zinc appears to be small and to have a rapid turnover rate, as shown by the prompt appearance of deficiency signs in laboratory animals. No single enzyme function has yet been identified that could explain the rapid onset of physiological and biochemical changes that follow the induction of zinc deficiency, but the requirement for zinc by many enzymes involved in gene expression (Chesters, 1982) could explain the immediate effect of deficiency on cell growth and repair.

The signs and symptoms of dietary zinc deficiency in humans include loss of appetite, growth retardation, skin changes, and immunological abnormalities. Studies in laboratory and domestic animals have shown that zinc deficiency during pregnancy may lead to developmental disorders in the offspring (Hurley and Baly, 1982).

The zinc content of typical mixed diets of North American adults has been reported to furnish between 10 and 15 mg day^{-1}. Pennington et al. (1984), in a survey of U.S. foods, found 13.2 mg of zinc in a 2,850 kcal diet. Infant and toddler diets containing 880 and 1,300 kcal contained 5.5 and 8.5 mg zinc, respectively. Elderly people generally have been found to consume 7 to 10 mg zinc daily (Greger, 1989). An evaluation of the most reliable balance studies indicates that at least 12 mg of zinc in a mixed U.S. diet is required to maintain the existing zinc status of healthy young men (Sandstead, 1985, Table 3).

These study results show that hazelnut can be important for completing the deficiency of Zn in dairy intake about 32% levels. The highest Zn element concentrations of hazelnut were found in the Cakildak (35.5 mg kg^{-1}), the lowest was Sivri (23.5 mg kg^{-1}) in 18 cultivars, respectively.

Manganese has been shown to be an essential element in every animal species studied. Signs of deficiency include poor reproductive performance, growth retardation, congenital malformations in the offspring, abnormal formation of bone and cartilage, and impaired glucose tolerance (Hurley and Keen, 1987). Several enzymes, such as decarboxylases, hydrolases, kinases, and transferases, are nonspecifically activated by manganese *in vitro*. There are two known manganese metalloenzymes: pyruvate carboxylase and superoxide dismutase, both localized in the mitochondria.

Whole grains and cereal products are the richest dietary sources of manganese, and fruits and vegetables are somewhat less so. Dairy products, meat, fish, and poultry are poor sources. Tea is a rich source of manganese, but typical drinking water consumed at the rate of 2 liters daily contributes only about 40 to 64 μg, or about 2 to 3% of the amount furnished by diet (NRC, 1980).

The Total Diet Study conducted in the United States between 1982 and 1986 indicated that the mean daily dietary manganese intake was 2.7 and 2.2 mg for adult men and women, respectively (Pennington et al. 1989). Teenage boys consumed an average of 2.8 mg day^{-1}, whereas girls consumed only 1.8 mg day^{-1}. Mean manganese intakes were 1.1 and 1.5 mg day^{-1}for 6- to 11-month-old babies and 2-year-old toddlers, respectively.

In humans, toxicity has been observed only in workers exposed to high concentrations of manganese dust or fumes in air, but not as a consequence of dietary intake by people consuming 8 to 9 mg of manganese per day in their food (WHO, 1973). In view of the remarkably steady tissue concentrations of manganese in the U.S. population (Schroeder et al., 1966) and the low toxicity of dietary manganese, an occasional intake of 10 mg day^{-1}by adults can be considered safe. To include an extra margin of safety, however, the subcommittee recommends a range of manganese intake from 2 to 5 mg day^{-1}for adults (Table 3).

The highest Mn element concentrations of hazelnut were found in the Cakildak (116.8 mg kg^{-1}), the lowest was Mincane (41.1 mg kg^{-1}) in 18 cultivars, respectively. According to our study result 18 hazelnut cultivar has got high Mn potential and may supply all of human daily needs.

Several abnormalities have been observed in copper-deficient animals, including anemia, skeletal defects, demyelination and degeneration of the nervous system, defects in pigmentation and structure of hair or wool, reproductive failure, myocardial degeneration, and decreased arterial elasticity. There are a number of important copper-containing proteins and enzymes, some of which are essential for the proper utilization of iron (Davis and Mertz, 1987).

Under normal circumstances, dietary copper deficiency is not known to occur in adults, but it has been observed in malnourished children in Peru; its manifestations are anemia, neutropenia, and severe bone demineralization (Cordano et al., 1964). In the early 1970s in the United States, similar findings were recognized in a few very small premature infants who were hospitalized for long periods and exclusively fed modified cow's milk formula or received prolonged parenteral alimentation. Presumably, these aberrations reflected a deficient dietary intake of copper (Cordano, 1974). More recently, copper deficiency has been shown to impair the growth of Chilean infants recovering from malnutrition (Castillo-Duran and Uauy, 1988).

Older analytical data indicating that most U.S. diets provide a daily copper intake between 2 and 5 mg are now being reexamined and questioned (Klevay, 1984). The Total Diet Study, based on the extensive dietary analyses performed by the U.S. Food and Drug Administration, showed that the daily intake of copper for adult males and females averaged about 1.2 and 0.9 mg, respectively, (Pennington et al., 1989). The intakes for infants 6 to 11 months old and toddlers 2 years old were 0.45 and 0.57 mg daily (Table 3).

In the past, estimates of the copper requirement for humans were derived from metabolic balance studies. However, the balance technique can lead to false estimates of nutritional requirements because the efficiency of copper absorption is increased or decreased in response to low or high copper intakes, respectively (Turnlund et al., 1989). Older balance studies suggested that the adult requirement for copper ranged from 2.0 to 2.6 mg/day, whereas later studies indicated that intakes less than 2.0 mg day^{-1}, and often not much more than 1.0 mg day^{-1}, could maintain positive copper balance (Mason, 1979).

The average Cu levels in varieties range from 12.8 to 21.7 mg/kg. We observed the highest amount of Cu in Mincane (21.7 mg/kg), but K-24 (12.8 mg/kg) hazelnuts had the lowest. Hazelnut varieties of 100 g amount can be source of Cu to supply 100% of Cu daily human intake.

Selenium is recognized as an essential trace element for human body at low concentrations (NRC, 1989). This element may be toxic if excess amounts are found in food and water (Bermejo- Barrera et al., 2002). This nutrient is also an important part of antioxidant enzymes that protect cells against the effects of free radicals that contribute to the development of some chronic diseases (Combs et al., 1998; Reilly, 1998; Rayman, 2000).

Average Se content of the varieties varied from 0.02 to 0.04 mg kg^{-1}. Se contents of hazelnuts were found lower to those reported (2.0-5.1 mg/100 g) in the Food Composition and Nutrition Tables (Souci et al., 2000), but similar those reported (0.47-1.69 mg 100 g^{-1}) in the literature (Dundar and Altundag, 2004). Results show that the contents of Se of our samples are higher than oilseeds walnuts (0.005-0.04 mg kg^{-1}), pistachio nuts (0.006-0.02 mg kg^{-1}), almonds (0.003-0.01 mg kg^{-1}) and peanuts (0-0.054 mg kg^{-1}) (Alarcon et al., 1996; Reilly, 1998; Souci et al., 2000). The present study show that hazelnuts and their products can also be a significant natural source of dietary Se instead of Se-fortified food products, and recommend to prevent diseases (about 40) linked with Se deficiency in some countries where the soil is extremely low in Se (Reilly, 1998). Moreover, hazelnuts have an advantage due to the fact that the bioavailability of Se in vegetable products is higher than that in animal foods such as meat, fish, seafood and offal (Alaejos et al., 2000).

Direct evidence of a requirement for selenium in human nutrition was lacking until 1979, when Chinese scientists reported an association between low selenium status and Keshan disease, a cardiomyopathy that affects primarily young children and women of childbearing age (Keshan Disease Research Group, 1979a). Sea foods, kidney, and liver, and to a lesser extent other meats, are consistently good sources of selenium, whereas grains and other seeds are more variable, depending on the selenium content of the soils in which they are grown (WHO, 1987; Keshan Disease Research Group, 1979b). Fruits and vegetables generally contain little selenium. Drinking water usually makes only a small contribution to selenium intake (WHO, 1987). Studies in animals have shown that the bioavailability of selenium in certain fish is less than that in other foods (Mutanen, 1986). Few data exist on the bioavailability of selenium in foods consumed by humans. However, bioavailability trials conducted in subjects with poor selenium status indicated that organically bound forms of se-tested caused similar increases in glutathione peroxidase activity (Levander et al., 1987; Thomson et al., 1982).

Analyses of national food composites in the United States indicate that the overall adult mean dietary selenium intake was 108 µg day^{-1} (Pennington et al., 1984), but 35-55 mg day according to Table 3. Several balance studies have been conducted to investigate selenium requirements of humans. However, humans apparently can adjust their selenium homeostatic mechanisms to remain in balance over rather wide ranges of dietary intakes. Therefore, the balance technique is of little help in delineating the selenium requirements of humans (Levander, 1987).

The level of dietary selenium exposure needed to cause chronic poisoning in humans is not known with certainty, but approximately 5 mg day^{-1} from foods resulted in fingernail changes and hair loss in a seleniferous zone of China (Yang et al., 1983). The Chinese investigators also reported that a person who had consumed 1 mg of selenium daily as sodium selenite for more than 2 years had thickened but fragile nails and a garlic odor in dermal excretions.

In this study, most common hazelnut cultivars in the Turkey were analyzed and found to contain significant quantities of variety of essential nutrients. All of the hazelnut varieties may be served as an excellent source of Se. Consuming the RDA of 100 g hazelnut supplies 21% of Se.

Conclusion

According to variance analysis, the effect of cultivars on microelements is found to be significant. The Mn concentration in hazelnut varieties was determined to be the highest level among the microelements, followed by Fe, Zn, B, Cu, Al, Ni, Co, Se, Mo, Pb, As, Cd,and Cr. The highest Se (0.05 mg kg^{-1}), Al (18.1 mg kg^{-1}), As (0.19 mg kg^{-1}), B (29.9 mg kg^{-1}), Cd (0.03 mg kg^{-1}) and Mo (0.31 mg kg^{-1}) element concentrations of hazelnut were found in the Uzunmusa cv, the highest Co (0.61 mg kg^{-1}), Cr (0.05 mg kg^{-1}), Pb (0.05 mg kg^{-1}), and Ni (2.9 mg kg^{-1}) in the Kargalak cv., and the highest Cu (21.5 mg kg^{-1}), Fe (49.4 mg kg^{-1}), Mn (166.8 mg kg^{-1}) and Zn (35.5) in the Cakildak cv. The levels of microelement changes according to varieties may have resulted from differences in use of fertilizers, the methods of cultivation and irrigation, soil composition of the production zone, the climate and the geographical region. Data obtained from hazelnut plants show that they had a very high nutritional potential, and their mineral content of 100 g hazelnut was greater than that of human daily consumption for this required level. Our results show that an important part of daily Co, Pb, Mo, Al, Cd, Mn, Pb, Cr, Cu, B, Zn, Fe, Ni and Se requirement for human nutrition can be supplied by the consumption of 100 g Uzunmusa cv, Kargalak cv., Cakildak cv. hazelnut daily. Hazelnut consumption with fruit juices containing particularly high ascorbic acid levels may be useful to enhance Fe absorption. Besides, hazelnuts are a good source to enrich dairy products for Fe. According to the daily microelement requirements, 100 g hazelnut provided about 50% for Fe and Cd, 41% for Mo, 32% for Zn, 21% for Se, 10% for Cr, 5% for B and 1% Ni of the RDA. On other hand, Mn, Cu, As, Pb, Al, and Co levels of 100 g hazelnuts are higher than the respective daily requirements, but slight overdoses of these elements are non-toxic for human health.

REFERENCES

Ackurt F, Ozdemir M, Biringen G, Loker M (1999). Effects of geographical origin and variety on vitamin and mineral composition of hazelnut (*Corylus aVellana* L.) varieties cultivated in Turkey. Food Chem., 65: 309-313.

Alaejos MS, Romero FJD, Romero CD (2000). Selenium and cancer: some nutritional aspects. Nutrition, 16: 376-383.

Alarcon JPD, Alarcon MN, Serrana LG, Martine MCL (1996). Determination of selenium in cereals, legumes and dry fruits from southeastern Spain for calculation of daily dietary intake. Sci. Total Environ., 184: 183-189.

Alasalvar C, Shahidi F, Amaral JS, Oliveira BPP (2009). Compositional characteristics and health effects of hazelnut (*Corylus avellana* L.): An overview. In Alasalvar C, Shahidi F (Eds.), Tree nuts: Compositions, phytochemicals, and health effects (pp. 185–214). Boca Raton, FL: CRC Press Taylor and Francis Group.

Anderson RA, Kozlovsky AS (1985). Chromium intake, absorption, and excretion of subjects consuming self-selected diets. Am. J. Clin. Nutr., 41:1177-1183.

Anonymous (2010). http://www.apps.fao.org/page.

Berger CE, Kroner A, Kluger R, Baron R, Steffan I, Engel A (2002). Effects of marathon running on the trace minerals chromium, cobalt, nickel and molybdenum. J. Trace Elem. Exp. Med., 15: 201-209.

Bermejo-Barrera P, Moreda-Pineiro A, Bermejo-Barrer AA (2002). Study of ammonium molybdate to minimize the phosphate interference in the selenium determination by electrothermal atomic absorption spectrometry with deuterium background correction. Spectrochimica Acta Part B., 57: 327-337.

Bonvehi JS, Coll FVA (1997). Chemical study of the mineral fraction of Tarragona hazelnuts (*Corylus aVellana* L.). Acta Aliment., 26: 243-253.

Bothwell TH, Charlton RW, Cook JD, Finch CA (1979). Iron Metabolism in Man. Blackwell, Oxford.

Bothwell TH, Finch CA (1968). Iron losses in man. Pp. 104-114 in Occurrence, Causes and Prevention of Nutritional Anaemias. Symposia of the Swedish Nutrition Foundation, VI. Almquist and Wiksell, Uppsala.

Brown RO, Forloines-Lynn S, Cross RE, Heizer WD (1986). Chromium deficiency after long-term total parenteral nutrition. Dig. Dis. Sci., 31: 661-664.

Castillo-Duran C, Uauy R (1988). Copper deficiency impairs growth of infants recovering from malnutrition. Am. J. Clin. Nutr., 47: 710-714.

Chesters JK (1982). Metabolism and biochemistry-of zinc. Pp. 221-238 in Prasad AS, ed. Clinical, Biochemical, and Nutritional Aspects of Trace Elements. Alan R. Liss, New York. Curr. Topics Nutr. Dis., p. 6.

Combs Jr. GF, Gray WP (1998). Chemopreventive agents: selenium. Pharmacol. Therap. 79: 179-192.

Cordano A (1974). The role played by copper in the physiopathology and nutrition of the infant and the child. Ann. Nestle, 33: 1-16.

Cordano A, Baertl JM, Graham GG (1964). Copper deficiency in infancy. Pediatrics, 34: 324-336.

Dahiya S, Karpe R, Hegde AG, Sharma RM (2005). Lead, cadmium and nickel in chocolates and candies from suburban areas of Mumbai, India. J. Food Compos. Anal., 18: 517-522.

Davis GK, Mertz W (1987). Copper. in W. Mertz, ed. Trace Elements in Human and Animal Nutrition, 5th ed., Academic Press, Orlando, Fla., 2: 301-364,

Dundar MS, Altundag H (2004). Selenium content of Turkish hazelnut varieties: Kara Findik, Tombul and Delisava. J. Food Compos Anal., 17: 707-712.

FAO (Food and Agriculture Organization of The United Nations) (2004). Agricultural statistics. Available online at: http://faostat.fao.org/faostat/collections?version.

Finch CA, Monsen ER (1972). Iron nutrition and the fortification of food with iron. J. Am. Med. Assoc., 219: 1462-1465.

Flyvholm M, Nielsen GD, Andersen A (1984). Nickel content of food and estimation of dietary intake. Eur. Food Res. Technol., 179: 427-431.

Greger JL (1989). Potential for trace mineral deficiencies and toxicities in the elderly. in C.W. Bales, ed. Mineral Homeostasis in the Elderly. Alan R. Liss, New York. Curr. Topics Nutr. Dis., 21: 171-200

Hallberg L, Hogdahl AM, Nilsson L, Rybo R (1966). Menstrual blood lossa population study. Variation at different ages and attempts to define normality. Acta Obstet. Gynecol. Scand., 45: 320-351.

Hambidge KM, Casey CE, Krebs NF (1986). Zinc. in Mertz W, ed. Trace Elements in Human and Animal Nutrition, 5th ed. Academic Press, Orlando, Fla., 2: 1-137.

Hunt CD (1996). Biochemical effects of physiological amounts of dietary boron. J. Trace Elem. Exp. Med., 9: 185-213.

Hurley LS, Keen CL (1987). Manganese. in Mertz W, ed. Academic Press, Orlando, Fla. Trace Elem. Hum. Anim. Nutr., 1: 185-223.

Hurley LS, Baly DL (1982). The effects of zinc deficiency during pregnancy. in Prasad AS, ed. Clinical, Biochemical, and Nutritional Aspects of Trace Elements. Alan R Liss, New York. Curr. Topics Nutr. Dis., 6: 145-159

IPCS (International Programme on Chemical Safety) (1988). Chromium. Environmental Health Criteria 61. World Health Organization, Geneva.

Jeejeebhoy KN, Chu RC, Marliss EB, Greenberg GR, Bruce-Robertson A (1977). Chromium deficiency, glucose intolerance and neuropathy reversed by chromium supplementation in a patient receiving long-term total potential nutrition. Am. J. Clin. Nutr., 30: 531-538.

Keshan Disease Research Group (1979a). Epidemiologic studies on the etiologic relationship of selenium and Keshan disease. Chin. Med. J., 92: 477-482.

Keshan Disease Research Group (1979b). Observations on effect of sodium selenite in prevention of Keshan disease. Chin. Med. J., 92: 471-476.

Klevay LM (1984). The role of copper, zinc, and other chemical elements in ischemic heart disease. Pp. 129-157 in O.M. Rennert and W.-Y. Chan, eds. Metabolism of Trace Metals in Man, Vol. 1. Developmental Aspects. CRC Press, Boca Raton, Fla.

Koksal I, Artik N, Simsek A, Gunes N (2006). Nutrient composition of hazelnut (*Corylus avellana* L.) varieties cultivated in Turkey. Food Chem., 99: 509-515.

Ladipo OA (2000). Nutrition in pregnancy: mineral and vitamin supplements. Am. J. Clin. Nutr., 72: 280-290.

Ladipo OA (1982). Selenium: biochemical actions, interactions and some human health implications. in Prasad AS, ed. Alan R. Liss, New York. Clinical, Biochemical, and Nutrititional Aspects of Trace Elements. Curr. Topics Nutr. Dis., 6: 345-368.

Levander OA (1987). A global view of selenium nutrition. Annu. Rev. Nutr., 7: 227-250.

Mason KE (1979). A conspectus of research on copper metabolism and requirements of man. J. Nutr., 109: 1979-2066.

Mehlenbacher SA (1990). Hazelnuts (Corylus), Generic resources of temperate fruit and crops 1. In: Moore JN, Ballington JR, editors. Acta Horticulturae 290 Wagenigen, The Netherlands: ISHS., pp. 791-820.

Mertens D (2005a). AOAC Official Method 922.02. Plants Preparation of Laboratuary Sample. Official Methods of Analysis, 18th edn. Horwitz,W., and G.W. Latimer, (Eds). AOAC-International Suite 500, 481. North Frederick Avenue, Gaitherburg, Maryland 20877-2417, USA. 3: 1-2.

Mertens D (2005b). AOAC Official Method 975.03. Metal in Plants and Pet Foods. Official Methods of Analysis, 18th edn. Horwitz, W, Latimer GW, (Eds). AOAC-International Suite 500, 481. North Frederick Avenue, Gaitherburg, Maryland 20877-2417, USA. 3: 3-4.

Mills CF, Bremner I (1980). Nutritional aspects of molybdenum in animals. in Coughlan MP, ed. Molybdenum and Molybdenum-Containing Enzymes. Pergamon Press, Oxford. pp. 517-542.
Mutanen M (1986). Bioavailability of selenium. Ann. Clin. Res., 18: 48-54.
NRC (National Research Council) (1980). The contribution of drinking water to mineral nutrition in humans. in Drinking Water and Health, Safe Drinking Water Committee, Board on Toxicology and Environmental Health Hazards, Assembly of Life Sciences. National Academy Press, Washington, D.D. Vol. 3: 265-404.
NRC (National Research Council) (1989). Food, Nutrition Board, Recommended Dietary Allowances, 10th Edition. National Academy Press, Washington, DC.
Pennington JA, Wilson DB, Newell RF, Harland BF, Johnson RD, Vanderveen JE (1984). Selected minerals in foods surveys, 1974 to 1981/82. J. Am. Diet. Assoc., 84: 771-780.
Pennington JAT, Jones JW (1987). Molybdenum, nickel, cobalt, vanadium, and strontium in total diets. J. Am. Diet. Assoc., 87: 1644-1650.
Pennington JAT, Young BE, Wilson DB (1989). Nutritional elements in U.S. diets: results from the Total Diet Study, 1982-86. J. Am. Diet. Assoc., 89: 659-664.
Rajagopalan KV (1988). Molybdenum an essential trace element in human nutrition. Annu. Rev. Nutr., 8: 401-427
Rayman MP (2000). The importance of selenium to human health. Lancet, 356: 233-241.
Reilly C (1998). Selenium: a new entrant into the functional food arena. Trends Food Sci. Technol., 9: 114-118.
Sandstead HH (1985). Are estimates of trace element requirements meeting the needs of the user? in Mills CF, Bremner I, Chesters JK, eds. Trace Elements in Man and Animals, TEMA-5. Commonwealth Agricultural Bureaux, Farnham Royal, United Kingdom. pp. 875-878.
Schroeder HA (1966). Chromium deficiency in rats: a syndrome simulating diabetes mellitus with retarded growth. J. Nutr., 88: 439-445.
Simsek A, Aslantas R (1999). Findigin bilesimi ve insan beslenmesi acisindan onemi. Gida 24, 209-216.
Simsek A, Aykut O (2007). Evaluation of the microeelemnt profile of Turkish hazelnut (*Corylus avellana* L.) varieties fro human nutrition and health. Int. J. Food Sci. Nutr., 58: 677-688.
Simsek A, Korkmaz S, Velioglu YS, Ataman OY (2003b). Determination of boron in hazelnut (*Corylus avellana* L.) varieties by inductively coupled plasma optical emission spectrometry and spectrophotometry. Food Chem., 83: 293-296.
Simsek A, Velioglu S, Coskun AL, Sayli BS (2003a). Boron concentrations of selected foods from borate regions in Turkey. J. Sci. Food Agric. 83: 586-592.
Soil Survey Staff (1992). Keys to Soil Taxonomi. 5th Ed. SMSS Technical Monograph No: 19, Blacksburg. Pocahontas Pres. Inc.
Souci SW, Fachmann W, Kraut H (2000). Food composition and nutrition tables. 6th rev. Stuttgart: Medpharm Scientific, p. 1182.
Tannenbaum SR, Young VR, Archer MC (1985). Vitamins and minerals. In: Fennema OR, editor. Food chemistry 2nd edn. New York: Marcel Dekker. 7: 477-543.
Thompson LU (1994). Antioxidant and hormone-mediated health benefits of whole grains. Crit. Rev. Food Sci. Nutr., 34: 473-497.
Thomson CD, Robinson MF, Campbell DR, Rea HM (1982). Effect of prolonged supplementation with daily supplements of selenomethionine and sodium selenite on glutathione peroxidase activity in blood of New Zealand residents. Am. J. Clin. Nutr., 36: 24-31.
Turkish Hazelnut Exporter's Union (2008). The Turkish Hazelnut. Giresun, Turkey: Turkish Hazelnut Exporter's Union.
Turnlund JR, Keyes WR, Anderson HL, Acord LL (1989). Copper absorption and retention in young men at three levels of dietary copper using the stable isotope. 65Cu. Am. J. Clin. Nutr., 49: 870-878.
U.S. Department of Agriculture (USDA) (2008). Food and nutrition information center. National Agricultural Library, Beltsville, MD: USDA.
Walker ARP, Arvidson UB (1953). Iron "overload" in the South African Bantu. Trans. R. Soc. Trop. Med. Hyg., 47: 536-548.
WHO (World Health Organization) (1973). Trace elements in human nutrition. Report of a WHO Expert Committee. WHO Technical Report Series. World Health Organization, Geneva. p. 532.
WHO (World Health Organization) (1987). Selenium, Envionmental Health Criteria 58: A Report of the International Programme on Chemical Safety. World Health Organization, Geneva.
Yang GS, Wang RZ, Sun S (1983). Endemic selenium intoxication of humans in China. Am. J. Clin. Nutr., 37: 872-881.
Yurttas HC, Schafer HW, Warthesen JJ (2000). Antioxidant activity of nontocopherol hazelnut (*Corylus spp.)* phenolics. J. Food Sci., 65: 276-280.

Different land use types in the semi-arid rangelands of Kenya influence soil properties

Kevin Z. Mganga[1]*, Nashon K. R. Musimba[1], Dickson M. Nyariki[1], Moses M. Nyangito[2], Wellington N. Ekaya[3], William M. Muiru[4] and Agnes, W. Mwang'ombe[4]

[1]Department of Range & Wildlife Sciences, South Eastern University College (SEUCO), P. O. BOX 170-90200, Kitui, Kenya.
[2]Department of Land Resources Management & Agricultural Technology, University of Nairobi
[3]RUFORUM, P. O. BOX 7062, Kampala, Uganda.
[4]Department of Plant & Crop Science, University of Nairobi, P. O. BOX 29053-00625, Nairobi, Kenya

Rangelands in semi-arid Kenya have recently witnessed extensive land use changes. These changes can mainly be attributed to increased livestock populations and the response of the increased human population to both local and exogenous opportunities and constraints. This study was carried out in Kibwezi district of Kenya mainly inhabited by agro-pastoralists. The main objective of this study was to establish how different land use types influence soil properties in tropical semi-arid rangelands. Disturbed and undisturbed soil samples from three land use types namely cultivated land (CL), grazing land (GL) and fallow land (FL) at a depth of 15-30cm were collected and analyzed using standard laboratories for soil nutrients and soil physical properties. Results show open grazing lands (CEC 19.59 meq/100g) to be more fertile than cultivated (13.88 meq/100g) and fallow (6.40 meq/100g) lands. This was attributed to the continuous dropping of faecal material by grazing livestock. Higher bulk density in grazing land (1.36 g/cm^3) compared to cultivated and fallow lands with 1.29 g/cm^3 and 1.33 g/cm^3 was attributed to the hoof action of livestock. These results suggest that different land use types in the semi-arid rangelands contribute immensely to soil properties.

Key words: Bulk density, cation exchange capacity (CEC), drylands, grazing lands, Kibwezi district, livestock.

INTRODUCTION

Kenyan rangelands are home for millions of pastoralists and agro-pastoralists. The semi-arid lands form part of the vast rangelands of Kenya, which occupy over 80% of the country's total land area (Mwang'ombe et al., 2011). The semi-arid rangelands in Kenya are characterized by a number of habitat structures ranging from open grasslands to closed woody or busy vegetation with varying amounts and composition of grass cover and grass species respectively. The composition of the grass species and abundance of vegetation cover are dependent on a number of ecological conditions including; thermal regimes, rainfall amounts and duration of wet versus dry seasons, soil moisture content and the phenology of the shrub and tree canopies in the areas they grow. Livestock is the major user of primary production in these semi-arid rangelands (Nyangito et al., 2009). Livestock production is partly dependant on the cultural/economic interest of the main ethnic group occupying the area. The herd size (livestock numbers) and structure (composition of goats, sheep and cows) varies from place to place and ethnic group to ethnic group, but they are not dependent on ecological productivity potential.

The semi-arid African rangelands are often characterized by threshold dynamics and alternate stable states that are highly resilient (Kinyua et al., 2010). However, the interaction of heavy grazing and climatic variability can cause dramatic ecological degradation (Wessels et al., 2007). Nevertheless, communities

*Corresponding author. Email: kzowe@yahoo.com or kmganga@seuco.ac.ke.

inhabiting semi-arid rangelands in Kenya have been in a dynamic equilibrium with their environment and ecological degradation has been, for the most part, non-existent. Over the last few decades, much has changed. The changes are a response to a mix of factors including land privatisation and other government policies, population growth and migration, and changing national and international markets for crops and livestock products (LUCID, 2006).

Human and livestock populations within and outside the semi-arid rangelands in Kenya have greatly increased (Ngugi and Nyariki, 2003). Increased livestock numbers has resulted to heavy grazing which has altered vegetation composition and decreased primary productivity. This has led to the reduction of species diversity and increased exposure to bare ground, leading to increased runoff and increased erosion, which in turn has led to reduced water availability, nutrient retention and plant establishment. In Eastern and Southern Africa, many rangelands are now pockmarked by large bare areas with minimal organic matter and a smooth, sealed surface crust (Kinyua et al., 2010). Additionally, the rapid human population growth has put intense pressure on the semi-arid lands leading to the increased conversion of pastureland to cropland for subsistence crop production.

The destruction of pasture ecosystems and conversion to cropland can reduce soil productivity, because of increased erosion, decline in fertility, changes in aeration and moisture content, salinization or change in soil flora and fauna (Emadi et al., 2008). Land use induced changes in nutrient availability may influence secondary succession and biomass production (Foster et al., 2003) and reduce soil organic carbon (SOC) which plays a crucial role in sustaining soil quality, crop production and environmental quality. Such changes directly affect soil physical, chemical and biological properties such as soil water retention and availability, nutrient cycling, gas influx, plant root growth and soil conservation (Emadi et al., 2008).

Identifying and monitoring changes in soil quality is important in counteracting ecological degradation in the fragile semi-arid rangelands. The objective of this study was to establish the contribution of different land use types in a semi-arid rangeland in Kenya to chemical and physical soil properties.

MATERIALS AND METHODS

This research was conducted in the year 2008 in semi-arid Kibwezi district, Kenya. Kibwezi lies between latitudes 2º 6′S and 3ºS and longitude 37º 36′E and 38º 30′E respectively and has a total area of 3400 km^2 (Mwang'ombe et al., 2011). The climate is typical semi-arid characterised by low and unreliable supply of enough moisture for plant growth (Mganga et al., 2010). The average annual rainfall is 600 mm (Musimba et al., 2004). Kenya receives a bi-modal rainfall pattern with the long rains expected between April-May and short rains between November-December. Soils in the semi-arid environment in Kenya are considered problematic because of their physic-chemical properties that limit their use for cultivation of crops (Biamah, 2005). They generally have low organic matter contents and unstable structure. Problems associated with the soils are high levels of salinity and sodicity, poor drainage, soil erosion, soil compaction, soil crusting and low soil fertility. Such soils are generally very vulnerable to physical erosion, chemical and biological degradation (El Beltagy, 2002).

The largest community in the area are the Kamba who practice agro-pastoralism as their mainstream economic activity. Crops grown include a variety of drought tolerant grains like maize, sorghum, millet, beans and pigeon peas. Livestock kept consist of local breeds mainly the Small East African Shorthorn Zebu cattle, Red Maasai sheep and the Small East African Goat (Nyangito et al., 2009).

Soil samples were collected from the three land use types: cultivated land (CL), open grazing land (GL) and fallow land (FL). The three land use types were located in close proximity within the same physiographic unit and with similar slope and topography. Soil samples: disturbed and undisturbed, were taken from 15-30 cm at each of the three locations in each land use type. Eighteen sampling points were randomly selected and composite soil samples made for analysis.

Disturbed soil samples were taken to determine soil moisture, soil texture and soil nutrients. Soil chemical analysis for nitrogen, phosphorus, potassium, carbon and cation exchange capacity (CEC) were done using standard soil laboratory analysis procedures. Organic carbon (OC) was determined by the Walkley and Black method (1934). Total Nitrogen (TN) was determined with the Kjeldahl method (McGill and Figueiredo, 1993) available Phosphorus (P) was measured by the Olsen method (Olsen et al., 1954). These tests were done to establish and compare the nutrient contents of the sampled soils under the different land use systems.

Soil moisture content of the samples was determined by the gravimetric method (Rowell, 1994). Soil texture was determined following the hydrometer method as described by Gee and Baunder (1986). The fine fraction of soil passing through a 2 mm sieve was taken for texture analysis using the Buoyoucos hydrometer. The textural class was determined using the standard USDA Triangle (USDA, 1975). Undisturbed soil core samples were used to determine soil bulk density and saturated hydraulic conductivity. Bulk density was determined by the core method (Blake and Hartge, 1986). Constant head permeameter as described by Klute and Dirksen (1986) was used to determine saturated hydraulic conductivity (Ksat).

Soil physical and chemical properties in the three land use systems were compared using One-Way Analysis of Variance (ANOVA) and means separated using Tukey's-B. Mean comparison were performed at $p < 0.05$ level. The Statistical Package for Social Sciences (SPSS) (Einstein and Abernethy, 2000) computer programme was used to analyze data.

RESULTS

Results from this study showed that there was significant difference ($p<0.05$) in the cation exchange capacity (CEC) in the different land use types (Table 1).

Soils in the open grazing lands (GL) showed the highest cation exchange capacity (CEC) of 19.59 me/100g. Cultivated land and fallow land were ranked second and third with 13.88 me/100 g and 6.40 me/100 g respectively. Results also showed that the soils in all the land use types were deficient of nutrients mostly potassium and nitrogen. The most abundant nutrient in all the three sites occurrence was phosphorus. Grazing land (GL) recorded the highest amounts of phosphorus 15.38

Table 1. Soil chemical properties in the study sites.

Land Use	C (%)	N (%)	P (ppm)	CEC (me/100g)	K (me/100g)
Cultivated	1.56 ± 2.02	0.37^{ab} ± 0.06	13.90 ± 8.42	13.88^{a} ± 7.45	2.03 ± 0.31
Grazing Land	0.75 ± 0.12	0.32^{a} ± 0.04	15.38 ± 4.63	19.59^{a} ± 3.63	2.21 ± 0.78
Fallow	0.92 ± 2.20	0.43^{b} ± 0.04	13.72 ± 5.47	6.40^{b} ± 6.91	1.92 ± 0.45

Column means with different superscripts are significantly different at $p < 0.05$

ppm. Cultivated land (CL) and fallow land (FL) were ranked third and fourth with 13.90 and 13.72 ppm respectively. However, there was no significant difference in the amounts of phosphorus in all the three land use types. Results from this study also showed that there was no significant difference in amounts of potassium in the soils sampled from the three land use types.

Percentage of carbon and nitrogen were very low in all the three land use types. Fallow land (FL) had the highest percent of nitrogen (0.43%), followed by cultivated land (CL) and open grazing land (GL) with 0.37 and 0.32% respectively. Percentage of nitrogen in the open grazing land was significantly different ($p<0.05$) from percentage of nitrogen in the cultivated land and fallow land. However, percentage of nitrogen was not significantly different between the site under cultivation and fallow land. Soils under cultivation had the highest percent carbon (1.56%). Fallow land and grazing land were ranked second and third with 0.92 and 0.75% respectively. There was no significant difference in percent carbon in all the three land use types.

Saturated hydraulic conductivities in the three land use systems were significantly different ($p< 0.05$) (Table 2). Soils under fallow land, which had a sandy clay loam soil texture showed the highest rates of hydraulic conductivity. Cultivated land and grazing land which had the same soil texture of sandy clay were ranked second and third respectively. The percentage of soil moisture in all the three sites was generally low. There was significant difference ($p< 0.05$) in percentage of moisture content between soils sampled from fallow land and those sampled from grazing and cultivated land. Grazing land had the highest percent moisture followed by cultivated and fallow lands which were ranked second and third respectively. Bulk densities in the three land uses showed a no significant difference ($p> 0.05$). However, grazing land recorded the highest bulk density. Fallow land and cultivated land were ranked second and third respectively.

DISCUSSION

Cation exchange capacity (CEC) is a measure of the ability of a soil to retain cations, some of which are plant nutrients. Higher CEC values in cultivated and grazing lands than fallow land can be attributed to the soil texture. Sandy clay soils have a higher clay content compared to sandy clay loams. Soils with higher clay content tend to have higher CEC. These results compare with those of Githae et al. (2011) who found sandy clay loams to have a higher CEC of 18.78 meq/100 g compared to sandy loams and loamy sands with 18.73 and 12.75 meq/100g respectively. Soils that have a low CEC hold few cations and may require more frequent applications of fertilizers and amendments than soils that have a high CEC. These results suggest that cultivated and grazing land have greater water holding capacity thus a higher capacity to hold cations compared to fallow land.

Moreover, CEC in the soil is a function of organic matter. The continuous faecal droppings in the open grazing lands could have also contributed to a higher CEC values in the open grazing land as compared to cultivated land and fallow land. Addition of organic manure in the soil has residual effects on the field (Kihanda et al., 2007). According to Gachene and Kimaru (2003), the CEC measurements indicate overall assessment of the potential fertility of a soil and possible response to fertilizer application. Soils with a CEC of <16 meq/100 g are considered not to be fertile and such soils are highly weathered while fertile soils have a CEC of >24 meq/100 g. Results from this study suggest that soils in the fallow land (CEC 6.40 meq/100 g) and cultivated land (13.88 meq/100 g) are not fertile, whereas the fertility of the soils in the grazing land (CEC 19.59 me/100g) is average. The poor fertility in the fallow and cultivated lands can be attributed to the continuous mining and depletion of soil nutrients as a result of continuous cultivation of crops and soil erosion.

The difference in the saturated hydraulic conductivities (Ksat) in the different land use types can be explained by the difference in soil types and the texture of the soils. Higher saturated hydraulic conductivity in the fallow land can be attributed to its sandy clay loam texture. Saturated hydraulic conductivity is influenced by grain size, which is reflected in the texture of the soil. Soils from the fallow land had a higher percentage of sand, which has larger soil grains, thus higher hydraulic conductivity. The cultivated and open grazing lands had sandy clay textured soils. This explains the lower hydraulic conductivities due to a higher percentage of clay, which has smaller grains. This trend of results is comparable to those found by Clapp and Hornberger (1978) who found higher saturated hydraulic conductivities in sandy clay

Table 2. Soil physical properties in the study sites.

Land Use	K sat (cm/hr)	Texture	Moisture (%)	Bulk density (g/cm^3)
Cultivated	$2.8^{a} \pm 1.31$	Sandy Clay	$6.9^{a} \pm 0.88$	1.29 ± 0.09
Grazing land	$1.9^{ab} \pm 1.39$	Sandy Clay	$7.4^{a} \pm 1.69$	1.36 ± 0.11
Fallow land	$4.9^{b} \pm 3.64$	Sandy Clay Loam	$2.5^{b} \pm 0.82$	1.33 ± 0.07

Column means with different superscripts are significantly different at $p< 0.05$.

loam textured soils (Ksat 1.99×10^2 m/yr) compared to sandy clay textured soils (Ksat 6.84×10^1 m/yr).

Low levels of percentage of carbon in all the three land use types suggest low levels of organic matter content in the soils. The soil organic carbon content observed in the present study ranged from 0.75 to 1.56%. Results from a study conducted in a semi-arid environment in Kenya (Githae et al. 2011), the percent soil organic carbon ranged from 0.9 to 3.2%. The higher lower and upper limits in this study can be attributed to a higher litter fall from the *Acacia senegal* species, thus increase in levels of organic matter on the upper soil horizons. Organic matter is the reservoir that holds the nutrients which can be released to the soil. The low amounts of soil nutrients in all the soils can therefore be attributed to the low amounts of organic matter. Moreover, soils with low organic matter have low water holding capacity and poor structure making them prone to erosion and vulnerable to degradation. This premise is supported by the findings of El Beltagy (2002) who concluded that soils in the semi-arid environment are generally very vulnerable to degradation through physical erosion and to chemical and biological degradation. Frequent soil erosion through wind and water in form of run-off has also contributed to the low nutrient levels. Overgrazing by livestock which depletes vegetation cover is a major contributor to soil degradation in the semi-arid ecosystems. Reduced vegetation cover leads to increased runoff and increased erosion, which in turn leads to reduced water availability, nutrient retention and plant establishment (Jones and Elser, 2004).

Higher amounts of phosphorus in the open grazing land compared to cultivated and fallow lands can be attributed to the continuous addition of organic manure in form of faecal droppings by the free grazing livestock. A study by Kihanda et al. (2007) on the effects of manure application on crop yield and soil chemical properties in a long-term field trial in semi-arid Kenya showed an increase in phosphorus after continued application of goat manure. The sites under cultivation and fallow lands are often excluded from livestock interference, thus no addition of animal manure. Animal manure contains significant amounts of phosphorus in organic forms. Organic manure has residual effects on the field (Kihanda et al., 2007). The higher amounts of potassium in the cultivated and grazing land can also be attributed to higher clay mineral content in the soil. As the clays weather, the potassium ions sand-witched between the layers are released. Furthermore, the amount of this exchangeable potassium in a soil depends on the cation exchange capacity (CEC) of the soil. Low levels of nitrogen in all the land use types can be attributed to nitrogen losses due to number of factors notably; low levels of organic matter, runoff and soil erosion. Soil erosion is the most visible form of land degradation in the area.

Higher bulk density in the grazing land as compared to cultivated and fallow lands could be attributed to the long term soil compaction caused by the grazing cattle. These results are similar to those found by Nyangito et al. (2009) who found higher soil bulk densities in grazed lands compared to ungrazed lands, while investigating soil physical properties of various perennial grass swards in grazed and ungrazed sites in the semi-arid rangelands of Kibwezi district, Kenya. Once bare ground is exposed, grazing livestock can cause soil compaction and reduce soil aggregate stability (Kinyua et al., 2010). Eventually, such structural degradation can lead to a formation of a surface seal that further reduces infiltration and hinders seed germination (Beukes and Cowling, 2003). Trampling by grazing animals is a primary reason for lower water infiltration in heavily grazed land thus increased rates of erosion leading to loss in soil nutrients. Long-term trampling by grazing livestock, especially in the semi-arid rangelands often characterised by limited amounts of vegetation coverage, can create a hard-pan in the sub-soils of open grazing areas. Lack of enough vegetation cover and shortage of litter being deposited on the soil surface, not only makes the soils more vulnerable to wind and water erosion but also accelerates the formation of the hardpan layer (Chaichi et al., 2005).

Soil moisture content (SMC %) in the semi-arid rangelands is a function of soil type and rainfall. The differences in soil moisture contents in the three land use types could be attributed to percentage of clay content in the soils. Higher soil moisture content in the cultivated and grazing lands can be attributed to the sandy clay soil textures with a higher percentage of clay content compared to sandy clay loam textured soils in the fallow land. Munaifu and Kinyamario (2007) also found similar results while working in the same study area. Soil moisture content (SMC %) increased with an increase in percent clay in the soil. Additionally, the lower soil moisture content in the fallow land can partly be attributed to the high rates of hydraulic conductivity characteristic of the sandy clay loam soil textured soils, compared to the sandy clay soils in the cultivated and grazing lands. High rates of hydraulic conductivity facilitate the free movement of water beyond the sampling horizon.

Conclusion

The semi-arid rangelands in Kenya were predominantly used as pastureland for both domestic livestock and wildlife. However, as a result of the increased livestock and human populations in these semi-arid rangelands of Kenya in recent past, these ecosystems have experienced changes in land use. This has lead to increased pressure on the fragile ecosystems as a result of cultivation of crops mainly for subsistence and overgrazing by livestock. Our results suggest that the land use types currently practised in the semi-arid rangelands of Kenya influence soil physical and chemical properties.

ACKNOWLEDGEMENTS

The authors acknowledge the Department of Land Resource Management and Agricultural Technology, University of Nairobi for allowing us use their laboratories for soil analysis. We gratefully acknowledge the Agricultural Innovations for Drylands Africa (AIDA) Project (Grant Number 043863-SSA Africa-2006) for funding this research. The lead author would sincerely like to thank the Tropical Biology Association (TBA) for sponsoring and inviting him to participate in the TBA-NK-NMK-DRECA Specialist Training Workshop on Communicating and Disseminating Research Results, 2011.

REFERENCES

Beukes PC, Cowling RM (2003). Evaluation of restoration techniques for the succulent Karoo, South Africa. Restoration. Ecol., 11: 308-316.

Biamah EK (2005). Coping with drought: Options for soil and water management in semi-arid Kenya. Tropical Resource Management Papers, No. 58. Wageningen University and Research Centre (Wageningen UR).

Blake GR, Hartge KH (1986). Bulk density. In. Kiule, A. (Ed) Methods of soil analysis. Agron No.9. Amer. Soc. Agron. Inc. and Soil Sci. Amer. Inc. Madison, Wisconsin, U.S.A.

Chaichi MR, Saravi MM, Malekian A (2005). Effects of livestock trampling on soil physical properties and vegetation cover (case study: Lar Rangeland, Iran). Int. J. Agri. Biol., 7(6): 904-908.

Clapp RB, Hornberger GM (1978). Empirical equations for some soil hydraulic properties. Water Resources. Res., 14: 601-604.

Einstein G, Abernethy K (2000). Statistical package for the social sciences (SPSS) Version 12.0. Furman University.

El Beltagy A (2002). ICARDA' Experience in the Rehabilitation of Degraded Drylands of Central and Western Asia and Northern Africa: In the proceedings of the International Workshop on Combating Desertification. Rehabilitation of Degraded Drylands and Biosphere Reserves, Aleppo, Syria.

Emadi M, Emadi M, Baghernejad M, Fathi H, Mahboub S (2008). Effect of Land Use Change on Selected Soil Physical and Chemical Properties in North Highlands of Iran. J. Appl. Sci., 8(3): 496-502.

Foster D, Swanson F, Aber J, Burke I, Brokaw N, Tilman D, Knapp A (2003). The importance of land-use legacies to ecology and conservation. Bioscience, 53: 77-88.

Gachene CKK, Kimaru G (2003). Soil fertility and land productivity. A guide for extensive workers in the Eastern Africa region. Nairobi, Kenya, p. 146.

Gee GW, Bauder JW (1986). Particle size analysis. In: Kiule, A. (Ed.). Methods of soil analysis. Part 1. 2nd Edition. Agron. No.9. Amer. Soc. Agron. Inc. and Soil Sci. Amer. Inc. Madison, Wisconsin, U.S.A.

Githae EW, Gachene CKK, Njoka TJ (2011). Soil physicochemical properties under *Acacia senegal* varieties in the dryland areas of Kenya. Afr. J. Plant. Sci., 5(8): 475-482.

Jones FE, Esler KJ (2004). Relationship between soil-stored seed banks and degradation in eastern Nama Karoo rangelands (South Africa). Biodiversity and Cons. 13: 2027-2053.

Kihanda FM, Warren GP, Micheni AN (2007). Effects of manure application on crop yield and soil chemical properties in a long term field trial in semi-arid Kenya. Advances in Integrated Soil Fert. Mgt. in Sub-Saharan Africa: Challenges and Opportunities.pp. 471-485.

Kinyua D, McGeoch LE, Georgiadis N, Young TP (2010). Short-term and long-term effects of soil ripping, seeding and fertilization on the restoration of a tropical rangeland. Restoration. Ecol., 18: 226-233.

Klute A, Dirksen C (1986). Hydraulic conductivity and diffusivity: Laboratory methods, pp. 687-734. In: Methods of soil analysis. Part 1. 2nd Edition. Klute, A. (Ed.). Agron. Ser. No.9. Amer. Soc. Agron. Inc., Madison, Wisconsin.

LUCID (2006). Sustainable intensification of mixed crop-livestock systems. LUCID Working Policy Brief No. 1. Land Use Change, Impacts and Dynamics.

McGill WB, Figueiredo CT (1993). Total Nitrogen. In: Soil Sampling and Methods of Analysis, Carter, M.R. (Ed.). Canadian Society of Soil Science/Lewis Publishers, pp. 201-211.

Mganga KZ, Musimba NKR, Nyangito MM, Nyariki DM, Mwang'ombe AW (2010). Improving Hydrological Properties of Degraded Soils in Semi Arid Kenya. J. Environ. Sci. Tech., 4(3): 217-225.

Munaifu M, Kinyamario JI (2007). Soil nutrient content, soil moisture and yield of Katumani maize in a semi-arid area of Kenya. Afr. J. of Environ. Sci. Tech., 1(4): 081-085.

Musimba NKR, Nyariki DM, Ikutwa CN, Teka T (2004). Dryland Husbandry for sustainable development in the southern rangelands of Kenya. OSSREA, Addis Ababa.

Mwang'ombe AW, Ekaya WN, Muiru WM, Wasonga VO, Mnene WM, Mongare PN, Chege SW (2011). Livelihoods under climate variability and change: an analysis of the adaptive capacity of rural poor to water scarcity in Kenya's drylands. J. of Environ. Sci. and Tech., 4: 403-410.

Ngugi RK, Nyariki DM (2003). Rural livelihoods in the arid and semi-arid environment of Kenya: Sustainable alternatives and challenges. Agri. Hum., 22(1): 65-71.

Nyangito MM, Musimba NKR, Nyariki DM (2009). Hydrological Properties of Grazed Perennial Swards in Semiarid Southeastern Kenya. Afr. J. Environ. Sci. Tech., 3(2): 026-033.

Olsen SR, Cole CW, Watanabe FS, Dean LA (1954). Estimation of available phosphorus in soils by extraction with sodium bicarbonate, US Department of Agriculture, Circular 939.

Rowell DR (1994). Soil science methods and application. Longman Group, UK.

USDA (1975). Soil taxonomy. Agric. Handbook No. 436 Soil Conservation Service, USDA, Washington DC.

Walkley A, Black CA (1934). An examination of the Degtjareff method of determining soil organic matter and a proposed modification of the chronic acid titration method. Soil. Sci., 37: 29-38.

Wessels KJ, Prince SD, Carroll M, Malherbe J (2007). Relevance of rangeland degradation in semi-arid northeartern South Africa to the non-equilibrium theory. Ecol. Applications., 17: 815-827.

Status and distribution of extractable micronutrients in Haplustults in Yamaltu-Deba Local Government Area, Gombe state, Nigeria

S. Mustapha[1], H. K. Mamman[1] and N. A. Abdulhamid[2]

[1]Crop Production Programme, Abubakar Tafawa Balewa University, Bauchi, Nigeria.
[2]Department of Agricultural Engineering, Faculty of Engineering, University of Maiduguri, Nigeria.

A study was conducted to determine the status and distribution of extractable Zn, Cu, Fe and Mn in Haplustults in Yamaltu-Deba Local Government Area (LGA) of Gombe State, Nigeria. Composite soil samples representative of the soils in the LGA were collected from 0 - 15 and 15 - 30 cm depths of five locations (Hinna, Kanti, Jangargari, Gidan Waya and Dadin Kowa) and analyzed. Results indicate that the soils were dominantly sandy loam, very strongly to moderately acidic (pH = 4.7 - 5.9) and low in CEC (mean = 7.21 cmolkg^{-1}), exchangeable bases (Means for Ca = 3.34, Mg = 0.71, K = 0.21, Na = 0.10 cmolkg^{-1}) and organic carbon (mean = 7.27 g kg^{-1}). Extractable Zn, Cu, Fe and Mn (in mgkg^{-1}) were 0.48 - 0.75 (mean = 0.58), 0.18 - 0.26 (mean = 0.21), 18.40 - 21.91 (mean = 19.96), and 30. 54 - 38.58 (mean = 33.00), respectively. The micronutrients were neither significantly (P = 0.05) influenced by the depths nor locations. The Fe and Mn contents were above the critical limits for crop production in all the locations. Copper was low in soils from Hinna and Jangargari and will consequently benefit from its application. Zinc was, however, generally low and its application in all the locations are recommended for successful crop production in the area.

Key words: Haplustults, micronutrients, distribution.

INTRODUCTION

Micronutrients refer to a number of metals and their ions which are mostly of high density (usually > 5.0 mgm^{-3}) and belong largely to the group of "transition elements' of the periodic table. They are usually of relatively low abundance in soils but play key roles in the growth and development of crop plants, hence, essential for plant growth.

In Nigeria, micronutrient deficiencies were quiet rare, owing in part to the extensive system of agriculture practiced that permitted the recuperation of soils; hence, replenishing its macro- and micro-nutrients that were hitherto lost. The increasing human and animal population in Nigeria in general and in Gombe State in particular, and the country's drive to attain food security, has necessitated the abandonment of the traditional extensive agricultural system, to a more scientific intensive one. This, coupled with the use of new high yielding crop varieties which are nutrient demanding, and the realization of the concept of balanced nutrition by farmers have unraveled micronutrient deficiencies in some Nigeria Savanna soils (Mustapha and Loks, 2005).

The drive towards self sufficiency in food production through the adoption of more scientific intensive agricultural systems has necessitated the evaluation of the nutrient status of soils; most especially the micronutrients which had hitherto been neglected (Mustapha, 2003). This study was conducted to evaluate the status and distribution of zinc, copper, iron and manganese in the Haplustults of Yamaltu-Deba LGA in Gombe State, Nigeria.

MATERIALS AND METHODS

The study area

The study was conducted in Yamaltu-Deba LGA, Gombe State, Nigeria, situated about 609.5 m above the sea level. It is in the

*Corresponding author. E-mail: shafiumusty@yahoo.com.

Table 1. Distribution of particle-size fractions, pH and Organic Carbon in some soils in Yamaltu-Deba LGA, Gombe State.

	Sand (%)	Silt (%)	Clay (%)	Texture*(%)	pH (in water)	Org. C. (gkg^{-1})
Location						
Hinna	40.9	27.4	31.7	CL	5.18	10.18
Kanti	68.2	13.4	18.3	SL	4.77	7.11
Jangargari	59.6	23.6	16.2	SL	5.93	7.11
Gidan Waya	59.6	22.1	18.3	SL	5.10	5.23
Dadin Kowa	56.9	18.8	24.3	SCL	5.75	6.55
Mean	57.0	21.1	21.9		5.5	7.27
LSD (P=0.5)	21.6	11.47	13.52		0.89	2.84
Depth (cm)						
0 – 15	57.3	22.6	20.1	SL	5.27	7.81
15 - 30	57.0	17.5	23.5	SCL	5.42	6.73
Mean	57.2	21.0	21.8		5.35	7.27
SE ±	3.88	2.06	2.43		0.15	0.89

Texture: CL = Clay loam, SL = Sandy loam, SCL = Sandy clay loam.

northern guinea savanna agro-ecological zone of Nigeria (BSADP, 1982). Its geomorphology comprises of greatly undulating plains and pediments.

The climate is characterized by high temperatures and seasonal rainfall. The mean temperature ranges from 30 - 32°C. The rainfall pattern is unimodal, ranging from 700 - 1250 mm and is characterized by distinct dry (October - May) and rainy (June - September) seasons.

Soil sampling and handling

Six representative composite soil samples each were collected from two depths (0 - 15 and 15 – 30 cm) from five locations (Hinna, Kanti, Jangargari, Gidan waya, and Dadin kowa) in Yamaltu-Deba LGA; making a total of 60 samples. Each soil sample was a composite of five sub-samples. The collected soil samples were stored in properly labeled polythene bags and taken to the laboratory for analyses. In the laboratory, each sample was separately dried in air, ground using a porcelain pestle and mortar and passed through a 2 mm sieve. The sieved samples were used for all laboratory analyses.

Laboratory analyses

The processed soil samples were analyzed for some physico-chemical properties including the micronutrient (Zn, Cu, Fe and Mn) under investigation following procedures described by Page et al. (1982). Particle size distribution was determined by the hydrometer method after dispersing in sodium hexametaphosphate solution (Bouyoucos, 1951). The soil pH was determined in 1:1 soil/water suspension using a glass electrode pH meter while organic carbon in the soil was determined by the wet combustion method of Walkley and Black (1934). Cation exchange capacity was estimated using the NH_4OAc saturation (pH 7) method, while the leachate was used to determine the exchangeable bases. The extractable micronutrients: Zn, Cu, Fe and Mn were extracted with 0.1M HCl solution (Osiname et al., 1973) and determined on an atomic absorption spectrophotometer at appropriate wave lengths.

For the purpose of micronutrient fertility ratings, the limits given by Esu (1991) were employed. For Zn, values < 0.8, 0.81 – 2.0 and >2.0 $mgkg^{-1}$ were respectively rated 'low", 'medium' and 'high' while for Cu, the respective fertility category rating limits were < 0.2, 0.21 – 2.0 and > 2.0 $mgkg^{-1}$. Iron was regarded as 'low' if < 2.5, 'medium' if 2.51 – 5.0 and 'high' if > 5.0 $mgkg^{-1}$. Manganese was low if the values were < 1.0, 'medium' if 1.1 - 5.0, and 'high' if > 5.0 $mgkg^{-1}$.

Data analysis

Data obtained were subjected to statistical analysis using the analysis of variance (Harry and Steven, 1995). Means that were significantly different were separated using the least significant difference (LSD).

RESULTS AND DISCUSSION

The data on the physico–chemical properties of the soils studied are presented in Tables 1 and 2. Table 1 shows the particle-size distribution of the soils in the study area. The sand, silt and clay fractions ranged, respectively, from 40.9 - 68.2 (mean = 57.0), 13.4 - 27.4 (mean = 21.1), and 18.3 - 31.7(mean = 21.9)%, giving the soils a generally sandy loam to clay loam texture. The fractions varied significantly (P < 0.05) between locations but not with depths.

The soils were generally acidic with soil pH varying widely from 4.77 (very strongly acid) to 5.93 (moderately acid) between the locations. Organic carbon content was generally low (mean = 7.27 gkg^{-1}) except at Hinna (10.18 gkg^{-1}) where it was medium (Esu, 1991). Gidan Waya (5.23 gkg^{-1}) and Dadin Kowa (6.55) recorded the lowest soil organic carbon (matter) contents. Depths considered did not significantly (P = 0.05) influence the organic

Table 2. Distribution of exchangeable bases and CEC (in cmol (+) kg^{-1}) in some soils in Yamaltu–Deba Local Government Area Gombe State, Nigeria.

	Ca^{2+}	Mg^{2+}	K^{+}	Na^{+}	CEC
Location					
Hinna	3.17	0.79	0.22	0.07	6.74
Kanti	3.96	0.81	0.23	0.08	7.64
Jangargari	3.68	0.68	0.21	0.16	7.69
Gidan Waya	2.79	0.64	0.17	0.07	6.19
Dadin Kowa	3.08	0.62	0.22	0.12	7.79
Mean	3.34	0.71	0.21	0.10	7.21
LSD(P=0.05)	NS*	NS	NS	NS	NS
Depth (cm)					
0 – 15	3.71	0.64	0.19	0.08	7.54
15 - 30	2.95	0.77	0.23	0.13	6.88
LSD (P= 0.05)	NS	NS	0.04	NS	NS

*NS = Not significant.

Table 3. Distribution of some extractable micronutrients (in $mgkg^{-1}$) in soils of Yamaltu – Deba Local Government Area, Gombe State, Nigeria.

	Zn	Cu	Fe	Mn
Location				
Hinna	0.51	0.18	19.74	31.34
Kanti	0.45	0.21	21.91	38.58
Jangargari	0.75	0.19	20.42	30.54
Gidan Waya	0.48	0.21	19.31	31.62
Dadin Kowa	0.69	0.26	18.40	32.62
Mean	0.58	0.21	19.96	33.00
SE ±	0.15	0.06	3.51	4.51
LSD(p=0.05)	NS	NS	NS	NS
Depth (cm)				
0 – 15	0.54	0.18	22.12	33.62
15 - 30	0.61	0.24	17.80	32.38
SE ±	0.09	0.04	2.22	2.85
LSD(P=0.05)	NS	NS	NS	NS

carbon content.

The exchangeable bases were generally low in the soils (Table 2). Their distribution was neither influenced by the locations nor by the depths considered; except for K^{+} that was significantly ($P < 0.05$) higher in the lower 15 – 30 cm (0.23 cmol(+)kg^{-1}) than the upper 0 to 15 cm (0.19 cmol(+)kg^{-1}). The CEC was generally low (<10 cmol (+) kg^{-1}) probably indicative of the presence and dominance of 1:1 kaolinitic and/or Fe ad Al oxide clays.

Zinc status

Table 3 shows that Zn content in the soils ranged from 0.48 to 0.78 (mean = 0.58) $mgkg^{-1}$. This, in line with the ratings provided by Esu (1991), is "low". The values recorded for there soils fall below the 1.2 – 4. 0 $mgkg^{-1}$ obtained by Kparmwang and Malgwi (1997) for the soils in the Northern guinea Savanna of Nigeria, 0.81 to 1.34 (mean = 1.13) $mgkg^{-1}$ for Ustults in Galambi District in

Bauchi State, Nigeria (Mustapha and Singh, 2003) and 1.1 to 6.9 $mgkg^{-1}$ obtained for volcanic ash soils elsewhere in Tanzania (Kamasho and Singh, 1982). It is pertinent to note that the Zn values obtained in this study fall below the critical 0.90 $mgkg^{-1}$ given by Lombin (1983). It, therefore, follows that for successful and sustainable crop production in all the locations studied, Zn application will prove beneficial.

Copper status

With Cu ranging from 0.81 to 0.26 (mean = 0.21) $mgkg^{-1}$ (Table 3), it is rated, according to Esu (1991), as generally "medium". It is noted, however, that soils from the locations: Hinna (0.18 $mgkg^{-1}$) and Jangargari (0.19 $mgCukg^{-1}$) fall below the critical 0.2 $mgkg^{-1}$ suggested by Esu (1991) and Lindsay and Norvell (1978) for optimal crop growth. These soils, falling in the "low" fertility category may hence, benefit from Cu application.

It is noteworthy (Table 3) that soils in the upper 0 -15 cm contained Cu (0.18 $mgkg^{-1}$) below the critical value of 0. 20 $mgkg^{-1}$ while those at the lower 15 – 30 cm depth contained higher Cu contents (0.24 $mgkg^{-1}$). While advocating for Cu application, light turning of the soils could also prove beneficial for crop production in the study area.

Iron status

Table 3 shows that Fe in the soils studied ranged from 18.40 to 21.91 (mean = 19.96) $mgkg^{-1}$. These values are much higher that the critical 2.5 $mgkg^{-1}$ reported by Esu (1991), but similar to the 12.40 - 45.1 $mgkg^{-1}$ reported by Mustapha and Singh (2003) for Ustults in similar agro-ecology in Nigeria. It is, therefore, unlikely that Fe deficiency is experienced in these soils. This is true especially when viewed against the report (Chen and Barak, 1982; Sakal et al., 1984; Mengel and Geurtzen, 1986) that Fe deficiency is unlikely in acid soils; as Fe is known to be soluble under relatively acidic and reducing conditions (Chesworth, 1991).

The presence of high concentrations of Fe in soils could lead to its precipitation and accumulations and upon complex chemical reactions, lead to the formation of soft unindurated plinthite (laterite). This, upon alternate wetting and drying, could irreversibly form hard indurated material called petroplinthite (Ironstone) which would restrict rooting depth and drainage, amongst others.

Manganese status

Available Mn in the area studied ranged from 30.54 to 38.58 (mean = 33.00) $mgkg^{-1}$ (Table 3). It is rated "high" in its status (Esu, 1991). This implies that the soils contain Mn above the critical available range of 3 to 5 $mgkg^{-1}$ reported by Lindsay and Norvell (1978) and 1 - 5 $mgkg^{-1}$ reported by Esu (1991). The values obtained for these soils are above those obtained for some Ustults in Bauchi, Nigeria (7.89 – 12.00; mean = 9.10 $mgkg^{-1}$) but averages about the 19.0 – 69 .3 $mgkg^{-1}$ reported by Kparmwang (1996) in similar Nigerian soils. The high content of available Mn in the soils may be related to the acidic nature of the soils. It has been reported (Sillanpaa, 1982) that above soil pH of 7.5, the availability of Mn is very low because of the formation of hydroxides and carbonates.

Concluding remarks

The results of the present study have indicated that soils in the study area ranged from moderately to extremely acidic (pH range = 4.7 – 5.9; mean = 5.4), low in CEC (mean = 7.2 cmol kg^{-1}) and exchangeable bases. The available Fe and Mn in the soils studied were found to be above the critical ranges suggested by Esu (1991). This means that deficiency symptoms of these plant nutrients on crops grown on these soils is unlikely. For Cu (except at Hinna and Jangargari that were low and will thus benefit from Cu application), the soils contained sufficient quantities of the plant nutrients and may not be supplemented. Zinc in the soils was low indicating that crops grown on these soils will benefit from its application.

REFERENCES

Bouyoucos CA (1951). A re-calibration of the hydrometer for making mechanical analysis of soils. Agron. J., 31: 510-513.

BSADP (1982). Bauchi State Agricultural Project. Zonal and Development Area Statistics, Bauchi, Nigeria. pp. 90-100.

Chen Y, Barak P (1982). Iron nutrition of plants in calcareous soils. Adv. Agron., 35: 217 -240.

Chesworth W (1991). Geochemistry of nutrients. In: J. J. Mortredt, L. M., Shuman and R. M. Welch (eds). Micronutrients in Agriculture, 2nd Edition Soil Science Society of America Inc., Madison, Wisconsin, USA. Pp. 427 -476.

Esu IE (1991). Detailed Soil Survey of NIHORT farm at Bunkure, Kano Sate, Nigeria. Institute for Agricultural Research, Ahmadu Bello University, Zaria, Nigeria. p. 32.

Harry F, Steven CA (1995). Statistics: Concepts and Applications. Cambridge University Press, Great Britain. p. 853.

Kamasho JA, Singh BR (1982). Available copper and zinc status of some Tanzanian volcanic ash soils: A case study. Pedologie., 32: 209-224.

Kparmwang T (1996). Inherent fertility status of upland and fadama soils in Bauchi State, Nigeria. Noma. 12: 1-7.

Kparmwang T, Malgwi WB (1997). Some available micronutrients in profiles of Ultisols and Entisols developed from sandstone in north-western Nigeria. In: B.R. Singh (Ed.). Management of Marginal Lands in Nigeria. Proceedings of the 23rd Annual Conference of Soil Science Society of Nigeria, UDU, Sokoto, Nigeria. pp. 245-255.

Lindsay WL, Norvell WA (1978). Development of DTPA soil test for zinc, iron, manganese and copper. Soil Sci. Soc. Am. J., 42: 421-428.

Lombin G (1983). Evaluating the micronutrient fertility of Nigeria's semi-arid savanna soils. 2. Zinc. Soil Sci., 136: 42-47.

Mengel K, Geurtzen G (1986). Iron chlorosis on calcareous soils. Alkaline nutritional conditions as the cause of chlorosis. J. Plant Nutr.

9: 161-173.
Mustapha S (2003). Fertility status and productivity constraints of the basement complex-derived Ustults in Bauchi State,Nigeria. J. Agric. Technol., 11: 11-19.
Mustapha, S. and Singh, B.R. (2003). Available zinc, copper, iron and Manganese status of the basement – complex rock–derived Ultisol in Bauchi State, Nigeria: A case study. Nig. J. Soil Res., 4: 35-40.
Mustapha S, Loks NA (2005). Distribution of available zinc, copper, iron and manganese in the fadama soils from two distinct agroecological zones in Bauchi State, Nigeria. J. Environ. Sci., 9(2): 22-28.
Osiname OA, Schults BT, Corey RB (1973). Soil tests for available copper and zinc in soils of Western Nigeria. J. Sci. food Agric., 24: 1341-1349.
Page ALP, Miller RH, Keeny DR (1982). Methods of Soil Analysis. Agron. ASA, Madison, Wisconsin, USA. 9: 2.
Sakal R, Singh BP, Singh AP (1984). Determination of threshold value of iron in soils and plants for the response of rice and lentil to iron application in calcareous soils. Plants and Soils., 82: 141-148.
Sillanpaa M (1982). Micronutrients and the nutrient status of soils. A global study FAO Soils Bulleting, 48, Rome, Italy.
Walkley A, Black CA (1934).Acid extractable Zn in soil in relation to the occurrence of Zn deficiency symptoms of corn: A method of analysis. Soil Sci. Soc. Am. Proceedings. 12: 143-148.

5

Effects of crop rotation on soil macronutrient content and pH in potato producing areas in Kenya: Case study of KARI Tigoni station

Jane Muthoni[1]* and Jackson N. Kabira[2]

[1]Kenya Agricultural Research Institute (KARI), National Potato Research Centre, Tigoni, Kenya.
[2]Kenya Agricultural Research Institute, National Potato Research Centre, Tigoni, P. O. Box 338 00217, Limuru, Kenya.

Proper crop rotation might help to conserve soil fertility among small scale potato farmers in Kenya. A study was carried out at KARI Tigoni station between April 2007 and May 2010 to evaluate the effect of crop rotation on soil pH and macro nutrient levels. Rotations involved maize, potatoes and cabbages. Results showed that, soils at KARI Tigoni station are generally acidic with pH ranging from 4.3 (extreme acidity) to 5.59 (medium acidity). Cultivation reduced pH further probably due to increased organic matter decomposition. Percent organic carbon was low and was significantly (P = 0.05) increased by the rotation. Percent total nitrogen was adequate for potato production. Phosphorus was generally low. Proper crop rotation should be employed which should include a legume. This study should be carried out over a longer period of time to get more comprehensive results.

Key words: Crop rotation, soil pH, soil macronutrients; potato production, KARI Tigoni.

INTRODUCTION

Crop rotations are effective in improving soil physical, chemical and biological characteristics (Verma and Shekhawat, 1991). The physical characteristics include increased soil aggregate stability, decreased crusting of soil surfaces, increased granular structure and friable consistence (Bullock, 1992). Rotations that include sod, pasture, or hay crops also help to decrease bulk soil density, which can greatly impede root growth and nutrient flow (Tian et al., 2000). The effect of crop rotations on soil nitrogen (N), phosphorous (P), potassium (K) and carbon (C) is very complex (Bullock, 1992). It has been reported that including deep-rooted cover crops in rotations helps to distribute phosphorous and potassium from deep within the soil profile to the soil surface, where plant roots have better access to them (Marschner, 1990; Clark et al., 1998).

A shallow-rooted crop like onions or carrots, may be followed by a deeper-rooted crop like maize to recover nutrients that were unused by the shallow feeders and may have leached by irrigation or rainfall to lower depths in the soil profile. Conversely, a deep-rooted heavy feeder may be followed by a shallow-rooted light feeder to scavenge nutrients that may remain after heavy applications of nutrients (Clark et al., 1998; Bullock, 1992). Examples of heavy feeders are maize, potatoes, vegetables and soybeans while light feeders include grass sods, legume sods, wheat, barley and oats (Clark et al., 1998). Table 1 shows the effective root zone depth (in inches) of some commonly grown crops.

In addition, legumes in crop rotations supply biologically fixed atmospheric nitrogen to the soil thus as a replacing or supplementing inorganic nitrogen fertilizer. The amount of nitrogen in legume cover crops varies among species, but legumes generally contribute 50 to 200 pounds of nitrogen per acre (Flint and Roberts, 1988; Clark et al., 1998). This nitrogen is mineralized over an extended period of time, so that any surplus of it does not readily run off into streams and underground water supplies. Researchers estimate that from 40 to 75 percent of the total nitrogen contained in a legume cover crop is available in the soil for subsequent crops,

*Corresponding author. E-mail: jayney480@yahoo.com.

Table 1. Effective root zone depth (in inches) of some commonly grown crops.

Field crops	Vegetable crops	Forage crops	Fruit crops
Barley, 24	Asparagus, 24	Alfalfa, 24	Apples, 24
Maize (field), 24	Beets, 12	Bluegrass, 18	Blueberries, 18
Cotton, 24	Broccoli, 12	Bromegrass, 24	Cane fruits and grapes, 18
Flax, 24	Cabbage, 12	Ladino clover, 18	Peaches, 18
Oats, 24	Cantaloupes, 18	Orchardgrass, 24	Pears, 18
Peanuts, 24	Carrots, 12	Red and sweet clovers, 24	Strawberries, 6
Rye, 24	Cauliflower, 12	Sudan grass, 24	
Sorghum, 24	Celery, 12	Ryegrass, 24	
Soybeans, 24	Corn (sweet), 24	Bermuda grass, 18	
Sunflower, 24	Cucumbers, 18	Tall fescue, 18	
Tobacco, 18	Kale, 18		
Wheat, 24	Lettuce, 6		
	Lima beans, 18		
	Onions (bunch), 6		
	Onions (dry),12		
	Peas, 18		
	Peppers, 18		
	Potatoes, 18		
	Radish, 6		
	Beans, 18		
	Spinach, 6		
	Squash, 18		
	Tomatoes, 18		
	Watermelons, 24		

Source: Flint and Roberts, 1988.

depending on environmental conditions (Clark et al., 1998; Utomo et al., 1992). Legume cover crops may contribute from 30 to 60 kgN/ha (Utomo et al., 1992) to 110 kgN/ha (Marschner, 1990; Tian et al., 2000) to the subsequent maize compared to continuously grown maize. Thus proper crop rotation especially with inclusion of a legume might help to conserve soil fertility in small scale farms managed by resource poor potato farmers in Kenya. Low soil fertility among small scale potato farmers in Kenya is mainly caused by continuous cultivation without a fallow period (Kiiya et al., 2006). This is worsened by inadequate crop rotation due to small farm sizes (Kaguongo et al., 2008).Crop rotations practiced in potato growing parts of Kenya mainly involve maize, cabbages and potatoes in that order or an intercrop of maize and beans or potatoes and beans (Kanyanjua and Agaya, 2006). Besides this rotation, fertilizers are usually applied below the recommended rate for potato production (90 kg N ha^{-1} + 230 kg P_2O_5 ha^{-1}) (Kaguongo et al., 2008).

Furthermore, application of organic matter to the fields is limited since crop residues are used as fodder; where cattle manure is used, the quantities applied are below the recommended rates and the quality is questionable Kanyanjua and Agaya, 2006). Nitrogen and phosphorus are the major nutrients limiting potato production in Kenya (Recke et al., 1997). Soil phosphorus in major potato growing parts of Kenya is as low as 2.9 ppm (modified Olsen) while total nitrogen may be lower than 0.15% (Recke et al., 1997). In addition to low soil fertility, high soil acidity limits potato production in Kenya. The high soil acidity is mainly caused by acidic parent rock and the situation maybe worsened by continuous application of Di-ammonium phosphate (DAP) ($18N:46P_20_5$) fertilizer (Kiiya et al., 2006). Most of potato growing areas in Kenya have a soil pH of less than 5.5 while values of 4 to 5 are most common (Recke et al., 1997; Kiiya et al., 2006). A pH of less than 5.5 severely limits availability of potassium, nitrogen, phosphorus, sulphur, calcium and magnesium, and excessive levels of aluminum, manganese, boron, iron copper and zinc (Recke et al., 1997; Kanyanjua and Agaya, 2006).

At the Kenya Agricultural Research Institute (KARI) Tigoni station, crop rotation involves potatoes, maize, and cabbages in any order and the fields are rarely fallow. In addition, DAP ($18N:46P_20_5$) is the only fertilizer used in all these crops. Cattle manure is not used because the quantities needed would be too high and are unavailable (Kaguongo et al., 2008). Continuous use of DAP might have partly contributed to high acidity at the station.

The situation might have been worsened by the fact that the soil type at KARI Tigoni is nitisol which is inherently acidic (Recke et al., 1997; Kanyanjua and Agaya, 2006; Jaetzold et al., 2006). Guided by this background, a study was initiated at KARI Tigoni station in 2007. The objective of the study was to evaluate the effect of crop rotation on soil pH and macro nutrient levels at KARI Tigoni.

MATERIALS AND METHODS

The KARI-Tigoni station is located 40 km North-west of Nairobi city center, at an altitude of 2131 masl, latitude of 1°15' S and longitude 23° 46' E (Jaetzold et al., 2006). The average annual rainfall is 1096mm with a bimodal distribution. Long rains season occurs between March and May while short rains season is between October and December (Jaetzold et al., 2006). The mean annual air temperature is 18℃ ranging between 12 and 24℃. The soil type is humic nitosol (alfisol) derived from quartz trachyte (Jaetzold et al., 2006). The soil is very deep and well drained with a pH range of 5.5 to 6.5. The soil is of medium inherent fertility with organic carbon content of 1.65%. Exchangeable bases of potassium, calcium and magnesium are moderate to high with available potassium being about 21.2 ppm (Jaetzold et al., 2006).

The experimental field at the station which is (approx. 15 hectares) was divided into 20 rotation blocks based on the crop rotations that were planted. Crop rotation program which is normally employed at KARI Tigoni was monitored for six consecutive seasons. This includes maize, potatoes and cabbages. The DAP ($18N:46P_2O_5$) was applied on these crops at the recommended rates. Long rains season crop was planted in April and harvested in August while short season crop was planted in October and harvested in February. The uncultivated blocks served as the control. The experiment ran from April 2007 to May 2010. Soils were sampled and analyzed for macronutrients and pH after every two seasons on March 2008, March 2009 and March 2010. In each block, 15 to 20 soil cores were taken depending on block size. These soil cores were then bulked, thoroughly mixed and a sample of about 0.25kg was taken for analysis to represent the block. Soil samples were taken from 0 to 25cm depth using a hand hoe. The samples were air dried and sieved with a 2mm sieve and then sent for laboratory analysis within the same day. Soil pH was determined in water (1:2.5 soil to water ratio) with the ELL glass-electrode pH meter (Maclean, 1965). Organic carbon (OC) was determined using dichromate wet oxidation method (Walkey and Black, 1934) as modified by Piper (1942). Total nitrogen (N) was determined using the macro Kjeldahl method as described by Jackson (1958). Available phosphorus (P) was determined using Bray II method as described by Bray and Kurtz (1945). Exchangeable acidity was determined by the titration method (Anderson and Ingram, 1993). Exchangeable cations (K and Ca) were extracted with ammonium acetate and then determined with a flame photometer (Anderson and Ingram, 1993).

Trends in change of pH and macronutrients were assessed between the two sampling interval i.e. 2008 to 2009 and 2009 to 2010. In all the rotation blocks that showed an increase in pH and macronutrient in the two sampling intervals, the % increase in pH and macronutrients was calculated and subjected to analysis of variance. Analysis of variance was done in a randomised complete block design where the two sampling intervals were the blocks and the rotation blocks were the treatments.

RESULTS AND DISCUSSION

Crop rotation pattern in the station for three years is shown in Table 2. This rotation did not include a legume. The soils are generally acidic (Table 3). This was expected since the inherent soils, the nitisols, are acidic. The pH ranged from 4.3 (extreme acidity) to 5.59 (medium acidity). There was no definite trend in change in pH over time although the uncultivated area (F9 and F10) had higher pH than the cultivated area (Table 3). Cultivation is likely to increase soil acidity due to increased oxidation of organic matter (Bullock, 1992). The rotation program as well the sampling intervals did not significantly ($P = 0.05$) increase soil pH (Table 5a) although F5A gave the highest percent increase (Table 5b). Percent organic carbon was low although there was a general increase over time (Table 3). Low organic carbon was probably due to continuous cropping coupled with removal of crop residues and no use of organic manures.

However, the general increase in organic carbon in this study was probably due to increase in organic matter, probably from increases in plant roots resulting from continuous application of DAP. The rotation program used as well as sampling time led to a significant ($P = 0.05$) increase in soil organic carbon (Table 6). Percent total nitrogen was adequate and there was a general increase over time (Table 4). The seemingly adequate and increasing levels of nitrogen were probably due to continuous use of DAP in high doses on short duration crops leading to build-up of residual effect. The % increase in nitrogen was significantly different ($P = 0.05$) for the various rotation blocks (Table 8). There was no definite trend in phosphorus changes over time although there was a general decline (Table 4). Where there was an increase, the percent increase was also not significant (Table 7a and b). This was unexpected because it was assumed that continuous application of DAP, as practiced in the experimental field, would lead to an accumulation of phosphorus. Maybe this will be achieved in the long run. There was a general increase in potassium and calcium over time (Table 4). Percent increase in potassium was not significant for the rotation blocks (Table 9). Percent increase in Calcium was significant for the rotation blocks (Table 10).

Nitrogen and phosphorus are the two nutrients limiting potato production in the Kenyan highlands (Recke et al., 1997). The rotations used in this study did not show any significant increase in either nitrogen or phosphorus. Maybe the study needs to be carried out over a longer period of time to come up with conclusive results.

RECOMMENDATIONS

Proper crop rotation should be employed which should include a legume. Less acidifying fertilizers such Single Superphosphate, Triple Super Phosphate or NPK should be used instead of DAP since the soil is inherently acidic. Well decomposed manure or compost should be applied. This study should be carried out over a longer period of

Table 2. Crop rotations at KARI Tigoni station.

Block	Crop rotation					
	Long rains crop 2007	Short rains crop 2007	Long rains crop 2008	Short rains crop 2008	Long rains crop 2009	Short rains crop 2009
F1A	Cabbages	Potatoes	Maize	Cabbages	Potatoes	Maize
F1(a)	Maize	Potatoes	Maize	Potatoes	Maize	Potatoes
F1B	Cabbages	Potatoes	Maize	Cabbages	Potatoes	Maize
F1(b)	Potatoes	Maize	Maize	Potatoes	Maize	Potatoes
F1C	Cabbages	Potatoes	Potatoes	Cabbages	Potatoes	Maize
F1c	Fallow	Fallow	Fallow	Fallow	Potatoes	MAIZC Cabbages
F2	Fallow	Potatoes	Maize	Potatoes	Maize	Maize
F3	Potatoes	Maize	Fallow	Cabbages	Fallow	Potatoes
F4	Cabbages	Fallow	Potatoes	Flowers	Flowers	Flowers
F5A	Flowers	Flowers	Flowers	Flowers	Flowers	Flowers
F5B	Flowers	Flowers	Flowers	Flowers	Flowers	Flowers
F5C	Flowers	Flowers	Flowers	Flowers	Flowers	Flowers
F6A	Flowers	Flowers	Cabbages	Cabbages	Potatoes	Maize
F6B	Cabbages	Potatoes	Cabbages	Cabbages	Potatoes	Maize
F7A	Cabbages	Potatoes	Potatoes	Cabbages	Potatoes	Potatoes
F7B	Cabbages	Potatoes	Potatoes	Potatoes	Fallow	Potatoes
F7C	Potatoes	Fallow	Potatoes	Fallow	Fallow	Potatoes
F8	Fallow	Fallow	Fallow	Potatoes	Maize	Maize
F9	Uncultivated	Uncultivated	Uncultivated	Uncultivated	Uncultivated	Uncultivated
F10	Uncultivated	Uncultivated	Uncultivated	Uncultivated	Uncultivated	Uncultivated

Table 3. Soil pH and % organic carbon concentrations at KARI Tigoni station.

Blocks	Soil pH				% organic carbon			
	2008	2009	2010	Direction of change	2008	2009	2010	Direction of change
F1A	4.81	4.76	4.64	Decrease	1.12	1.28	1.5	Increase
F1(a)	4.82	4.86	4.89	Increase	1.10	1.28	1.51	Increase
F1B	4.71	4.67	4.41	Decrease	1.08	1.27	1.48	Increase
F1(b)	4.71	4.67	4.3	Decrease	1.09	1.27	1.51	Increase
F1C	4.51	4.41	4.73	No trend	0.89	1.17	1.47	Increase
F1c	4.48	4.44	4.41	Decrease	0.92	1.26	1.45	Increase
F2	4.71	4.76	4.9	Increase	1.08	1.29	1.47	Increase
F3	5.77	5.72	5.59	Decrease	0.86	1.27	1.66	Increase
F4	4.65	4.81	5.01	Increase	0.76	1.12	1.59	Increase
F5A	4.72	4.81	5.54	Increase	0.98	1.31	1.6	Increase
F5B	4.84	4.81	4.78	Decrease	1.02	1.31	1.69	Increase
F5C	5.32	5.38	5.52	Increase	0.66	1.31	1.86	Increase
F6A	4.80	4.82	4.84	Increase	0.91	1.29	1.56	Increase
F6B	4.52	4.65	4.92	Increase	1.02	1.31	1.49	Increase
F7A	4.65	4.75	5.03	Increase	1.12	1.38	1.64	Increase
F7B	4.90	4.94	4.96	Increase	1.21	1.4	1.65	Increase
F7C	5.30	5.02	4.71	Decrease	1.23	1.49	1.65	Increase
F8	4.42	4.46	4.66	Increase	1.02	1.25	1.39	Increase
F9	5.71	5.65	5.6	Decrease	0.96	1.55	2.03	Increase
F10	6.22	6.1	5.82	Decrease	0.98	1.58	1.94	Increase

Table 4. Soil macronutrient concentration at KARI Tigoni station.

	Phosphorus (ppm)				Total N %				Potassium (me %)				Calcium (me %)			
Block	2008	2009	2010	Direction of change	2008	2009	2010	Direction of change	2008	2009	2010	Direction of change	2008	2009	2010	Direction of change
F1A	12	23	52	Increase	0.15	0.25	0.31	Increase	1.33	1.39	1.43	Increase	3.8	4.2	4.7	Increase
F1(a)	26	23	13	Decrease	0.15	0.25	0.32	Increase	1.19	1.39	1.52	Increase	4.3	4.2	3.7	Increase
F1B	29	27	24	Decrease	0.14	0.25	0.31	Increase	1.25	1.35	1.52	Increase	3.9	4.0	4.1	Increase
F1(b)	31	27	26	Decrease	0.22	0.25	0.28	Increase	1.15	1.35	1.62	Increase	4.2	4.0	3.7	Decrease
F1C	26	21	13	Decrease	0.33	0.36	0.38	Increase	1.09	1.29	1.68	Increase	3.2	3.2	4.9	Increase
F1c	25	19	11	Decrease	0.33	0.33	0.38	Increase	1.11	1.31	1.74	Increase	2.6	3.4	4.3	Increase
F2	31	22	11	Decrease	0.36	0.34	0.32	Decrease	1.03	1.23	1.46	Increase	2.9	3.4	4.7	Increase
F3	38	29	22	Decrease	0.33	0.33	0.35	Increase	1.19	1.59	2.06	Increase	6.3	6.6	6.9	Increase
F4	19	18	18	No trend	0.30	0.31	0.35	Increase	1.15	1.35	2.02	Increase	3.5	4.8	7.1	Increase
F5A	12	18	61	Increase	0.29	0.31	0.33	Increase	1.07	1.37	2.1	Increase	4.2	5.4	7.9	Increase
F5B	17	18	18	No trend	0.28	0.31	0.35	Increase	1.17	1.37	1.6	Increase	4.4	5.4	6.1	Increase
F5C	13	34	72	Increase	0.30	0.31	0.35	Increase	1.27	1.37	2.22	Increase	3.2	5.4	9.1	Increase
F6A	39	27	18	Decrease	0.30	0.31	0.31	Increase	1.03	1.03	1.2	Increase	4.2	4.0	4.7	No trend
F6B	37	22	13	Decrease	0.28	0.31	0.32	Increase	0.91	0.97	1.08	Increase	3.0	3.2	3.3	Increase
F7A	11	16	28	Increase	0.30	0.33	0.37	Increase	1.13	1.43	1.72	Increase	2.1	3.4	5.7	Increase
F7B	22	19	15	Decrease	0.32	0.35	0.37	Increase	1.19	1.49	1.82	Increase	4.3	4.8	5.1	Increase
F7C	13	17	21	Increase	0.30	0.32	0.33	Increase	1.29	1.59	1.84	Increase	5.2	5.5	6.3	Increase
F8	23	21	20	Decrease	0.30	0.31	0.31	Increase	1.21	1.21	1.12	Decrease	4.0	4.0	4.6	Increase
F9	11	16	24	Decrease	0.52	0.52	0.66	Increase	1.81	1.91	2.02	Increase	4.1	5.4	6.9	Increase
F10	17	23	31	Decrease	0.44	0.49	0.55	Increase	1.49	1.69	1.92	Increase	4.3	5.2	6.1	Increase

Table 5a. Analysis of variance for % pH increase.

Variate: % pH increase Source of variation	d.f.	s.s.	m.s.	v.r.	F pr.
Sampling intervals	1	36.367	36.367	4.51	0.063
Rotation blocks	9	108.092	12.010	1.49	0.281
Residual	9	72.605	8.067		
Total	19	217.064			

Table 5b. Table of means for % pH increase.

Grand mean 2.09							
Rotation blocks	F1 (a)	F4	F22	F5A	F5C	F6A	F6B
	0.72	3.80	2.00	8.54	1.87	0.42	4.34
Rotation blocks	F7A	F7B		F8			
	4.02	0.61		2.69			

Table 6. Analysis of variance for % increase in % organic carbon content.

Variate: % increase in % carbon					
Source of variation	**d.f.**	**s.s.**	**m.s.**	**v.r.**	**F pr.**
Sampling intervals	1	1508.0	1508.0	12.72	0.002
Rotation blocks	19	7686.3	404.5	3.41	0.005
Residual	19	2252.4	118.5		
Total	39	11446.7			

Table 7a. Analysis of variance in % increase in phosphorus.

Variate: % increase in phosphorus					
Source of variation	**d.f.**	**s.s.**	**m.s.**	**v.r.**	**F pr.**
Sampling intervals	1	2922	2922	1.18	0.31
Rotation blocks	7	36502	5215	2.10	0.174
Residual	7	17355	2479		
Total	15	56779			

Table 7b. Tables of means in % increase in phosphorus.

Grand mean 72							
Rotation blocks	F1A	F5A	F5B	F5C	F7A	F7C	F9
	109	144	14	137	60	27	48
Rotation blocks	F10						
	35						

Table 8. Analysis of variance in % increase in nitrogen.

Variate: % increase in nitrogen					
Source of variation	**d.f.**	**s.s.**	**m.s.**	**v.r.**	**F pr.**
Sampling intervals	1	284.4	284.4	1.55	0.23
Rotation blocks	18	8413.8	467.4	2.54	0.027
Residual	18	3306.5	183.7		
Total	37	12004.7			

Table 9. Analysis of variance for % potassium increase.

Variate: % increase in nitrogen					
Source of variation	**d.f.**	**s.s.**	**m.s.**	**v.r.**	**F pr.**
Sampling intervals	1	398.9	398.9	3.07	0.098
Rotation blocks	17	3661.5	215.4	1.66	0.153
Residual	17	2206.2	129.8		
Total	35	6266.5			

Table 10. Analysis of variance for % increase in Calcium.

Variate: % increase in nitrogen					
Source of variation	**d.f.**	**s.s.**	**m.s.**	**v.r.**	**F pr.**
Sampling intervals	1	39.31	39.31	1.01	0.33
Rotation blocks	14	12134.18	866.73	22.30	<.001
Residual	14	544.11	38.86		
Total	29	12717.60			

time to get more comprehensive results.

REFERENCES

Anderson JM, Ingram JSI (1993). Tropical Soil Biology and Fertility: a Handbook of Methods. CAB International, Wallingford, UK.

Bray RH, Kurtz LT (1945). Determination of total organic and available forms of phosphorus in soil. Soil Sci., 59: 39-45.

Bullock DG (1992). Crop rotation. Critical Rev. Plant Sci., 11: 309-326.

Clark MS, Horwath WR, Shennan C, Scow KM (1998). Changes in soil chemical properties resulting from organic and low-input farming practices. Agro. J., 90: 662-671.

Flint ML, Roberts PA (1988). Using crop diversity to manage pest problems: Some California examples. Am. J. Alt. Agric., 3: 163-167.

Jackson ML (1958). Soil Chemical Analysis. Eagle Wood Cliff. N. Y. p. 498.

Jaetzold R, Schmidt H, Hornetz B, Shisanya C (2006). Farm Management Handbook of Kenya. Vol. II. Natural Conditions and Farm Management Information. 2nd Edition. Part B. Central Kenya. Subpart B2. Ministry of Agriculture, Nairobi, Kenya.

Maclean IO (1965). Aluminum. In: Black LA (ed.) Method of Soil Analysis. (Part II) Am. Soc. Agron. Madison, NI, pp. 976-979.

Kaguongo WP, Gildemacher P, Demo P, Wagoire W, Kinyae P, Andrade J, Forbes G, Fuglie K, Thiele G (2008). Farmer practices and adoption of improved potato varieties in Kenya and Uganda. Social Sciences Working Paper 2008-5. International Potato Center (CIP), Lima, Peru. ISSN 0256-8748.

Kanyanjua SM, Agaya GO (2006). A guide to choice of mineral fertilizers in Kenya. KARI Technical Note No. 17. KARI, Nairobi, Kenya. ISBN: 9966-879-66-8.

Kiiya WW, Mureithi JG, Kiama JM (2006). Improving production of irish potato (Solanum tuberosum, L.) in Kenya: The use of green manure legumes for soil fertility Improvement. *In* J. G. Mureithi et al. (ed.) Development and up scaling of green manure legumes technologies in Kenya. KARI, Nairobi, Kenya.

Marschner H (1990). Mineral Nutrition of higher plants.4th edition. Academic Press Limited, London 1986.

Piper CS (1942). Soil and Plant Analysis. Int. Sci. Publ. Inc. N. Y., p. 368.

Recke H, Schnier HF, Nabwile S, Qureshi JN (1997). Responses of Irish potatoes (*Solanum tuberosum* L.) to mineral and Organic fertilizer in various agro-ecological environments in Kenya. Exp. Agric., 33: 91-102.

Tian G, Kolawole GO, Kang BT, Kirchhof G (2000). Nitrogen fertilizer placement indexes of legume cover crops in the derived Savanna of West Africa. Plant Soil, 224: 287-296.

Utomo VH, Sitompul SM, Noordwijk MV (1992). Effects of leguminous cover crops on subsequent maize and soybean crop son an ultisol in Lampang. Agrivita, 15: 44-53.

Verma RK, Shekhawat GS (1991). Effect crop rotation and chemical soil treatment on bacterial on bacterial wilt of potato. Indian Phytopathol., 449: 5–8.

Walkey A, Black CA (1934). An examination of the Degtjareff method for determining soil organic carbon in soil and proposed modification of the chromic acid titration method. Soil Sci., 37: 29-38.

Effects of different sources of nitrogen on potato at Tigoni, Kenya

Jane Muthoni* and Jackson N. Kabira

Kenya Agricultural Research Institute (KARI), National Potato Research Centre, Tigoni, P. O. Box 338 00217, Limuru, Kenya.

A field experiment was set up to investigate the effects of different sources of nitrogen on potato at Tigoni in Kenya. The treatments consisted of ten fertilizer materials and two potato *Solanum tuberosum* varieties, namely, 'Tigoni' and 'Asante'. The experiment was conducted for two consecutive seasons: October, 2008 to February, 2009 (first season) and April, 2009 to August, 2009, (second season). The experiment was a factorial laid in randomized complete block design with three replications. The potato yields were high with variety 'Tigoni' giving an average of 81.0 tons/ha in the first season and 86.8 tons/ha in the second season. Variety 'Asante'yielded an average of 59.5 tons/ha in the first season and 62.1 tons/ha in the second season. The fertilizer materials, the varieties, and the interaction between fertilizers and varieties were all significant (P = 0.05) in both seasons. The experiment need to be repeated on farmers' fields where the masking effects of the residual fertilizers can be eliminated. In addition, the cost of different fertilizer combinations needs to be evaluated so as to make the most profitable recommendations to potato growers in Kenya.

Key words: Asante, Kenya, nitrogen sources.

INTRODUCTION

Potato is the second most important food crop after maize in Kenya (MoA/GTZ, 1998; CIP, 2006; MoA, 2008). It is grown by about 500,000 farmers, cultivating 120,000 hectares per season with an annual production of about 1 million tonnes in two growing seasons (MoA, 2008). Potato in Kenya is produced both as a subsistence and commercial crop, mainly by small scale farmers (MoA, 2005). Yields on farmers' fields vary from 4.4 to 15 t ha^{-1} with an average of 6.7 t ha^{-1}. However, yields of 40 t ha^{-1} had been attained from research fields (Lung'aho et al., 1997; MoA, 2008). Potato yields in the country are generally low (Kaguongo et al., 2008); Muthoni et al., 2010). One of the contributing factors to low potato yields is low soil fertility, occasioned by continuous cultivation without adequate replenishment of mined nutrients (Kaguongo et al., 2008. In addition, fertilizers are usually applied below the recommended rate (90 kg N ha^{-1} + 230 kg P_2O_5 ha^{-1}) for potato production in Kenya (Kaguongo et al., 2008).

Furthermore, application of organic matter to the fields is limited. This is because crop residues are used as fodder, and cattle manure when used, the quantity and quality are often below the recommended rates (Muthoni et al., 2010). The most limiting soil nutrients to potato production in Kenya are nitrogen and phosphorus (Recke, 1997).

Against this background, an experiment was set up whose objective was to investigate the effects of different sources of nitrogen on potato.

MATERIALS AND METHODS

The experiment was conducted at Kenya Agricultural Research Institute (KARI), National Potato Research Centre, Tigoni. It was conducted for two consecutive seasons: October, 2008 to February, 2009 (first season) and April, 2009 to August, 2009 (second season). The KARI-Tigoni station is located 40 km North-west of Nairobi city centre, at an altitude of 2131 masl, latitude of 1°15' S and longitude 23° 46' E (Jaetzold et al., 2006). The average annual

*Corresponding author. E-mail: jayney480@yahoo.com

Table 1. Fertilizer treatments.

No.	Fertilizer materials	Application rate/ plot
1	Di ammonium phosphate (DAP) (18:46:0)	600 g
2	DAP + Manure	300 g + 3 kg
3	Nitrogen Phosphorus Potassium (NPK) (20:20:20)	540 g
4	NPK + Manure	270 + 3 kg
5	Manure (1.8% N)	6 kg
6	Manure + Calcium ammonium nitrate (CAN) (26:0:0)	3 kg + 173 g
7	Triple super phosphate (TSP) (0:46:0) + CAN (26:0:0)	600 g + 346 g
8	NPK + TSP	540 g + 3 65 g
9	DAP (1/2 Rate)	300 g
10	Control	Nil

rainfall is 1096 mm with a bimodal distribution. Long rains season occurs between March and May while short rains season is between October and December (Jaetzold et al., 2006). The mean annual air temperature is 18°C and ranges between 12 and 24°C. Rainfall, temperature, and relative humidity during the study period were recorded. The soil type is humic nitosol (alfisol) derived from quartz trachyte (Jaetzold et al., 2006). The soil is very deep and well drained with a pH range of 4.3 to 5.82 (Muthoni and Kabira, 2010). The soil is of medium inherent fertility with organic carbon content of about 1.65%. Exchangeable bases of potassium, calcium and magnesium are moderately high with available potassium being about 1.7 % milliequivalent (Muthoni and Kabira, 2010). The treatment combinations consisted of ten fertilizer materials and two potato varieties, 'Asante' and 'Tigoni'. The manure was collected from open paddocks and was found to contain 1.8% nitrogen. The experiment was laid out in a factorial arrangement with potato varieties and fertilizer materials as factors giving a total of twenty treatment combinations. These treatment combinations were laid in a Randomized complete block design replicated three times. Blocking was introduced due to gradient /slope of the experimental site.

Each plot measured 4 m long x 3 m wide and contained four rows. Potatoes were planted at a spacing of 75 x 30 cm giving a total of 53 plants per plot. Fertilizer materials application rates were calculated on the basis of % nitrogen content so as to give the recommended nitrogen rate for potato production in Kenya which is 90 kgN/ha (Table 1). The half rate of Di-ammonium Phosphate (DAP) used was to simulate farmers' field practice. All the fertilizer materials were mixed with soil in the furrow during planting except CAN which was top dressed 21 days after planting. Plants were weeded and irrigated regularly as need arose. The plants were also sprayed against late blight using Ridomil MZ 68^{R} WP (Metalaxyl-M 40 g/kg + Mancozeb 640 g/kg) (Syngenta Crop Protection, AG basele, Switzerland) at 30 gm/15 L of water and against insect pests using Actellic Super (Syngenta Crop Protection, AG basele, Switzerland) at 20 mls/15 L of water. This was done three times as was dictated by the prevailing weather conditions and insect pest infestation.

Data on the number of leaves and primary branches per stem were obtained from five plants in the middle two rows. This was done at 60 days after planting when the varieties had reached 50% flowering. At maturity, tubers were harvested from six hills from the two middle rows per plot and weighed out. The yields were converted into tons/ha. Data were subjected to analysis of variance (ANOVA) using Genstat statistical package (Genstat, 2009). The treatment effects were tested for significance using F-test at 5%. Where the F-test showed significance effects, treatment means were separated using least significant difference (LSD) test at 5% (Gomez and Gomez, 1984).

RESULTS AND DISCUSSION

The mean air temperature was below 18°C for most of the experimental period (Table 2).

Tuber yields

There were significant differences (P = 0.05) in tuber yields between varieties in the first season (Table 3). The fertilizer effects were also significant. However, the interaction between the variety and the fertilizer was not significant (Table 3).

Variety 'Tigoni' gave an average yield of 81.0 tons/ha while 'Asante' yielded 59.5 tons/ha on average. Although DAP + Manure and DAP treatments were not significantly different (LSD = 0.05) from each other, DAP + Manure gave higher yields than all the other treatments in the first season (Table 4). During the second season, the fertilizer treatments had significantly different effects (P = 0.05) on tuber yields (Table 5). The same applied to potato varieties. However, the interaction between potato varieties and fertilizer treatments was not significant (Table 5).

Variety 'Tigoni' yielded 86.8 tons/ha on average while 'Asante' yielded 62.1 tons/ha on average. The overall mean yield was 74.4 tons/ha. The NPK (20:20:20) gave the highest yields (Table 6).

The yields recorded in this experiment were very high. This could have been due to the high fertility of the experimental site because the fields are normally fertilized each season, and there is a likelihood of residual fertilizer effects. The high yields could also have been due to other nutrients in the fertilizers in addition to nitrogen because the fertilizer materials were not straight fertilizers. In addition, the cool weather that prevailed

Table 2. Weather conditions at Tigoni during the study period.

Year	Month	No. of rainy days	Monthly rainfall (mm)	Temperature (°C)	RH
2008	January	11	80.3	16.11	67.81
2008	Feb	6	72.0	16.01	64.85
2008	March	15	304.1	16.10	69.53
2008	April	11	323.1	15.43	69.23
2008	May	7	51.3	15.47	67.23
2008	June	3	8.3	12.94	70.77
2008	July	12	60.5	12.56	77.10
2008	August	4	14.0	18.97	66.84
2008	September	6	79.8	15.50	68.87
2008	October	13	194.2	15.39	73.37
2008	November	8	69.9	15.88	69.93
2008	December	-	88.3	16.74	88.31
2009	January	3	21.7	16.80	64.80
2009	February	3	21.0	16.70	61.50
2009	March	3	61.9	18.10	65.20
2009	April	13	212.8	17.00	71.20
2009	May	17	180.1	15.40	72.90
2009	June	7	66.5	14.90	70.30
2009	July	11	46.2	14.00	74.70
2009	August	-	-	14.10	51.60
2009	September	3	37.8	16.20	67.97
2009	October	11	138.4	15.42	74.10
2009	November	11	111.1	15.90	72.30
2009	December	12	142.8	16.02	73.90

Table 3. Analysis of variance for total tuber yields (ton/ha) for the first season (October, 2008 to February, 2009).

Source of variation	df	ss	ms	vr	Fpr
Block stratum	2	4240.8	2120.4	4.13	
Block.*Units* stratum					
Variety	1	41456.4	41456.4	80.83	<.001
Fertilizer	9	24534.3	2726.0	5.32	<.001
Variety. Fertilizer	9	3282.4	364.7	0.71	0.699
Residual	338	173357.1	512.9		
Total	359	246871.0			

during the experimental period could have contributed to these high yields. Cool temperatures are known to enhance tuberization in potatoes and to lengthen the growing period hence more yields (Acquaah, 2007).

Number of leaves per stem

In the first season, the fertilizer materials, the varieties, and the interaction between fertilizers and varieties were all significant (P = 0.05) (Table 7). In NPK + TSP and DAP half rate, there were no significant (P = 0.05) differences in the number of leaves per stem between the two potato cultivars (Figure 1).

During the second season, the fertilizer materials, the varieties, and the interaction between fertilizers and varieties were all significant (P = 0.05) (Table 8).

The greatest differences in leaves per stem between the two potato cultivars were found in NPK + Manure, Manure + CAN, and TSP + CAN fertilizers (Figure 2). In both seasons, variety 'Tigoni' had significantly (P = 0.05) higher number of leaves per stem than 'Asante' (Table 9).

The differences between cultivars 'Tigoni' and 'Asante' in terms of leaves per stem could be due to genetic differences between them. Generally, cultivar 'Tigoni' is taller and has more foliage than cultivar 'Asante'.

Table 4. Effects of fertilizers on mean tuber yield for the first season (October, 2008 to February, 2009).

Fertilizer	Mean tuber yields (tons/ha)
DAP + Manure	83.7a
DAP(18:46:0)	78.0ab
NPK (20:20:20)	77.0ab
TSP(0:46:0) + CAN (26:0:0)	75.8ab
NPK + Manure	68.7bc
Manure	68.3bc
Manure + CAN (26:0:0)	66.4bc
NPK + TSP	70.0bc
DAP(1/2 Rate)	59.4cd
Control	55.2d
Mean yield	70.3
% CV	6.0
LSD	10.5

Within a column, values followed by the same letter are not significantly different at LSD = 0.05.

Table 5. Analysis of variance for tuber yields during the second season (April, 2009 to August, 2009).

Source of variation	df	ss	ms	vr	Fpr
Block stratum	2	6036.1	3018.0	8.28	
Block.*Units* stratum					
Variety	1	54935.5	54935.5	150.75	<.001
Fertilizer	9	117673.9	1963.8	5.39	<.001
Variety. Fertilizer	9	4602.1	511.3	1.40	0.185
Residual	338	123174.1	364.4		
Total	359	206421.7			

Table 6. Effects of fertilizer on mean tuber yield for the second season (April, 2009 to August, 2009).

Fertilizer	Mean tuber yields (tons/ha)
NPK (20:20:20)	85.9 a
TSP (0:46:0) + CAN (26:0:0)	83.0 a
DAP + Manure	81.2 ab
DAP (18:46:0)	79.8 abc
NPK (20:20:20) + TSP (0:46:0)	72.5 bcd
NPK + Manure	70.8 cd
Manure	67.4 d
Manure + CAN (26:0:0)	68.9 d
DAP (1/2 rate)	69.4 d
Control	65.2 d
Mean yield	74.4
% CV	6.7
LSD	8.85

Within a column, values followed by the same letter are not significantly different from each other at LSD = 0.05.

Table 7. Analysis of variance. Leaves per stem for the first season (October, 2008 to February, 2009).

Source	df	ss	ms	vr	Fpr
Block	2	927.05	463.52	38.78	
Fertilizer	9	317.85	35.32	2.96	0.002
Variety	1	1795.85	1795.85	150.27	<0.001
Fertilizer x variety	9	406.41	45.16	3.78	<0.001
Residual	278	3322.42	11.95		
Total	299	6769.59			

LSD for interaction 2.485.

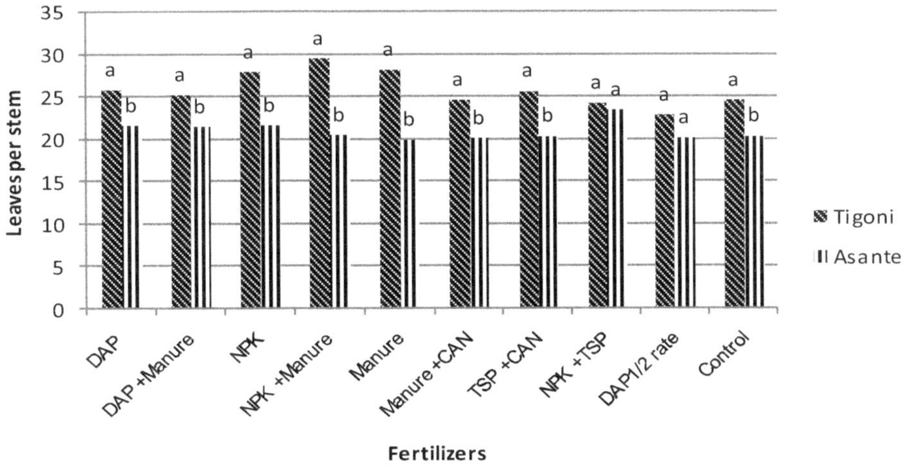

Figure 1. Leaves per stem during the first season.

Table 8. Analysis of variance: Leaves per stem during the second season (April, 2009 to August, 2009).

Source	df	ss	ms	vr	Fpr
Block	2	981.087	490.543	60.99	
Fertilizer	9	1726.350	191.817	23.85	<.001
Variety	1	2964.163	2964.163	368.56	<.001
Fertilizer x variety	9	457.470	50.830	6.32	<.001
Residual	278	2235.847	8.043		
Total	299	8364.917			

LSD for interaction 2.039.

Differences in the effects of the fertilizer materials could have been due to other nutrients in the fertilizers in addition to nitrogen because the fertilizer materials were not straight fertilizers. Nitrogen in the presence of adequate phosphorus and potassium stimulates canopy growth, leaves and branches. This is through production of extra leaves and branches, extension of leaf area duration and expansion of leaf area. Nitrogen needs to be present from emergence to flowering to promote rapid canopy growth (White et al., 2007).

Number of primary branches per stem

There were no significant differences among the fertilizer treatments on the number of primary branches per stem during the first season (Table 10). However, varieties showed significant differences (P = 0.05) while the interaction between fertilizer and varieties was not significant.

Generally, variety 'Tigoni' had significantly higher number of primary branches than 'Asante' (Table 11).

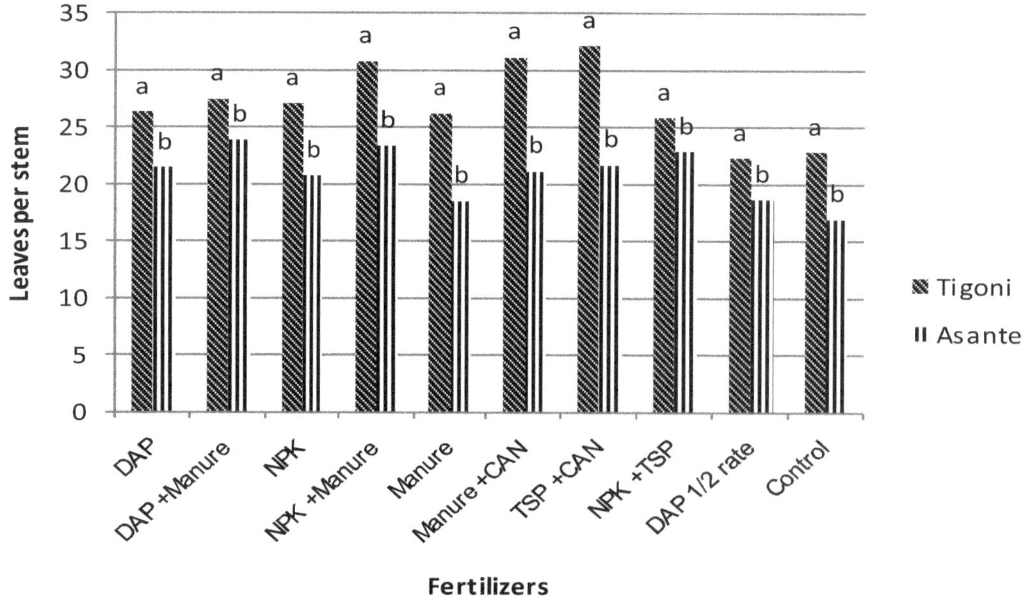

Figure 2. Leaves per stem during the second season.

Table 9. Mean leaves per stem for the two potato cultivars in both seasons.

Cultivar	Season 1	Season 2
Tigoni	25.72	27.23
Asante	20.83	20.94
LSD (0.05)	0.786	0.645

Table 10. Analysis of variance: Primary branches per stem during the first season (October, 2008 to February, 2009).

Source	df	ss	ss	vr	Fpr
Blocks	2	30.780	15.390	2.34	
Fertilizer	9	66.733	7.415	1.13	0.342
Variety	1	225.333	225.333	34.32	<.001
Fertilizer x variety	9	32.933	3.659	0.56	0.831
Residual	278	1825.220	6.566		
Total	299	2181.000			

Table 11. Varietal differences in primary branches per stem (first season).

Variety	Tigoni	Asante	LSD (0.05)
Branches/stem	14.37	12.63	0.582

There were significant differences among the fertilizer treatments on the number of primary branches per stem during the second season (Table 12). Varieties were also significantly different (P = 0.05). However, the interaction between fertilizer and varieties was not significant. Generally, variety 'Tigoni' had significantly higher number of branches than 'Asante' (Table 13).

The NPK + Manure treatment gave the highest number of branches per stem, although there were no significant differences between this treatment and DAP + Manure

Table 12. Analysis of variance: Branches per stem during the second season (April, 2009 to August, 2009).

Source	df	ss	ms	vr	Fpr
Block	2	63.447	31.723	8.88	
Fertilizer	9	242.187	26.910	7.53	<.001
variety	1	507.000	507.000	141.89	<.001
Fertilizer x variety	9	50.600	5.622	1.57	0.123
Residual	278	993.353	3.573		
Total	299	1856.587			

Table 13. Varietal differences in branches per stem during the second season (April, 2009 to August, 2009).

Variety	Tigoni	Asante	LSD (0.05)
Branches/stem	15.41	12.81	0.43

Table 14. Effects of fertilizer materials of branches per stem during the second season (April, 2009 to August, 2009).

Fertilizer materials	Mean branches per stem
NPK + Manure	15.17 a
DAP + Manure	15.13 ab
DAP	15.13 ab
NKP + TSP	14.60 abc
TSP + CAN	14.17 abcd
NPK	14.07 bcd
Manure + CAN	13.67 cd
Manure	13.57 cd
DAP(1/2 Rate)	13.33 d
Control	12.23 e
Grand mean	14.11
%CV	13.4
LSD (0.05)	0.961

and DAP alone (Table 14).

The differences between cultivars 'Tigoni' and 'Asante' in terms of number of primary branches per stem could be due to genetic differences between them. Generally, cultivar 'Tigoni' is taller and has more foliage than cultivar 'Asante'. Differences in the effects of the fertilizer materials could have been due to other nutrients in the fertilizers in addition to nitrogen because the fertilizer materials were not straight fertilizers.

ACKNOWLEDGEMENT

The authors acknowledge the assistance of technicians who assisted in field work. The authors are grateful to KARI seed Unit, Tigoni, who funded the study.

RECOMMENDATIONS

The experiment need to be repeated on farmers' fields where the masking effects of the residual fertilizers can be eliminated. In addition, the cost of different fertilizer combinations needs to be evaluated so as to make the most profitable recommendations to potato growers in Kenya.

REFERENCES

Acquaah G (2007). Principles of plant genetics and breeding. Blackwell Publishing Limited, Malden, MA. p. 538.

Centro Internacional de la Papa (CIP) (2006). World potato atlas [Online]. Available at https://research.cip.cgiar.org/confluence/display/wpa/Kenya. (Accessed on 25 March 2010). CIP, Lima, Peru.

GenStat (2009). GenStat for Windows. Lawes Agricultural Trust, Rothamsted Experimental Station, Harpenden, UK.

Jaetzold R, Schmidt H, Hornetz B, Shisanya C (2006). Farm Management Handbook of Kenya. Vol. II. Natural Conditions and Farm Management Information. 2nd Edition. Part B. Central Kenya. Subpart B2. Ministry of Agriculture, Nairobi, Kenya. pp. 61-65.

Kaguongo WP, Gildemacher P, Demo P, Wagoire W, Kinyae P, Andrade J, Forbes G, Fuglie K, Thiele G (2008). Farmer practices and adoption of improved potato varieties in Kenya and Uganda. CIP, Lima, Peru. 5: 42.

Lung'aho C, Mmakwa C, Kidanemariam H (1997). Effect of source of mother plant, variety, and growing conditions on the production of stem cuttings and subsequent yield of mini-tubers in the Kenyan potato programme. *In* Proceedings of the 4th Triennial Congress of the African Potato Association, Pretoria, South Africa. 23–28 February. African Potato Association, Kampala, Uganda. pp. 72–91.

Ministry of Agriculture (2005). National policy on potato industry. Policy and reforms in the industry to improve production, research, marketing, and regulatory framework. MoA, Nairobi, Kenya.

Ministry of Agriculture (2008). National policy on potato industry. Policy reforms to revitalize the potato industry. MoA, Nairobi, Kenya.

Ministry of Agriculture and Deutsche Gessellschaft fur Technische Zusammenarbeit (1998). Postharvest systems of potato and sweet potato in Kenya. Final report. MoA/GTZ, Nairobi, Kenya.

Muthoni J, Kabira JN (2010). Effects of crop rotation on soil macronutrient content and pH in potato producing areas in Kenya: A case study of KARI Tigoni station. J. Soil Sci. Environ. Manage, 1: 227-233.

Muthoni J, Mbiyu MW, Nyamongo DO (2010). A review of potato seed systems and germplasm conservation in Kenya. J. Agric. Food Inform., 11: 157–167.

Recke H, Schnier HF, Nabwile S, Qureshi JN (1997). Responses of Irish potatoes (*Solanum tuberosum* L.) to mineral and organic fertilizer in various agro-ecological environments in Kenya. Exp. Agric., 33: 91-102.

White PJ, Wheatley RO, Hammond JP, and Zhang K (2007). Minerals, soils, and roots. In D Vreugdenhil et al. (ed.) Potato biology and biotechnology: Advances and perspectives. Elsevier, Amsterdam. pp. 739-751.

Studies on sub-soil characteristics of sand deposits in some parts of Bayelsa State, Eastern Niger Delta, Nigeria

P. O. Youdeowei[1] and H. O. Nwankwoala[2*]

[1]Institute of Geosciences and Space Technology, Rivers State University of Science and Technology, P. M. B. 5080, Nkpolu-Oroworukwo, Port Harcourt, Nigeria.
[2]Department of Geology, University of Port Harcourt, P. M. B. 5323, Choba, Port Harcourt, Nigeria.

Sub-surface investigation was undertaken by drilling boreholes within and along the Nun River Channel covering three communities of Kaiama, Opokuma and Sabagreia, Bayelsa State, Nigeria. The route extent was over 5 km long and the study was necessitated to provide materials as structural fills for road construction purposes. Eight boreholes were drilled in the three locations to a maximum depth of 14 m from the river bed. Results reveal an overlying ligh grey to dark brown, very fine grained (anhedral), soft to stiff consistency, highly plastic silty clay soil in all the boreholes. This unit varies in depth vary in depth of 1.0 to 7.0 m and is proceeding thinner towards the Sabagreia axis from Kaiama. Beneath this overburden lie light grey, medium grained sands, in BH 2 to BH 6 covering the Kaiama and Opokuma axis. The sands are about 7 to 12 m in thickness to the maximum drilled depth of 14 m. In BH 7 and BH 8 in the Sabagreia axis, the underlying sands are coarse grained in texture and about 13 in thickness to the borehole depth. It is recommended that the sand may be derived at minimum depths of about 1.0 m below the river bed in the Sabagreia area, about 2.0 m in the Opokuma region and about 4.5 to 7.0 at the Kaiama axis. Medium to coarse grained; poorly graded (well sorted) sands which display uniform size gradation in the sand range were encountered below the river bed in the area.

Key words: Soils, grain size, engineering properties, stratigraphy, Niger Delta.

INTRODUCTION

The need for accurate information and adequate understanding of the geotechnical properties of the foundation of subsoil cannot be over emphasized. Geotechnical information are useful in ensuring that the effects of projects on the environment and natural resources are properly evaluated and mitigated where necessary (Nwankwoala et al., 2009).

Several environmental problems are expected to occur if no secure and environmental friendly measures associated with development are carried out, as this can prevent severe contamination in soil and groundwater (Bruno, 2007; Barends and Hagenaar, 2008; Depountis et al., 2009; Youdeowei and Nwankwoala, 2010a; Youdeowei and Nwankwoala, 2010b).

Bayelsa State is a fast growing state in the Southern Nigeria with associated environmental problems, which have been a subject of primary concern. There are severe drainage problems with seasonal and temporary flooding due to heavy rainfall and rise in groundwater table, resulting in almost total submergence during the wet season with exception of the natural levees (Allen, 1965).

In the study area, only little effort have been made in understanding the engineering and environmental properties of the shallow subsurface soils which are the foundation of civil engineering structures such as housing

*Corresponding author. E-mail: hyci2002@yahoo.com.

units, oil production facilities, waste management facileties etc that are directly or indirectly related to the oil and gas industry. This study therefore, provides geotechnical data on subsoil characteristics to guide future development in the area.

DESCRIPTION OF THE STUDY AREA

The study sites are characterized by a typical freshwater ecology of the upper reaches of the River Nun within the Niger Delta area. The area lies within the outcropping Benin Formation made up of continental deposits of Miocene to recent sediments. The areas are associated with freshwater swamps, backswamps and meander belts of flat to sub-horizontal elevation (Etu-Efeotor and Akpokodje, 1990; Allen, 1965). The river banks vary in heights of 4 to 6 m above the low water stage with gradients ranging from gentle to very steep (30 to 70°). The major soil types are light to dark grey, fine sand to silty clay. The mean annual rainfall exceed 2000 mm and rain falls all year round with over 85% during the wet season (Iloeje, 1972).

Geomorphologically, the Niger Delta has been described by a number of writers (NEDECO, 1961; Allen, 1965; Weber, 1971). The topography of the area is essentially flat, sloping very slightly seawards. The area is low lying and is drained and criss-crossed by an anastomosing network of distributaries. The area constitutes an extensive plain exposed to periodical inundation by flooding when the rivers and creeks overflow their banks. A prominent feature of the rivers and creeks is the occurrence of natural levees on both banks, behind which occur vast areas of back-swamps and lagoons/lakes where surface flow is negligible.

Although various types of morphological units and depositional environments have been recognized in the area (coastal flats, ancient/modern sea, river and lagoonal beaches, sand bars/flats, flood plains, seasonally flooded depressions, swamps, ancient creeks and river channels), the area can be sub-divided into five major geomorphological units, namely: (1) Active/ abandoned coastal beaches; (2) Saltwater, mangrove swamps; (3) Freshwater swamps, backswamps, deltaic plain alluvium and meander belt, (4) Dry deltaic plain with abundant freshwater swamps (Sombreiro- Warri deltaic plain), and (5) Dry flat land and plain. Table 1 shows the geologic units of the Niger Delta.

Our present knowledge of the geology of the Niger Delta was derived from the works of Reyment (1965), Short and Stauble (1967), Murat (1970), Merki (1970), and the exploration activities of oil and gas companies. The formation of the so-called proto-Niger Delta occurred during the second depositional cycle (Campanian-Maastrichtian) of the southern Nigerian basin. However, modern Niger Delta was formed during the third and last depositional cycle of the southern Nigerian basin which started in the Paleocene. Generally, the Niger Delta was formed during the Tertiary period as a result of the interplay between subsidence and deposition arising from a succession of transgressions and regressions of the area. These phenomena evolved the three main lithostratigraphic units of Akata, Agbada and Benin Formations, with an overall approximate thickness of over 5,000 m of sediment body (Short and Stauble, 1967).

The Akata Formation (outcropping as the Imo Shale Group) is made up of marine clay, shale and limestone and is estimated to be about 1,000 m in thickness. The Agbada (Eocene) which outcrops as the Ameki Formation overlies the Akata Formation (Paleocene) and comprises various alaternations of marine shales and sandstones. It has an estimated thickness of 1,700 m. The youngest of these formations, the Benin Formation (Miocene – Recent) basically consists of coarse sand with occasional clay and shales with a total thickness of over 2000 m. Quaternary deposits of about 100 m in thickness comprising recent deltaic sediments made up of sand, silt and clay beds, overlie the Benin Formation in the swampy areas of the delta and comprises rapidly alternating sequences of sand and silt/clay with the latter becoming increasingly more prominent seawards (Etu-Efeotor and Akpokodje, 1990).

STUDY TECHNIQUES

Field investigation

Subsurface investigation was undertaken by drilling boreholes within and along the Nun River Channel covering the three communities. The bathymetric sounding data of the river profile provided the necessary preliminary information for the borehole drilling programme. A 12' × 14' floating raft (pontoon) with a drilling spot (moon) at the centre was constructed and the drilling rig mounted on it for the drilling exercise. Soil boring was done with the use of a hand-operated light cable percussion rig of the pilcon way farrer type to maintain minimal disturbance of soil samples retrieved for logging and analysis.

Eight boreholes were drilled at the sites to a maximum depth of 14 m from the river bed. The boreholes were logged on site and soil samples recovered at intervals of 1 m for field and laboratory examination. Grain size distribution analysis was carried out on the sandy soils in accordance with British Standards (B.S, 1377, 1975) to delineate the textural characteristics.

RESULTS AND DISCUSSION

Grain size distribution of soil samples

Grain size analysis results show a range of poorly graded (well-sorted) sands with uniform gradation curve displaying very little or no fines (0.1 to 0.4% passing 0.075 mm sieves). The results are presented in Table 2 and the gradation pattern shown in the particle-size distribution curve (Figure 1). The sands are classified as SP (pooly graded) under the unified soil classification system

Table 1. Geologic units of the Niger Delta (after Akpokodje, 1989).

Geologic unit	Lithology	Age
Alluvium (General)	Gravel, sand, clay, silt	
Freshwater backswamp, meander belt	Sand, clay, some silt, gravel	
Saltwater mangrove swamp and backswamp	Medium-fine sands, clay and some silt	Quaternary
Active/abandoned beach ridges	Sand, clay, and some silt	
Sombreiro-warri deltaic plain	Sand, clay, and some silt	
Benin formation (Coastal Plain Sand)	Coastal to medium sand; subordinate silt and clay lenses	Miocene-Recent
Agbada formation	Mixture of sand, clay and silt	Eocene-Recent
Akata formation	Clay	Paleocene

Table 2. Grain size distribution of soil samples.

Sample no and depth (m)	Particle size distribution (percent passing sieves)				Unified soil classification system (U.S.C.S.)
	No. 4 (4.75 mm)	No. 10 (2.00 mm)	No. 40 (0.42 mm)	No. 200 (0.075 mm)	
BH1 (13)	100	100	90.8	0.2	SP
BH 2 (10)	100	100	92.4	0.2	SP
BH3 (8)	100	100	88.5	0.1	SP
BH 4 (6)	100	100	96.7	0.4	SP
BH 5 (2)	100	100	95.0	0.3	SP
BH 6 (12)	100	100	95.4	0.4	SP
BH 7 (12)	100	97.8	76.8	0.3	SP
BH 8 (9)	100	97.5	79.7	0.3	SP

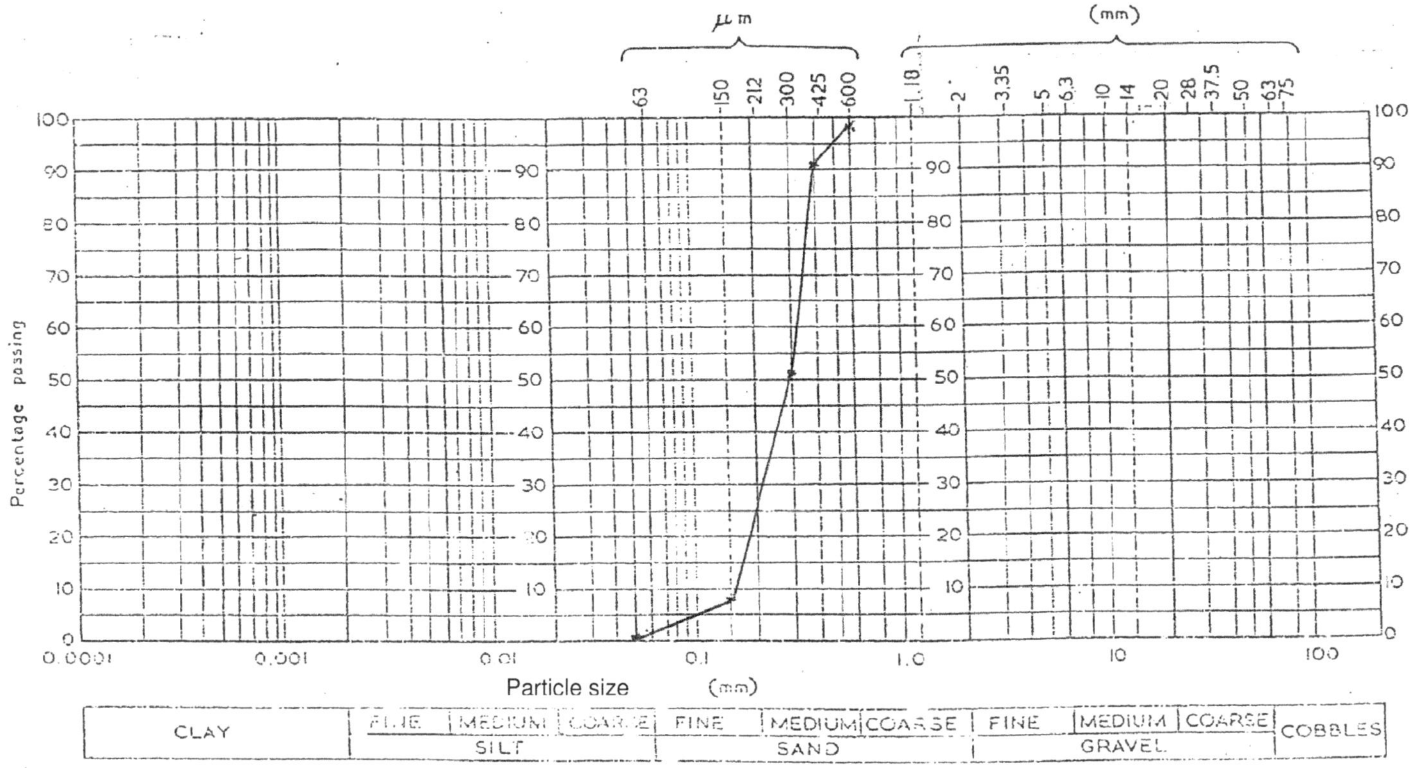

Figure 1. Particle size distribution curve.

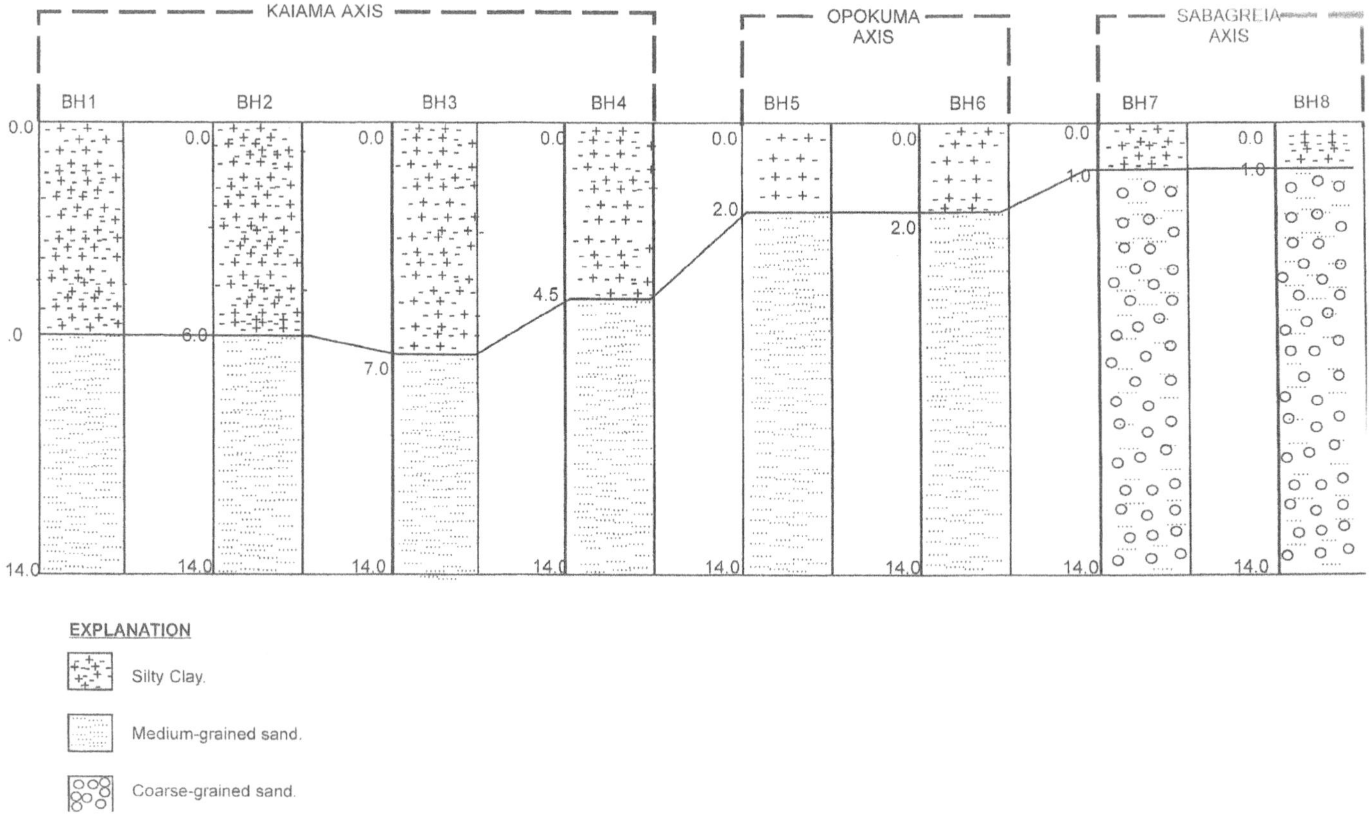

Figure 2. Cross-section and stratigraphic correlation of boreholes along the river channel.

(U.S.C.S).

Soil stratigraphy and physical properties

The sub-soil stratigraphic profile and correlation of the eight (8) boreholes drilled at the three (3) sites is shown in Figure 2. There is an overlying light grey to dark brown, very fine grained (anhedral), soft to stiff consistency, highly plastic silty clay soil in all the boreholes. The unit varies in depth of 1.0 to 7.0 m and proceeding thinner towards the Sabagreia axis from Kaiama. Beneath this overburden lies light grey, medium grained sand in BH 1 to BH 6 covering the Kaiama and Opukuma areas. The sands are about 7 to 12 m in thickness to the maximum drilled depth of 14 m. In BH 7 and BH 8 in the Sabagreia axis, the underlying sands are coarse grained in texture and about 13 m in thickness to the borehole depth.

GEOTECHNICAL IMPLICATIONS

The fine grained nature of the soils in the area implies a low permeability potential and the proportion of silty clay size fractions in the soils will have significant influence on the behaviour of the soils, whereby it may display a considerable amount of compressibility.

The strength properties of the soil should be evaluated for proper design of shallow foundation system that would bear the weight of structures. The foundation system may derive their bearing strength from the silty clay soil which has a stiff consistency, with attention given to the water table condition in the area.

Generally, considering the fact that the study area is within the Eastern Niger Delta (coastal zone), geotechnical considerations are very desirable. The coastal zone which comprises the beach ridges and mangrove swamps is underlain by alternating sequence of sand and clay with a high frequency of occurrence of clay within 10 m below the ground surface (Youdeowei and Nwankwoala, 2010c). Because of the nearness of the compressible clays to the surface, the influence of imposed loads results to consolidation settlement. The impact of the imposed loads is exacerbated by the thickness and consistency of the compressible layer. This in addition to other intrinsic factors contributes to the failure of civil engineering structures.

More importantly, it is not anticipated that the soil conditions as revealed in this study will vary significantly. However, the soil conditions may vary during construction, hence, it would be necessary to evaluate the engineering significance of such variations which could result in further investigations.

Conclusion

The lithostratigraphic profile along the river channel of the three (3) study sites reveal an overburden of very fine grained, soft to stiff consistency, silty clay overlying medium to coarse grained, plain sands. The thickness of the underlying coarse sand to the maximum drilled depth is greater at the Sabagreia area. The medium grained sands which cover the Opukuma and Kaiama axis preponderates over the coarse sand in areal extent. It is recommended that sand may be derived at minimum depths of about 1.0 m below the river bed in the Sabgreia are, about 2.0 m in the Opokuma region and about 4.5 to 7.0 m at the Kaiama axis. Medium to coarse grained, poorly graded (well sorted) sands which display uniform size gradation in the sand range were encountered below the river bed in all the locations.

ACKNOWLEDGEMENTS

The authors thank and appreciate the input of two anonymous reviewers for providing helpful and unbiased comments that improved the technical quality of this paper.

REFERENCES

Akpokodje EG (1989). Preliminary studies on the geotechnical characteristics of the Niger Delta Sub-soil. Engine. Geol., 26: 247-257.

Allen JRL (1965). Coastal geomorphology of eastern Nigerian beach ridges barrier islands and vegetated tidal flats. Geol. Mijabonw., 44: 1-21.

B.S 1377 (1975). Methods of Test for Soils for Civil Engineering Purposes. British Standard Institution (BSI), London.

Barends FBJ, Haganaar J (2008). Following the Geotechnical Studies of the West-Breaker at Sines: Hydrogeotechnical Stability of Breakwaters. ASCE Publications, Conference Proceeding, pp. 124-167

Bruno B (2007). Hydrogeotechnical properties of hard rock tailings from metal mines and emerging geo-environmental disposal approaches. Canadian Geotech. J., 44(9): 1019-1052

Depountis N, Koukis G, Sabatakakis N (2009). Environmental problems associated with the development and operation of a lined and unlined landfill site: A case syudy demonstrating two landfill sites in Patra, Greece. Engine. Geol., 56(7): 1251-1258.

Etu-Efeotor JO, Akpokoje EG (1990). Aquifer systems of the Niger Delta. J. Mining Geol, 26(2): 279-284.

Iloeje NP (1972). A New Geography of West Africa, Longman Group Limited, Nigeria.

Merki PJ (1970). Structural Geology of the Cainozoic Niger Delta. African Geology, University of Ibadan Press, pp. 251-268.

Murat RC (1970). Stratigraphy and paleogeography of the Cretaceous and Lower Tertiary in Southern Nigeria. In: African Geology, University of Ibadan Press, Ibadan, pp. 635-648

Nwankwoala HO, Youdeowei PO, Ngah SA (2009). Expanding Hydrogeotechnical Considerations in Environmental Impact Assessment Studies: A Case Study of Ogorode- Sapele, Delta State, Nigeria. J. Appl. Sci. Environ. Mange., 13(1): 67-71.

Reyment RA (1965). Aspects of the Geology of Nigeria. University of Ibadan Press, Ibadan, p. 133

Short KC, Stauble AJ (1967). Outline Geology of the Niger Delta. Am. Ass. Petroleun Geologists Bull., 51: 761-779.

Weber KJ (1971). Sedimentological aspects of oilfields in the Niger Delta. Environ. Geol., 50: 559-576.

Youdeowei PO, Nwankwoala HO (2010a). Subsoil characteristics and hydrogeology of the Export Processing Zone, Calabar-Southeatern Nigeria. J. Appl. Sci. Environ. Manage. 14(2): 11-15.

Youdeowei PO, Nwankwoala HO (2010b). Assessment of some geo-environmental problems associated with road construction in the Eastern Niger Delta. Afr. J. Environ. Pollut. Health, 8(1): 1-7.

Youdeowei PO, Nwankwoala HO (2010c). Suitability of soils as bearing media at a freshwater swamp terrain in the Niger Delta. (In Press).

Effects of tillage and manure application on soil physicochemical properties and yield of maize grown on a degraded intensively tilled alfisol in southwestern Nigeria

Adeyemo A. J. and Agele S. O.*

Department of Crop, Soil and Pest Management. Federal University of Technology, Akure, Nigeria.

Intensive tractorised tillage practices especially in the tropical and subtropical climate produces compaction in crop root zone (20 to 30 cm depth), degradation of soil quality, loss of organic matter and declining soil and crop productivity. The effects of tillage and manure application on soil physical and chemical properties, growth and yield of maize grown on an intensively mechanized and degraded Alfisol was investigated on the field in 2006 and 2007. The treatments consisted of 3 x 3 factorial combinations of tillage methods and manure application rates arranged in a split-plot design. The three tillage methods were: manual clearing (M_c), ploughing at 10 cm followed by harrowing (P_1H_1) and ploughing at 20 cm followed by harrowing (P_2H_1) while manure application at 10 t. ha^{-1} farm yard manure (FYM) and 200 kg.ha^{-1} NPK fertilizer (a compound mineral fertilizer which contain N, P and K in the ratio 20:10:10). The tillage constituted the main plot while farm yard manure and NPK fertilizer were the subplots. There was an unmanured control. Ploughing was carried out using mouldboard plough mounted on a tractor and harrowing was carried out after ploughing. There were no significant differences ($P < 0.05$) in soil moisture contents and bulk densities (0 to 60 cm depth) between tillage alone and tillage systems combined with manure application at 4 and 10 weeks after planting. The tillage-manure combination produced higher values of soil organic matter (1.3, 1.6%), total N (0.08, 0.11%) and available P (14.5, 18.9 mg/g) over tillage alone. Compared with tillage systems alone, application of manure significantly increased ($P = 0.05$) root and shoot dry weights, leaf area (0.11and 0.25 m^2), cob and seed yield of maize. The increases in grain yield due to manure (10 tha^{-1} FYM) and 200 Kgha^{-1} NPK fertilizers were 10, 13 and 12%, respectively.

Key words: Alfisol, degradation, tillage, manure, soil, physico-chemical, maize.

INTRODUCTION

Soil compaction is a major cause of soil degradation in most agricultural soils (Van Lynden, 2000). In a compacted soil, the particles are pressed together, thus reducing pore spaces which contain air and water necessary for good plant growth (Aina et al., 1985). Compaction can be caused by the use of heavy machinery, ploughing at the same depth for many years and trampling by animals, reduced use of organic matter, frequent use of chemical fertilizers (Ohu et al., 1994; Lampurlanes and Martinez, 2003).

Intensive soil tillage practices especially in the tropical and subtropical climate accelerate soil erosion, cause compaction (between 20 to 30 cm depth), degradation of soil and loss of organic matter (Lal, 1976; Ojeniyi, 1982). These may result in declining soil productivity (Sobulo and Osiname, 1986).

Tractorized multi-pass tillage practices destroys soil structure such as aggregates and cause compaction within the plough layer which further deteriorate soil conditions (Lupwayi et al., 2001). Any increased bulk density from such practices could explain the adverse effect of land ploughing and harrowing on crop performance

*Corresponding author. E-mail:ohiagele@yahoo.com.

(Ojeniyi, 1986; Agele et al., 2005). Compaction also results in soil aggregate fragmentation involving the break down of large structural units into smaller fragments which further reduces the air space (Kay, 1990; Ojeniyi and Agboola, 1995). Intensive soil compaction has adverse effect on soil water flow and storage, impedes root growth and therefore limits the volume of soil explored by roots. This in turn can decrease the plant's ability to take up nutrients and water (Stewart and Vyn, 1994; Zhang et al., 1997).

A wide range of special tillage operations involving soil inversion, chiseling, subsoiling or deep tillage (for soils with an impeding layer within rooting depth) have been found to be beneficial by minimizing soil hardening or bulk density, improving soil porosity, infiltration, soil water storage, and root development (Lal, 1984; Ogunremi et al., 1986). Soil tillage (seedbed preparation) methods in addition to manure use are basic practices involved in crop production. These practices are known to influence soil physical conditions, nutrient availability, growth and yield of crops (Ojeniyi and Agboola, 1995). Tillage operation tends to modify soil bulk density and pore size distribution, loosens, granulates, crushes and even compacts soil particles (Klute, 1982). Bennie and Botha (1986) found that tillage reduces soil bulk density by breaking the aggregates into smaller particles and increasing aeration. It enhances rapid root elongation with depth, induces a high root density in the subsoil and increases final yield.

Degraded soils are usually characterized by low organic matter contents, superficial effective rooting depth, high bulk densities and compaction and exposed subsoil (Agele, 2007). Such soils can be reclaimed via amendment or enrichment with commercial fertilizers, and/or organic waste materials such as domestic refuse and farmyard manure (Wong and Wong, 1989; Raymond and Roy, 1992; Chu and Bradshaw, 1996; Agele, 2000). Improved understanding is required on the efficacy of these materials when used on degraded lands (Kamamori and Yasuda, 1979; Agele 2007). In addition to the provision of essential plant nutrients to soils, organic manure improves soil structure through enhanced soil water holding capacity, aeration and drainage which encourage good root formation and plant growth (Cooke, 1975). Elmar and Wolfgang (1990) attributed the promotion of biological activities, nutrient exchange capacity and organic matter content that lessen soil erosion to the use of organic manure. Brady (1990) established the relationship of soil aggregate stability to its organic matter content. In crop production, soils must be loose enough to allow root penetration and seedling emergence. Appropriate soil tillage (seedbed preparation) methods in addition to manure use are basic practices required for maximum crop yield. These practices are known to influence soil physical conditions, nutrient availability, growth and yield of crops (Ojeniyi and Agboola, 1995).

Maize (*Zea mays* L.), is an important food crop grown in Nigeria. The crop is grown as food component of human diet as well as livestock feed, and has various industrials uses such as in the production of corn oil, sugar and alcohol. This study examines the effects of tillage methods, manure and mineral fertilizer application and their combinations on some soil properties, growth and yield performance of maize grown in an intensively tilled alfisol in Nigeria.

MATERIALS AND METHODS

Site for the experiment

This study was carried out at the Teaching and Research Farm of the Federal University of Technology (FUTO) Akure (lat.7º15' N, Long. 5º10'E) in a tropical rainforest zone of Nigeria. The trial was carried out during the late season (September-December) and early season (March to June) of 2006 and residual study conducted in 2007. Table 1 presents meteorological variables at the study site during maize growth in the respective late and early rainy seasons of 2006 and 2007. The sandy loam soil at the study site is an Alfisol classified as clayey skeletal Oxic-Paleustalf (USDA, 1999). The experiment was sited within a 20 ha agricultural field that has been mechanically tilled and cropped with tropical arable crops (maize, cowpea and cassava) continuously for over fifteen years.

Treatments

Treatments were a 3 x 3 factorial combination of seedbed types and manure application, arranged in a split-plot design. Tillage constitutes the main plot while manure application is split within the main plots as subplot and was randomized within the main plot. In each year, the experimental site was therefore divided into three blocks (5 x 20 m, main plot treatment; constituted by the seedbed types and were split into subplots (5 x 5 m, fertilizer treatments). There were 3 m guard rows between plots and 3 m between blocks. The tillage methods were manual clearing (Mc) and mechanised tillage involving ploughing at 10 cm depth plus one pass harrowing (P_1H_1) and ploughing at 20 cm depth plus one pass harrowing (P_2H_1). Ploughing and harrowing were accomplished using a 72 hp, 2 wheel drive, 2590 kg tractor (Marsey Ferguson MF, UK). Manual clearing (Mc) treatment consisted of a seedbed was manually cleared of existing vegetation using cutlass and the cleared residues were left in situ. The tillage treatments were amended with 10 tha^{-1} farm yard manure (FYM) and NPK (a compound fertilizer containing NPK in the ratio 20: 10:10) was applied at 200 kgha^{-1} and an unfertilized control (where neither farm yard manure nor NPK fertilizer was applied). Farmyard manure (FYM) consisted of partially decomposed cow dung and poultry droppings plus bedding materials. The main plot was constituted by tillage methods while fertilizer (farm yard manure and NPK fertilizer) were the subplots. Treatments were replicated three times and were arranged in a split- plot design. Table 1 presents some meteorological conditions at site of experiment during maize growth in 2006 and 2007. Maize (*Z. mays* L.) were planted at two seeds per hole at the spacing of 60 x 30 cm. Weed was controlled manually (hand weeding) at 3 and 7 WAP for all treatments.

Determination of soil properties

Pre-planting soil samples were collected using core samplers (steel corers -10 cm diameter and depth), from selected depths (0 - 20,

Table 1. Meteorological conditions at the site of the experiment during maize growth.

	2006					2007				
	Aug.	Sept.	Oct.	Nov.	Dec.	Mar.	Apr.	May	June	July
Rainfall (mm)	327	271	193	68	23	19	39	189	257	288
Min. Temp(°C)	20.8	21.5	21.8	20.9	19.5	22.3	22.9	21.3	20.8	21.2
Max. Temp(°C)	28.7	29.6	30.3	31.7	30.6	33.3	32.9	31.6	30.3	29.6
Rel. humidity (%)	82	63	52	48	43	48	55	63	67	73
Total sunshine (h)	93	138	219	235	193	238	206	183	177	128
Solar radiation (MJ/m^2/day)	9.8	12.4	13.8	15.8	14.1	17.8	17.1	17.2	15.9	12.9
Open water Evap. (mm)	86	97	102	130	148	238	132	108	98	91

Table 2. Selected profile characteristics of the soil before commencement of the experiment.

Horizontal depth (cm)	Sand (%)		Silt (%)		Clay (%)		Bulk density (g.cm^{-3})		Total porosity (%)	
	2006	2007	2006	2007	2006	2007	2006	2007	2006	2007
0-20	40	39	31	33	29	28	1.32	1.32	47	47
20-35	30	27	35	41	35	32	1.39	1.39	45	45
35-50	28	30	38	40	34	30	1.43	1.43	43	43
50-65	29	39	31	30	40	31	1.47	1.47	42	42

20- 35, 35-50 and 50-65 cm) from profile pits dug at two points within the experimental plot. The depth-wise samples were analysed for physical properties of bulk density and textural class (percent of sand, silt and clay) using standard methods. Starting from 2 weeks after planting (WAP), and at weekly intervals thereafter, measure-ments were made on soil temperature using soil thermometer. The soil thermometer was inserted to 5 cm depth and three determinations were made per treatment replicate. Soils were sampled at 4 and 10 WAP from three randomly selected sampling points from each treatment replicates within the field and from depths similar to those of pre-planting. From these sampling points, soil samples were collected using core samplers (steel corers – 10 cm diameter and depth) at soil depths of (0-20, 20-35, 35-50 and 50-65 cm) for the determination of soil moisture contents and bulk density. The field moist soil samples were oven-dried for 24 h and at 105°C for the determination of soil moisture contents (gravimetric method). Bulk density values were obtained from the weight of oven-dried soil samples and volume of the soil corer while total porosity was calculated from bulk density using particle density of 2.65 cm^{-3}. Volumetric soil moisture was obtained from the product of gravimetric moisture contents and bulk density. Particle size analysis was carried out using hydrometer method (Bouyoucos method).

Surface (0 to 20 cm) soil samples were collected from each site at start of experiment and at crop harvest from each treatment plot for chemical analysis. On each treatment replicate, five sites were randomly located as sampling points. Soil samples collected from sampling points in each replicate were bulked per treatment and were analyzed in the laboratory for chemical properties. Soil organic carbon (SOC) was analyzed by the Warkley and Black method (Warkley and Black, 1934) and soil organic matter was computed by multiplying the values of SOC by 2.72. Total N after Kjeldahl digestion. Phosphorus (P) was extracted using Bray-P1 solution and determined using molybdenum blue colorimetry. Exchangeable K, Ca and Mg were determined by EDTA titration method (Jackson, 1962). Soil pH was determined in soil-water medium at ratio 1:2 using the digital electronic pH meter.

Growth and yield parameters

Root and shoot biomass, plant height, days to 50% flowering, yield and yield components were monitored. Five plants were randomly sampled from 2 m^2 at the center of each plot at 12 weeks after planting (WAP) for the determination of root and shoot dry matter yields obtained from oven-dried plant samples at 80°C for 48 h. For root biomass determination, a cubic coring tool (10 x 10 x 10 cm) was drill inserted into the soil and to a depth of 40 cm within the circumference of 5 sampled plants per plot. The excavated roots from the soil samples were put in 2 mm sieves and were gently washed free of soil in the laboratory using moderate jets of water. The values of plant height, leaf area, number of leaves per plant and stem girth were determined at 12 weeks after planting. Leaf area was calculated as number of terminal leaflets (N) x 2.7 x 1/37 cm^2 (Nangju and Wanki, 1980). Ten maize plants were harvested per plot at harvest for yield components determination. The components were as follows: weight of cob per plants (g), number of cobs per plants, seed weight per plant, 100 seed weight and shoot weight. Afterwards, the entire plots were separately harvested and cobs were air-dried and weighed.

Statistical analysis

Data collected from the field were subjected to analysis of variance for a split-plot design (Steel et al., 1997) to determine the separate and interaction effects of the main plot and subplot treatments on some soil physical properties, growth and yield characteristics of maize.

RESULTS

Effects of seedbed types on soil physical properties

Table 2 presents the results of some selected profile

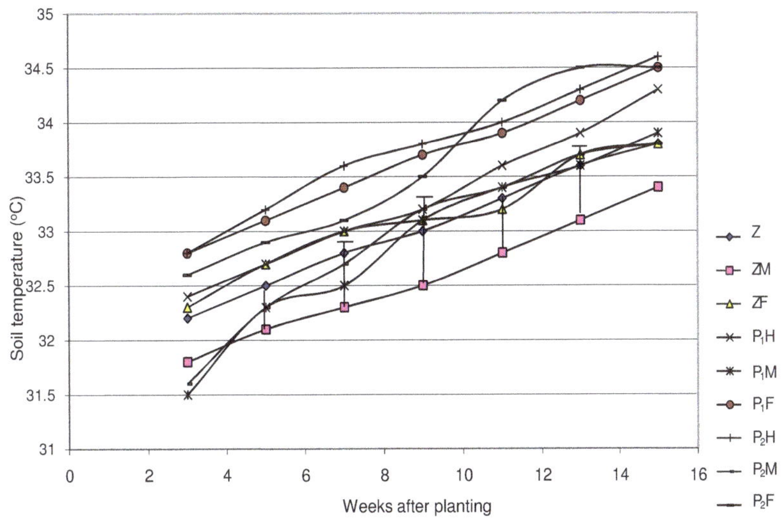

Figure 1a. Trends in soil temperature during growth of late season maize. Vertical lines are LSD bar. Z: manual clearing; P_1H: ploughing 10 cm depth; P_1HM: ploughing at 10 cm + farm yard manure; P_1HF: ploughing at 10 cm + fertilizer; P_2H: ploughing 20 cm depth; P_2HM: ploughing at 20 cm + farm yard manure; P_2HF: ploughing at 20 cm + fertilizer.

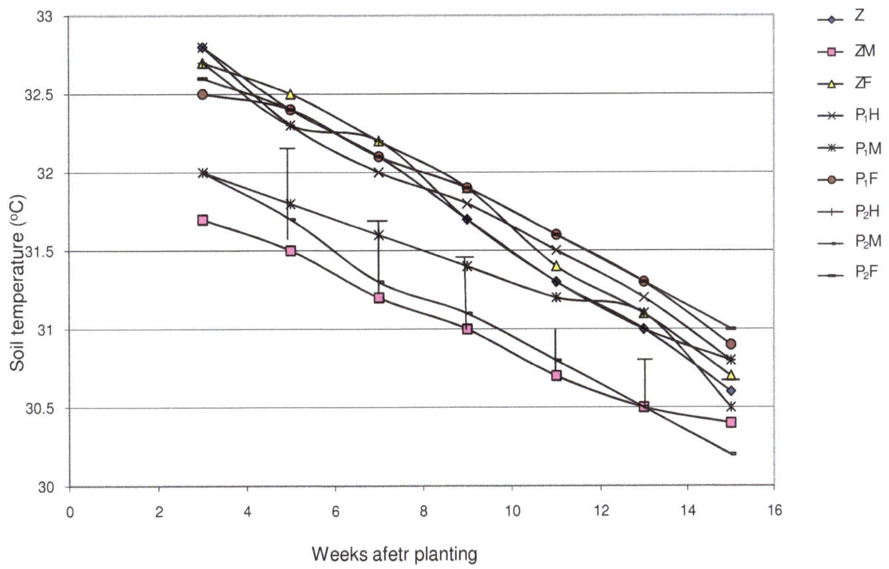

Figure 1b. Trends in soil temperature during growth of late season maize. Vertical lines LSD bar (P = 0.05). Z: manual clearing; P_1H: ploughing 10 cm depth; P_1HM: ploughing at 10 cm + farm yard manure; P_1HF: ploughing at 10 cm + fertilizer; P_2H: ploughing 20 cm depth; P_2HM: ploughing at 20 cm + farm yard manure; P_2HF: ploughing at 20 cm + fertilizer.

characteristics of the soil at the site of the experiment before the commencement of the experiment is presented. There were increases in bulk density and this possibly was as a result of the cofounding effects of natural and tillage-induced compaction. Increases were found for bulk density and this situation possibly resulted from cofounding effects of natural and tillage-induced compaction.

Effect of tillage methods and manure on soil physical properties

The dynamics of soil moisture and temperature regimes at the experimental site are shown in Figures 1a and b respectively. Tillage alone produced higher ranges of soil temperatures over tilled soil amended with FYM and NPK

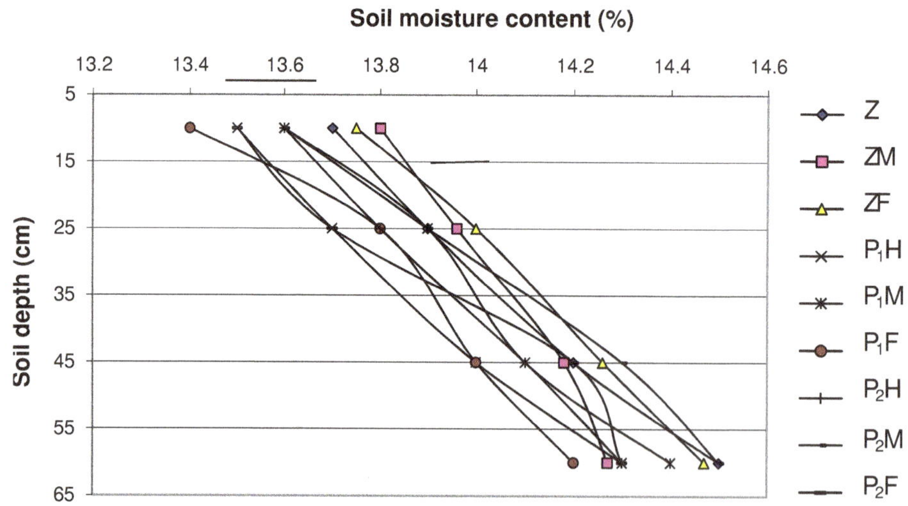

Figure 2a. Dynamics of gravimetric moisture contents (4 WAP). Horizontal lines are LSD bar (P = 0.05). Z: manual clearing; P_1H: ploughing 10 cm depth; P_1HM: ploughing at 10 cm + farm yard manure; P_1HF: ploughing at 10 cm + fertilizer; P_2H: ploughing 20 cm depth; P_2HM: ploughing at 20 cm + farm yard manure; P_2HF: ploughing at 20 cm + fertilizer.

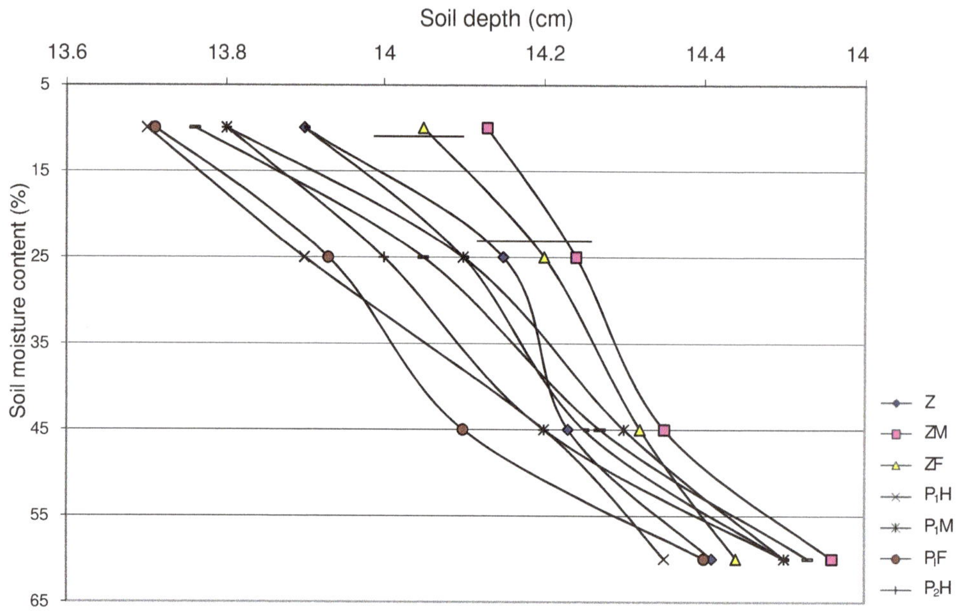

Figure 2b. Dynamics of gravimetric moisture contents (10 WAP). Horizontal lines are LSD bar (P = 0.05). Z: manual clearing; P_1H: ploughing 10 cm depth; P_1HM: ploughing at 10 cm + farm yard manure; P_1HF: ploughing at 10 cm + fertilizer; P_2H: ploughing 20 cm depth; P_2HM: ploughing at 20 cm + farm yard manure; P_2HF: ploughing at 20 cm + fertilizer.

fertilizers across all sampling dates. The lower soil temperatures under FYM incorporation could have stemmed from lower bulk density (improved total porosity) enhanced soil water infiltration. The differences at times attain significance level, were obtained in the values of soil temperatures in the respective tillage methods alone and tillage combined with manure application. There were differences in gravimetric moisture content among treatments and in the different sampling dates within 0 to 60 cm depth (Figures 2a and b). Mean soil moisture content was higher under manual clearing and deep tillage (P_2H_1), little variations in soil moisture

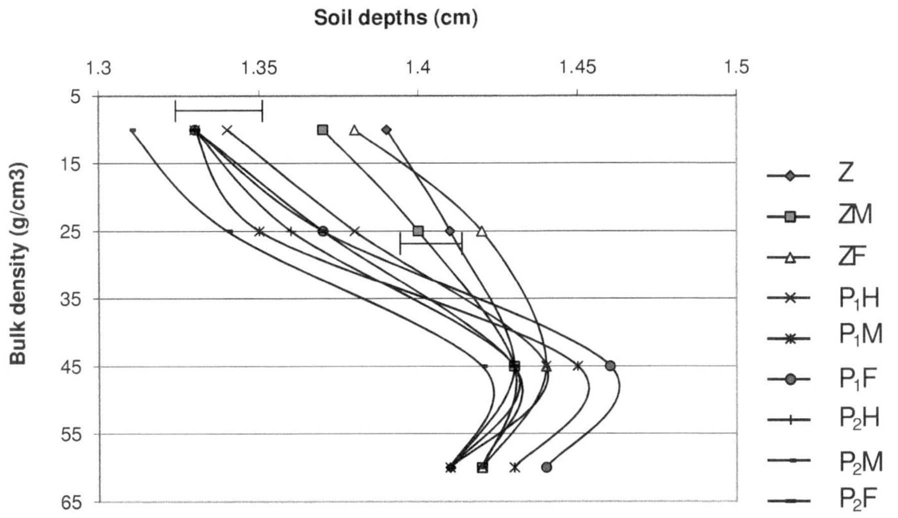

Figure 3a. Dynamics of bulk density as affected by tillage and organic waste incorporation (4 WAP). Horizontal lines are LSD bar (P = 0.05). Z: manual clearing; P_1H: ploughing 10 cm depth; P_1HM: ploughing at 10 cm + farm yard manure; P_1HF: ploughing at 10 cm + fertilizer; P_2H: ploughing 20 cm depth; P_2HM: ploughing at 20 cm + farm yard manure; P_2HF: ploughing at 20 cm + fertilizer.

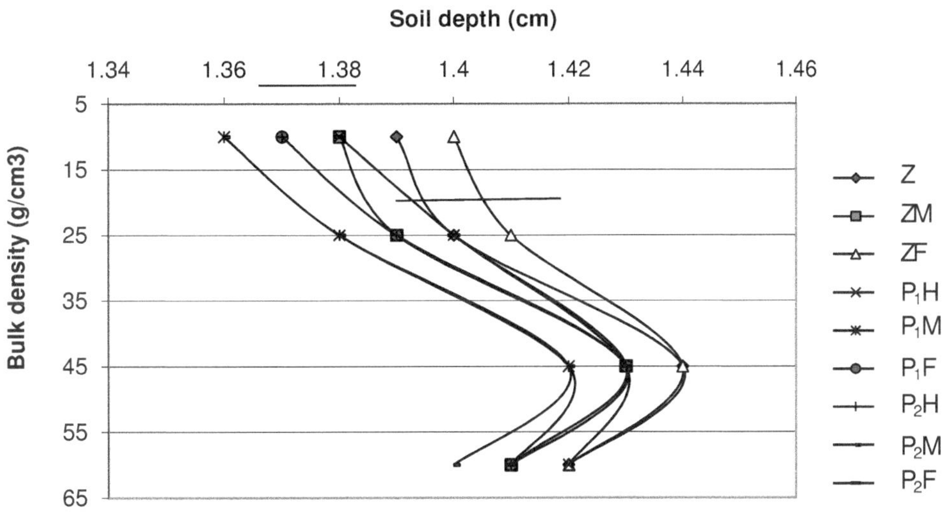

Figure 3b. Dynamics of bulk density as affected by tillage and organic waste incorporation (10 WAP). Horizontal lines are LSD bar (P = 0.05) Z: manual clearing; P_1H: ploughing 10 cm depth; P_1HM: ploughing at 10 cm + farm yard manure; P_1HF: ploughing at 10 cm + fertilizer; P_2H: ploughing 20 cm depth; P_2HM: ploughing at 20 cm + farm yard manure; P_2HF: ploughing at 20 cm + fertilizer.

storage in tillage alone and tillage-manure combinations were obtained. This trend connotes differences in the magnitudes of soil moisture storage within the root zone and at different phases of maize growth under the different treatments. Manure improved organic matter status and total porosity, this might have favourably impacted soil water holding capacity and modify rainfall infiltration. The values of bulk density taken at 4 and 10 WAP within the crop root zone (0 to 60 cm) varied among the treatments -ploughing depths and manure (Figures 3a and b), The results of soil bulk densities as affected by tillage methods and manure application show that the tillage-manure combinations had lower bulk density values compared treatments involving tillage alone within depth of soil sampled (0 to 60 cm) (Figures 2a and b).Bulk densities values were about half the values obtained under tillage alone treatment, however, bulk densities obtained under P_1H_1 and P_2H_1 in combination

Table 3. Effect of tillage methods and manure application on soil bulk density and moisture contents (0 to 60 cm) and temperature (0-5cm) at crop maturity. (The values of bulk density and moisture contents were mean of 0 to 60 cm sampling depths, soil temperature at 5 cm depth)

Treatments	Bulk density ($g.cm^{-3}$)		Total porosity (%)		Volumetric moisture (%)		Temperature (°C)	
	2006	2007	2006	2007	2006	2007	2006	2007
Z	1.38	1.39	41.9	40.0	15.8	23.9	29.8	28.2
Z_1M	1.27	1.27	46.0	47.0	19.8	25.4	27.0	26.0
Z_1F_1	1.36	1.40	42.0	41.0	16.1	24.6	29.2	28.0
P_1H_1	1.28	1.29	48.5	46.4	11.7	19.6	31.2	29.4
P_1HM	1.24	1.23	52.9	54.0	16.3	21.9	28.6	27.8
P_1HF	1.29	1.28	47.6	46.8	12.3	20.5	30.8	28.0
P_2H_1	1.36	1.38	43.7	41.0	11.7	19.9	32.5	30.1
P_2HM	1.25	1.27	48.0	48.0	14.8	21.7	30.0	28.6
P_2HF	1.35	1.41	44.0	40.1	12.2	20.9	32.0	29.0
LSD (0.05)	0.17	0.27	1.8	2.4	3.3	4.1	1.6	1.5
Unmanured	1.36	1.38	44.7	42.5	13.2	21.5	31.2	29.2
FYM	1.22	1.25	48.9	49.7	16.8	22.9	28.5	27.5
NPK	1.32	1.36	44.6	42.6	13.6	20.0	30.7	28.3
LSD (0.05)	0.11	0.17	0.9	1.0	1.2	1.4	0.9	0.9

Legend: Z (manual clearing); P_1H (ploughing 10 cm depth) plus harrowing; P_2H (ploughing 20 cm depth) plus harrowing; F (200 kg NPK); M (10 t FYM).

with FYM were similar. Lower bulk densities were obtained under tillage at 10 and 20 cm depths (P_1H_1 and P_2H_1) compared to the manually cleared plots. Bulk density values differed between sampling dates (4 and 10 WAP) within the sampling depth, especially for the manured plots. The lower soil strength could have enhanced rainfall infiltration, soil water storage and root development of maize.

Tillage methods combined with FYM application produced lower soil temperatures (Table 3). In all manured plots, lower soil temperatures were attained using 10 $t.ha^{-1}$ FYM compare to NPK application. The significantly lower soil temperatures in treatments where FYM was incorporated into the soil could have stemmed from livestock manure enhanced soil moisture storage. Bulk density, total porosity, moisture content and temperature of the soil are affected by tillage treatments. Ploughing at 10 cm and at 20 cm depth (in addition to one pass harrow) produced relatively lower bulk density, higher total porosity, lower moisture content and higher temperature compared with manual clearing. However, lower values of bulk density, higher total porosity and moisture content and soil temperature were produced by P_2H_1 compared with P_1H_1. In both years of experiment, manually cleared plots produced bulk density and soil moisture content which were significantly ($P = 0.05$) higher than P_2H_1 (Table 3). Application of manure in combination with tillage systems improved soil physical conditions as indicated by reduced soil bulk density and soil temperature and increased moisture content and total porosity in 2006 and 2007 respectively across the tillage treatments.

Effects of tillage methods and manure on soil chemical properties

The results of the chemical analysis of soil samples surface soil (0-15 cm) at the end of 2006 and 2007 experiments are shown in Tables 4a and b respectively. Manual clearing produced the highest values of soil organic matter (SOM), N, P and exchangeable K and Ca compared with mechanized tillage (P_1H_1 and P_2H_1). Ploughing 10 cm plus harrowing had higher values of soil organic matter (SOM), N, P, exchangeable K, Ca and Mg compared to ploughing at 20 cm. Tillage had no significant ($P = 0.05$) effect on soil pH in 2006 (Table 4a) but the effect of tillage on soil pH in 2007 (rainy season) was significant (Table 4b). The pH values decreased at the end of 2007 experiment (Table 4b). The decline was less for manual clearing plots compared with ploughing at 10 and 20 cm depth. Application of manure significantly increased ($P = 0.05$) soil organic matter, total N, available P, exchangeable K, Ca and Mg. Application of 10 t/ha FYM increased soil pH, statistically significant soil pH in both 2006 and 2007 experiments. In general, soil nutrient concentrations in FYM treated plots were higher than in no manure (control).

Among the tillage treatments, the soil nutrient status after harvest in 2006 and 2007 decreased in the order:

Table 4a. Effect of tillage methods on surface soil (0 to 20 cm) chemical properties (2006).

Tillage methods	pH (1:2)	SOM (%)	N (%)	P (mg/Kg)	K (cmol/kg)	Ca (cmol/Kg)	Mg (cmol/Kg)
Manual clearing (MC)	6.17	1.43	0.09	18.7	0.77	4.7	1.52
Ploughing 10 cm + harrowing (P_1H_1)	6.09	1.28	0.1	18.5	0.78	4.6	1.50
Ploughing 20 cm + harrowing (P_2H_1)	6.05	1.24	0.09	17.8	0.71	4.5	1.46
LSD (0.05)	0.1	0.1	0.01	0.2	0.06	0.09	0.02

Table 4b. Effect of tillage methods on surface soil (0 to 20cm) chemical properties (2007).

Tillage methods	pH (1:2)	SOM (%)	N (%)	P (mg/Kg)	K (cmol/kg)	Ca (cmol/Kg)	Mg (cmol/Kg)
Manual clearing (MC)	6.03	1.41	0.10	17.54	0.75	4.56	1.51
Ploughing 10 cm + harrowing(P_1H_1)	5.85	1.22	0.094	16.81	0.70	4.48	1.52
Ploughing 20 cm + harrowing (P_2H_1)	5.81	1.19	0.091	17.48	0.71	4.45	1.48
LSD (0.05)	0.05	0.06	0.004	0.19	0.06	0.07	0.01

MC, P_1H_1 and P_2H_1 (Tables 4a and b). Tillage alone reduced the values of organic matter, total N, available P and exchangeable K, Ca and Mg contents after the first and second seasons of maize cultivation, whereas, application of manure in combination with tillage significantly increased (P = 0.05) values of these parameters at crop maturity in both 2006 and 2007 experiments. The fertility of the soil in the tillage- manure combination increased significantly (P = 0.05), manual clearing plus 10 t/ha FYM recorded the highest values of soil chemical properties measured. Organic matter decreased by 37, 47 and 49.5% in manual clearing, P_1H_1 and P_2H_1 respectively, but percentage increase in organic matter contents under the various combinations involving tillage plus manure ranged from 76 to 45%. Total N decreased in both years under manual clearing, P_1H_1 and P_2H_1 contrasting the increase in values when FYM was applied (Tables 4a and b). Available P decreased by 12, 14 and 13% for manual clearing, P_1H_1 and P_2H_1 respectively, while increases under the various combinations involving the tillage plus manure ranged from 13 to 2%. Percentage decreases in exchangeable K under the different tillage systems were 27, 37 and 37% for manual clearing, P_1H_1 and P_2H_1, which was in contrast to increases ranging from 40-29 % for the various tillage-FYM combinations (Tables 4a and b).

Effect of tillage methods and manure on growth and yield components of maize

The tillage methods influence growth parameters of maize (Table 5). The highest values of plant height, leaf area and stem girth were obtained from manual clearing which was significantly higher (P = 0.05) than values obtained at 10 cm plus harrowing. But in 2007 experiment (rainy season crop), stem girth and 50% flowering recorded the highest values under manual clearing and this was not significantly higher (P = 0.05) from other tillage treatments. In both 2006 and early 2007 cropping seasons, neither number of leaves per plant nor 50% flowering date was significantly higher (P = 0.05) except in the late 2006 where ploughing 20 cm plus harrowing (P = 0.05) than manual clearing. Manure improved growth components of maize over unmanured plots and there were significant (P = 0.05) treatment differences for most of the parameters. The application of manure slightly reduced the number of days to attain 50% flowering and significantly increased (P = 0.05) plant height, stem girth over the unmanured plots in both experiments (Table 5). The values of maize growth components under the various methods of tillage alone were lower than under the various combinations of tillage and manure combinations. The combination of manure and tillage did produce significantly increases (P = 0.05) in number of leaves per plant and 50% days to flowering. Manual clearing plus manure gave the highest values of growth components at different levels of farmyard manure and NPK fertilizer followed by P_1H_1 and P_2H_1. However, in 2007 experiments, the highest values of plant height were obtained in ploughing at 10 cm depth plus manure.

The effects of tillage were significant on root and shoot dry weights, 100 seed weight and grain yield of maize in both 2006 and 2007 experiments (Table 6). Maize grown under manual clearing produced the highest grain yield in 2006 followed by ploughing at 20 cm depth plus harrowing. Manual clearing and ploughing at 20 cm depth plus harrowing out yielded ploughing at 10 cm depth plus harrowing by 17.4 and 2.1%, respectively. Application of manure significantly increased ($P < 0.05$) root dry weight, dry matter yield, 100 seed weight and grain yield of maize

Table 5. Effect of tillage methods and manure on growth characters of maize.

Tillage methods	Plant height (m)		Stem girth (cm)		No. of leaves		Flowering days	
	2006	2007	2006	2007	2006	2007	2006	2007
Z	1.43	1.63	4.5	6.8	3.5	3.5	55	54
ZM	1.81	1.86	5.3	7.1	3.6	3.8	55	52
ZF	1.84	1.92	5.5	7.3	3.7	3.9	57	53
P_1H_1	1.41	1.63	4.3	6.7	3.3	3.5	54	54
P_1HM	1.77	1.92	5.1	7.2	3.7	4.1	56	53
P_1HF	1.82	1.88	5.6	7.5	3.9	4.4	58	54
P_2H_1	1.25	1.55	3.6	6.4	3.3	3.4	56	55
P_2HM	1.51	1.75	4.2	6.6	3.6	3.8	53	55
P_2HF	1.49	1.84	4.8	7.0	3.8	4.1	58	54
LSD (0.05)	0.03	0.1	0.21	0.31	0.26	0.32	2.33	2.22
Unmanured	1.36	1.54	4.13	6.29	3.36	3.46	51	54
10 t/ha FYM	1.7	1.81	4.83	6.81	3.51	3.69	55	55
200 kg/ha NPK	1.69	1.89	4.97	6.90	3.52	3.73	55	56
LSD (0.05)	0.02	0.06	0.13	0.18	0.15	0.19	1.44	1.29

Legend: Z (manual clearing); P_1H (ploughing 10 cm depth) plus harrowing; P_2H (ploughing 20 cm depth) plus harrowing; F (200 kg NPK); M (10 t FYM).

Table 6. Effect of tillage methods and manure on yield components of maize.

Tillage methods	Root dry wt (g)		Shoot dry wt (g)		Leaf area (m^2)		100 grain wt (g)		Grain Yield (t/ha)	
	2006	2007	2006	2007	2006	2007	2006	2007	2006	2007
Z	6.8	8.4	48.6	54.8	0.12	0.12	19.9	20.6	4.11	4.32
ZM	8.0	9.3	58.4	65.7	0.19	0.15	21.3	20.9	4.42	4.51
ZF	7.6	10.2	54.5	68.3	0.28	0.22	20.59	22.1	4.34	4.60
P_1H_1	5.5	7.2	42.5	57.4	0.11	0.11	16.23	16.9	3.43	3.52
P_1HM	6.5	8.2	46.2	63.7	0.15	0.14	17.97	18.7	3.78	3.90
P_1HF	7.3	8.8	52.3	60.5	0.31	0.24	18.13	19.1	3.75	3.81
P_2H_1	6.5	7.8	44.1	59.6	0.08	0.10	17.18	17.5	3.58	3.67
P_2HM	7.7	9.4	50.8	67.3	0.12	0.13	18.63	18.83	3.93	4.01
P_2HF	7.4	9.7	47.2	70.5	0.15	0.17	18.20	18.90	3.73	3.82
LSD (0.05)	0.6	0.9	1.7	1.6	0.02	0.03	1.94	1.37	0.07	0.13
Unmanured	6.28	9.43	43.73	64.87	0.1	0.11	17.79	18.31	3.71	3.87
FYM	7.40	10.33	51.80	71.27	0.20	0.18	19.29	19.90	4.05	4.19
NPK	7.43	10.23	50.67	68.05	0.22	0.20	18.97	19.73	3.94	4.05
LSD (0.05)	0.38	0.63	1.03	0.71	0.01	0.02	0.8	0.51	0.03	0.06

Legend: Z (manual clearing); P_1H (ploughing 10 cm depth) plus harrowing; P_2H (ploughing 20 cm depth) plus harrowing; F (200 kg NPK); M (10 t FYM).

(Table 6). The increases between unmanured and manured tilled plots 10 t/ha FYM and 200 kg/ha NPK were 13 and 12% respectively. The increases in the values for grain yield were 6, 9 and 5 % respectively. Relative to unmanured plots (control), 10 t/ha FYM and 200 Kg/ha NPK treated plots increased grain yields by 8.7 and 5.3%, respectively. In both trials, application of manure in combination with tillage significantly affected the growth and yield components of maize (Table 6). Among tillage plus manure combination, manual clearing plus manure (10 t/ha FYM and 200 Kg/ha NPK) gave the highest values of root dry weight, shoot dry weight, dry cob weight, dry grain weight per cob, 100 grain weight and grain yield of maize in 2006 and 2007. This was followed by ploughing at 20 cm depth plus harrowing plus manure combinations (P_2H_1 + 10 t/ha FYM and P_2H_1+ 200 Kg NPK). Ploughing at 10 cm depth plus harrowing plus manure (P_1H_1 + 10 t/ha FYM and P_1H_1 + 200 Kg/ha NPK) gave the least values of yield components (Table 6).

Non-significant interactions between seedbed type and manure application were found for some soil physical and chemical properties, growth and yield of maize. The independent effects of these factors (treatments) also indicate that the seedbed types responded same way to types of manure applied.

DISCUSSION

In this trial the treatments applied produced the changes in soil chemical (fertility) and physical (hydrothermal, porosity and soil water storage) properties and the improvement in soil conditions enhanced maize performance. Tillage practices and soil amendment using organic wastes are known to lead to improvement in soil conditions (Stamatiadis et al., 1997; Wilson et al., 2000; Agele, 2007). The mechanized tillage resulted into disturbance of the soil, and this has been reported to lead to the destruction of soil aggregates (Hulugalle et al., 1985; Ohu et al., 1994). Application of FYM and NPK fertilizer improved the soil physical conditions and nutrients status presumably due to the supplied plant nutrients from the materials (Eaton, 2001; Agele et al., 2006).

The lower soil bulk density and high porosity produced by ploughing at 20 cm plus harrowing could be attributed to the loosening effects of tillage (Hulugalle et al., 1985; Lal, 1997). Although, higher soil bulk density produced by ploughing at 10 and 20 cm depth plus harrowing could be due to effects of tractor wheel-traffic and implement passes and lower macro-porosity and evaporation rate (Donahue et al., 1990; Agbede, 2006), however, Ojeniyi (1981) found that for sandy soils of southwest Nigeria that manual clearing had higher mean soil bulk density and lower total porosity at soil surface over tractorised tillage. Hulugalle et al. (1985) and Ike (1986) reported that higher bulk density for soils under zero tillage is due to soil compaction of untilled soil in southern Nigeria soils. This observation implies that in the tropics, continuous exposure of untilled soil to rainfall without mechanical tillage impacts the soil (Ojeniyi, 2003). Increased machinery traffic on agricultural soil associated with increased compaction has been attributed to increases in the number of wheel traffic passes (Hulugalle et al., 1985; Voorhees, 1987; Zhang et al., 1997).

Soil amended with FYM had lower bulk density and higher total porosity possibly due to increases in the proportion of macro-aggregates and soil organic (Agele et al., 2005). Bulk density is important to rainfall infiltration, root distribution and root function, which in turn affect plant water uptake and growth (Graecen and Gardner, 1982). Soil at the experimental site had high bulk density before planting maize. Meinke et al. (1993) reported the association of increasing dryness, high soil strength, and penetration resistance with increasing bulk density with depth. Application of manure reduced soil bulk density and temperature and increased moisture content and porosity in both years of experiment. The widely reported influence of manure application on soil physical conditions had been attributed to the enhancement of soil organic matter and possible improvements in soil porosity and structure (Hulugalle et al,. 1985; Agele et al., 2006).

The manually cleared soils had higher moisture content and lower soil temperature in both experiments compared with soils that were ploughed to at 10 cm or 20 cm depth. The highest moisture status of manual clearing could be adduced to the minimum soil disturbance with little exposure of the soil surface to the atmosphere and consequent reduction in water evaporation, therefore conserving soil temperatures and the available water in the soil (Maurya and Lal, 1979; Ghuman and Lal, 1984, Agele et al., 2000). Soil profile characteristics (moisture and density) are important to root distribution and root function, which in turn affect plant water uptake and growth (Graecen and Gardner, 1982).

Ploughing at 10 and 20 cm depth plus harrowing expose organic materials which were formerly inaccessible to microbial decomposition. Ploughing is known to accelerate mineralization rate of organic materials (Grace et al., 1993; De Neve and Holfman, 1997). The enhanced mineralization is necessary to sustain increased microbial activities due to the incorporated organic substrates especially plant residue of high C:N ratio (Grace et al., 1993; Mazzarino et al., 1993; Agele et al., 2006). The soil at the site of the experiment was deficient in organic matter, total N, available P, exchangeable K and Ca. Nottidge et al. (2005) also found that Alfisol was low in organic matter, N, P and exchangeable cations. Therefore, it is expected that maize grown on the soil at site of experiment would require amendments such as farmyard manure and inorganic fertilizer application in order to attain optimum productivity.

Lal (1979) in experiments conducted using different tillage techniques on an Alfisol in south western Nigeria, found that maize under non-tillage treatment with relatively high bulk density out yielded those planted under conventional tillage for two consecutive seasons. The range of bulk density recorded for the manual tillage methods (1.33 to 1.45 g/cm^3) was not limiting maize growth. Our results suggest that increase in soil bulk density up to 1.45 g/cm^3 did not reduce nutrient uptake, root penetration, dry matter and grain yields of maize. Combination of farmyard manure and NPK (a mineral fertilizer) with manual clearing (10 t/ha FYM and 200 kg/ha NPK) gave higher values for soil nutrient (N, P, K,

Ca and Mg) contents, growth and grain yield parameters compared with other tillage-manure treatments such as P_1H and P_2H (10 t/ha FYM and 200 kg/ha NPK). This finding attests to the positive cumulative effect of manual clearing methods and manure on soil productivity.

In this study, while the effects of tillage and manure were independently noticed, the tillage-manure combinations improved soil physical and chemical conditions and in addition affected the growth and yield of maize.

Conclusion

Field experiments were conducted to study the effect of tillage and manure application on soil physical and chemical properties, growth and yield of maize grown on a degraded Alfisol following several years of intensive mechanized tillage. The treatments brought about improvement in soil physical and chemical properties and enhanced growth and seed yield of maize in a humid zone of Nigeria. The manually cleared soil had higher moisture content and lower soil temperature in both experiments compared with soils that were ploughed to 10 or 20 cm depth (P_1H_1 and P_2H_1). The highest moisture status of manual clearing could be adduced to the minimum soil disturbance with little exposure of the soil surface to the atmospheric demand and consequent reduction in soil water evaporation could have conserved soil water. However, the mechanized tillage treatments (involving plough at 10 and 20 cm depth plus ploughing) reduced bulk density and enhanced porosity compared with manual clearing. However, the range of bulk density recorded for the manual tillage methods (1.33 to 1.45 g/cm^3) was not limiting to maize. Favorable bulk density and hydrothermal regime in the root zone and improved soil fertility can be advanced for the improved performance of maize under tillage combined with livestock manure application. Soil amendment with farmyard manure and inorganic fertilizer brought about improvement in soil fertility status of organic matter, N, P and exchangeable cations. Tillage depths and manuring appeared to have resulted in improved soil physical conditions (soil porosity and structure and reduced soil bulk density and temperature and increased moisture content) and enhanced growth and seed yield of maize in both years of experiment. The results of this study showed that manual clearing was the most productive soil in terms of performance of maize and soil fertility. The treatments also conserved soil water and reduced soil temperatures relatively to mechanized tillage systems.

REFERENCES

Agbede TM (2006). Effect of tillage on soil properties and yam yield on an Alfisol in Southwestern Nigeria. Soil Till. Res., 86:1-8.

Agele SO (2000). Effects of animal manure and NPK fertilizer on simulated erosion and maize yield. J. Environ. Educ. Inf., 19(2): 131-138.

Agele S O, Ewulo BS, Oyewusi IK (2005). Effect of some management systems on soil physical properties, microbial biomass and nutrient distribution under rainfed maize production in a humid rainforest Alfisol. Nutr. Cycling Agroecosyst., 72: 121-134.

Agele SO (2007). Efficacy of organic amendments on the establishment and biomass yield of grass species on adegarded alfisol. J. Plant Interactions, 2(3): 195-201.

Aina PO, Fapohunda H, Idowu J (1985). Compaction and moisture suction effects on soil strength and crop emergence. Ife J. Agric., 7: 16-25.

Bennie ATP, Botha FJP, (1986). Effect of deep tillage and controlled traffic on root growth, water use efficiency, and yield of irrigated wheat and maize. Soil Till. Res. 17: 85-95.

Brady NC (1990). Soil organic matter and organic soils. The nature and properties of soils. Macmillan Publishing Co. New York, p. 279-313.

Chu CM, Bradshaw AD (1996). The value of pulverized refuse fines (PRF) as a substitute for topsoil in land reclamation.1. Field studies. J. Appl. Ecol.. 33: 851-857.

De Neve S, Hofman G (1997). Modelling N mineralization of vegetable residues during laboratory incubations. Soil Biol. Biochem., 28: 1451-1457.

Eaton WD (2001). Microbial and Nutrient activity in soils of Ghana under different. Appl. Soil Ecol., 8: 19-24.

Elmar S, Wolfgang W (1990). Organic manure: No alternative to mineral fertilizers for developing countries. Fertilizer Int. pp. 7-12.

Grace PR, MacRae IC, Myers RJK (1993). Temporal changes in microbial biomass and N mineralization under simulated field condition. Soil Biol. Biochem., 25(12): 1745-1754.

Hulugalle NR, Lal R, Opara-Nadi OA (1985). Effect of tillage system and mulch on soil properties and growth of yam (*Dioscorea rotundata*) and cocoyam (*Xanthosoma sagillifolium*) on an ultisol. J. Root Crops, 11: 9-22.

Ike IF (1986). Soils and crop responses to tillage practices in a ferruginous soil in the Nigeria Savanna. Soil Till. Res. 6: 261-272.

Jackson ML (1962). Soil Chemical analysis. Prentice – Hall Inc. New Jersey, pp. 1-498.

Kanamori T, Yasuda T (1979). Immobilisation, mineralisation and the availability of the fertilizer nitrogen during the decomposition of organic matter applied to soil. Plant Soil, 52: 215-229.

Kay BD (1990). Rates of changes of soil structure under different cropping systems. Adv. Soil Sci., 12:1-52.

Lal R (1976). No-tillage effect on soil properties under different crops in western Nigeria. Soil Sci. Soc. Am. Proc., 40: 762-768.

Lal R (1979a). Influence of six years of no-tillage and conventional ploughing on fertilizer response of maize on an Alfisol in the tropics. Soil Sci. Soc. Am. J., 43: 399-403.

Lal R (1984). Mechanized tillage system effects on soil erosion from an Alfisol in water sheds cropped to maize. Soil Till. Res., 4:349-360.

Lal R (1997). Long-term tillage and maize monoculture effects on a tropical Alfisol in western Nigeria on crop yield and soil physical properties. Soil Till. Res., 42: 145-160.

Lampurlanes J, Martinez CC (2003). Soil bulk density and penetration resistance under different tillage and crop management systems and their relationship with Barley root Growth. Agron. J., 95: 526-536.

Lupwayi NZ, Arshad MA, Rice WA, Clayton GW (2001). Bacterial diversity in water stable aggregates of soils under conventional and zero tillage management. Appl. Soil Ecol., 16(3): 251-261.

Meinke H, Hammer GL, Wart P (1993). Potential soil water extraction by sunflower on a range of soils. Field Crops Res., 32: 59-81.

Maurya PR, Lal R (1980). Effect of no tillage and ploughing on roots of maize and leguminous crops. Exp. Agric., 16: 185-193.

Mazzarino MJ, Szott L, Jimenex JM (1993). Dynamics of total C and N, microbial biomass and water stable C in tropical agroecosystems. Soil Biol. Biochem., 25(2): 205-214.

Nangju D, Wanki SBC (1980). Estimating leaf area of cowpea and soybean using dry weights of terminal leaf lets. Exp. Agric., 16: 149-151.

Ogunremi LT, Lal R, Babalola O (1986). Effect of tillage and seedling method on soil physical properties and yield of upland rice for an ultisol in southeast Nigeria. Soil Till. Res., 6: 223-234.

Ohu JO, Ekwue EI, Folorunso OA (1994). The effect of addition of organic matter on the compaction of a vertisol from northern Nigeria. Soil Technol., 7(2): 155-162.

Ojeniyi SO (1982). Soil structure produced by tillage implements and their effects on soil water. Ghana J. Agric. Sci., 14-19: 45-49.

Ojeniyi SO (1986). Effect of zero tillage and disc ploughing on soil water, soil temperature and growth and yield of maize (*Zea mays*). Soil Till. Res., 7: 173-182.

Ojeniyi SO, Agboola AA (1995). A review of tillage requirement of crops in Africa Soils 28, 259-266.

Raymond WM, Roy LD (1992). Organic Matter and Container Media Soils: An Introduction to soils and Plant Growth, 6th edn. (Delhi: Prentice Hall), pp. 181-224.

Sobulo RA, Osiname OA (1986). Soil properties and Crop yields under continuous cultivation with different management system. In: R. Lal,P.A. Sanchez and R. W. Cunnings (eds). Land clearing and development in the tropics. A. A. Balkama, Rotterdam, Boston, pp. 363-370.

Stamatiadis S, Werner M, Buchanan M (1997). Field assessment of soil quality as affected by compost and fertilizer application in a broccoli field (San Benito Country, California). Appl. Soil Ecol., 12(3): 217-225.

Stewart GA, Vyn TJ (1994). Influence of high axle loads and tillage systems on soil properties and grain corn yield. Soil Till. Res., 29(2-3): 229-235.

Van Landen GWJ (2000). Soil Degradation in Central and Eastern Europe: The assessment of the status of human-induced degradation (ver. 1.0) Wageningen: FAO and ISRC-World Soil information, (ISRIC Report) 2000/5), p.15

Voorhees WD (1987). Assessment of soil susceptibility to compaction using soil and climatic base. Soil Till. Res. 10: 29-38.

Walkley TA, Black IA (1934). An estimation of the Degtjareff method for determining soil organic matter and a proposed modification of the chromic acid titration method. Soil Sci., 37: 29-38.

Wilson CW, Shaw PE, Knight Jr RS (1982). Analysis of Oxalic Acid in Carambola (Averrhoa Carambola L.) and Spinach by high performance liquid Chromatography J. Agric. Food Chem., 30: 1106-1108.

Wilson TC, Paul EA, Harwood RR (2000). Biologically active organic matter fractions in sustainable cropping systems. Appl. Soil Ecol. 16(1): 63-76.

Wong MH, Wong JWC (1989). Germination and seedling growth of vegetable crops in fly ash amended soil. Agric. Ecosyst. Environ., 26: 232-235.

Zhang HQ, Hartge HK, Ringe H (1997). Effectiveness of organic matter incorporation in reducing soil compactibility. Soil Sci. Soc. Am. J. 61(1): 239-245.

Spatial variability mapping of some soil properties in El-Multagha agricultural project (Sudan) using geographic information systems (GIS) techniques

M. Y. Mohamed[1]* and B. M. Abdo[2]

[1]Faculty of Agriculture, Alzaiem Al-Azhari University, Khartoum North, Box 1432, Sudan.
[2]Ministry of Agriculture and Forestry, Sudan.

Soil surveying and mapping is an important operation, since it plays a key role in the knowledge about soil properties and how it can be used for agriculture, irrigation, urbanization and other land uses. This study was carried out in order to map out some soil characteristics in the Northern State in resettlement projects associated with the newly established Merowe Dam. Some physical and chemical properties of the soil were determined to assess the current situation of the region, since no similar study had been conducted in the area to assess the soil status. The use of the geographic information systems (GIS) and sampling broad analysis through the use of GPS is found to be a very efficient tool. This study demonstrates the usefulness of these technologies in soil survey. The objective of this study is to conduct a soil survey in a new developing area using a nontraditional technique. GIS was used in this study to present the laboratory findings of the soil analysis; as such, this was achieved through a preparation of digital maps for soil properties. These digital maps can be used in precision farming studies with the application of variable rate technology. The obtained results showed the alkaline reaction of soil, where the range of pH was between 8.8 and 7.0. Moreover, the EC's range was found to be 2.34 to 3.30, where the value of 3.38 dS/m was found in depth of 30 to 60 cm. Nonetheless, the SAR values fall within the slight, moderate and strong degrees of salinity.

Key words: Geographic information systems (GIS), global positioning system (GPS), soil survey, soil properties mapping, spatial variability.

INTRODUCTION

Soils play a central role in the life and development of lands. They provide a vital substratum for people, animals, plants and micro-organisms. They are a key component of the ecosystems, in particular for water and nutrients cycling. In addition, they constitute a major genetically reservoir and are a place of major regulation and transformation processes. As a central element of the mountainous landscape, soils have always occupied a central position in the cultural and economic life of human communities. In particular, they constitute the basis for agriculture, including forestry and raising of livestock, and a major component for recreational and touristic activities. It is therefore essential to promote the conservation and sustainable use of the soil resource, taking into account the specific sensitivity of soils to degradation and alteration processes.

Soil survey can be of use to the agricultural producers, farm managers, and consultants; hence, it contains information that is the basis for conservation efforts aimed at erosion protection from wind and water (FAO, 1988). Soil survey evaluation is a tool to assess, manage and induce changes in the soil and to link existing resource concerns to environmentally sound land management practices and soil survey assessment, which is thus used to evaluate the effects of management on the soil. An

*Corresponding author. E-mail: Mustafayousif70@yahoo.com.

intimate knowledge about the types of soils and their spatial distribution is a prerequisite in developing rational land use plan for agriculture, forestry, irrigation and drainage etc.

Soil survey provides an accurate and scientific inventory of different soils, their kind, nature and extent of distribution so that one can make prediction about their characters and potentialities. It also provides adequate information in terms of land form, terraces and vegetation (Brown et al., 1978).

Reliable and timely information on soil is a vital issue conducted throughout the nation in order to gather spatial information and associated statistics on crops, rangeland and other related types of soil uses. Such types of information are most important for the implementation of effective management and decisions making at district levels.

Advent of GIS technology and its great potential in the field of soil have opened newer possibilities of improving soil statistic system as it offers accelerated, repetitive, spatial and temporal synoptic view. GIS is a potential tool for handling voluminous data and has the capability to support spatial statistical analysis, thus there is a great scope to improve the accuracy of soil survey through the application of GIS. It may be difficult to implement such programs without regulations to deal with information that enables researchers and decision-makers to benefit from the vast quantity of data and maps that can be used effectively by using methods of collection, storage and traditional or customary analysis; therefore, it is now important to use some systems such as GIS so as to create geographic information rules that will help in the accounting and analysis of the achievements and maintenance programs. The development of plans and processes of data acquisition and analysis is very fast through the use of GIS as compared to conventional methods.

The main objective of this study is to conduct a soil survey in a newly developing area using a nontraditional technique. GIS will be used in this study to represent and illustrate the geo-referenced the laboratory's analysis findings of the soil analysis, spatially to help in identification of areas suitable for the cultivation of crops in the study area and to develop a suitability map for the study area. This will be achieved through the preparation of a digital map for some soil properties. New tools allow both the scientists who implement soil survey and the decision makers who use them to better analyze and utilize the data. GIS has greatly improved the ability to analyze and spatially depict soil data. In addition, custom soil data queries can be made and the results can be presented spatially using real world locations.

MATERIALS AND METHODS

This work was conducted in a study area located in the Western bank of the River Nile, in the Northern State of Sudan, and the area was confined between longitudes 30º 00' 08" and 31º 15' 31"E and latitudes 17º 50' 11" and 18º 10' 11" N (Figure 1).

The total area of the project is about 147000 km^2. The project area is a true desert with extremely high temperatures and solar radiation in summer, low temperatures in winter, low rain fall and high wind speed. The diurnal range of temperature is wide all year round. The mean maximum and minimum temperatures are 37 and 19°C, respectively, whereas the climate is hyper-arid with low vapor pressure and relative humidity. The shine duration of the mean bright sun is 10.5 h; nevertheless, solar radiation is as high as 25.88 MJM in May. Rainfall is low with a mean annual amount of 21 mm, whereas the wind prevails from the North for most parts of the year. The winter season is characterized by the strong dry North direction winds which cause serious sand drift in the area. However, the average wind speed is 15.7 km/h (Ministry of Agriculture and Forestry (MAF), 1999).

The data required for this study were collected from different sources and were used in the database construction. The ground truth and field observations were obtained using a GPS navigator at a 500 m grid interval covering the study area. The field activities included an examination of the landscape set up and description of the soil mapping units and laboratory analysis of the collected soil samples from the field. The data were collected from the Ministry of Agriculture and Forestry for characterization of the standard chemical and physical properties and the general properties of the soil of the Northern State area.

RESULTS AND DISCUSSION

Soil surveying and mapping is an important operation, since it plays a key role in the knowledge about soil properties and how it can be used for agriculture, irrigation, urbanization and other land uses. This study was carried out to in order to map out some soil characteristics. As such, some physical and chemical properties of the soil were determined to assess the current situation of the region; thus the results from this study can be presented as follow.

Chemical properties of the soil

The values of soil analysis showed that pH has no significant difference ($P \leq 0.05$), in all depths. In the depth of 30 to 60 cm, alkaline soils have a pH of 7.5 to 8.2. The overall average of the pH for all depths is 8.0, while the average pH of the top and sub soils is 7.6 and 6.3, respectively. The pH values on saturated paste were almost invariably low reflecting the effect of exchange-able sodium, whereas the pH values for the soil samples in different depths were plotted and interpolated using kriging interpolation algorithm method to show the spatial distribution of the pH in 0 to 30 and 30 to 60 cm depths, respectively as shown in Figures 2 and 3.

From Figure 2, it can be observed that the alkaline soils are dominant in the study area and they occupy more than 80% of the area, while the acidic soils cover only about 8% of the area. The spatial distribution of the pH on the sub soil layer (30 to 60 cm) shows that the soils are alkaline.

On the other hand, the spatial analysis was used to

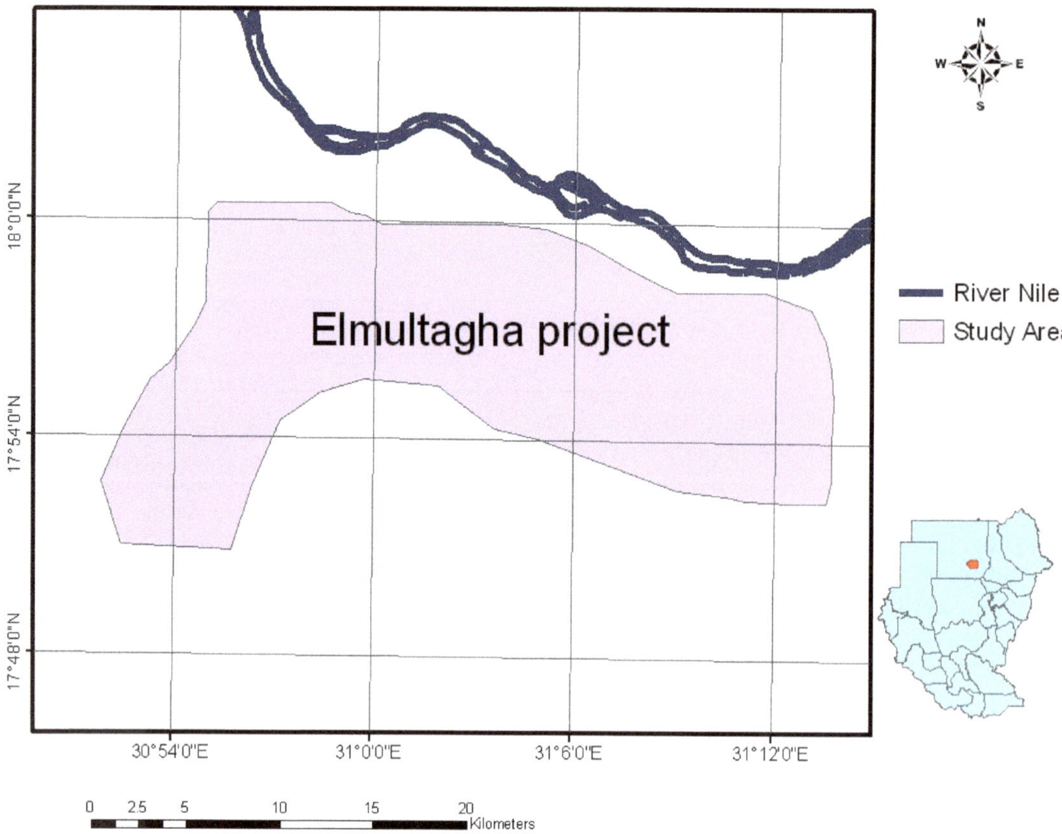

Figure 1. Location of El-Multagha project.

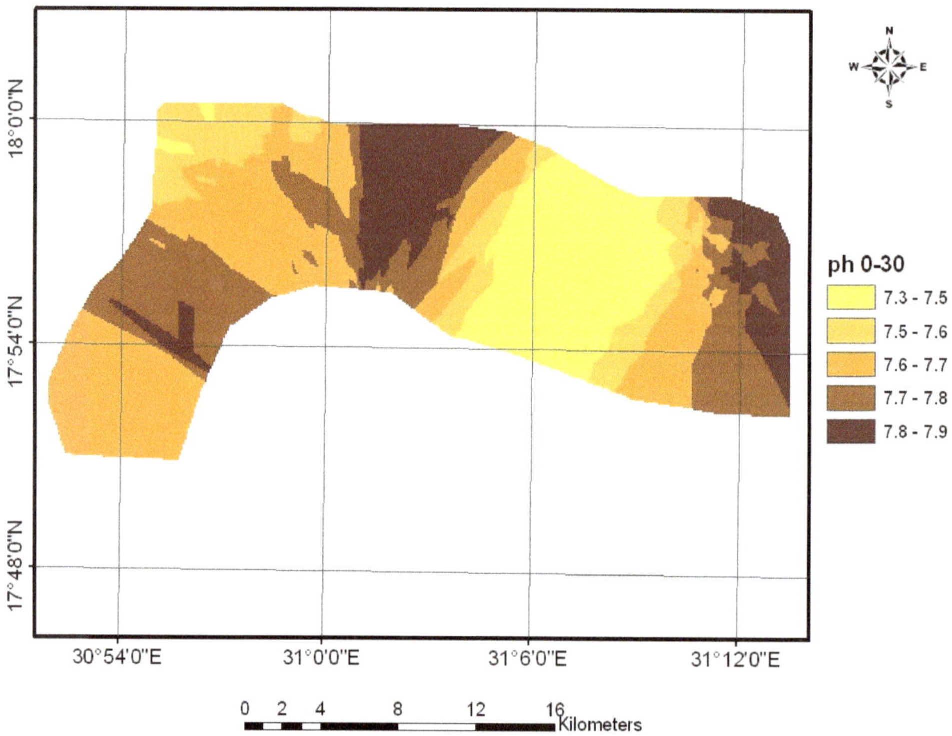

Figure 2. Spatial distribution of the pH in 0 to 30 cm depth.

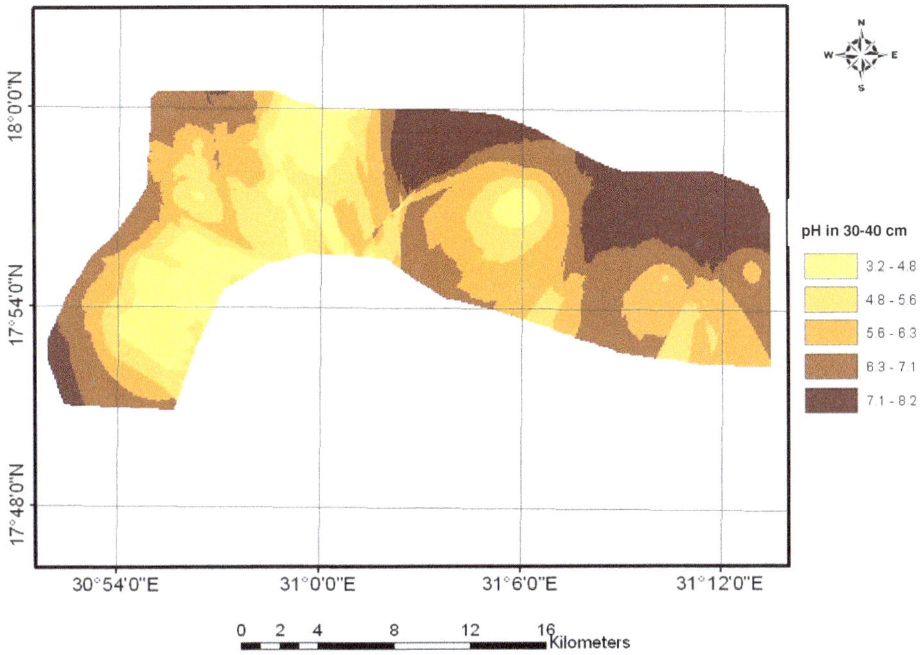

Figure 3. Spatial distribution of the pH in 30 to 60 cm depth.

show the distribution of the EC in the study area as shown in Figures 4 and 5 for the depths of 0 to 30 cm and 30 to 60 cm, respectively. From Figure 4, it can be observed that the non saline soils covered the majority of the soil, and this means the top layer of the soil (0 to 30 cm) in the study area is fertile and suitable for agriculture. From the same figure, the salty soil represent less than 5% of the soil and medium salinity covering 2%, while the high salt soils cover only 0.35% of the area.

In the depth of 30 to 60 cm, the non saline soil covers 90% of the area, while the saline soil represents 7% as shown in Figure 5. However, medium salinity soil represents 1.5% of the area and very saline soils covers only 0.50% of the area.

The results from soil samples analysis for sodium adsorption ratio (SAR) were plotted and spatially analyzed to show the distribution of SAR in the study area. Hence, Figures 6 and 7 show the spatial distribution of SAR.

From these figures, it can be observed that the majority of soils have less than 4 and less than 8 of the SAR in the depths of 0 to 30 cm and 30 to 60 cm, respectively. From Figure 6, the non sodic soils cover about 77% of the soils, while it covers 80% of the lower depth (Figure 7). Results showed that there were no statistical significant differences in SAR in the different depths.

Soil physical properties

Soil texture is one of the most important characteristic which influences the physical properties of the soil and has great significance in land use and management. The results show no significant difference in soil texture in different layers. Table 1 shows the soil texture in the different layers, while Figure 8 represents the spatial distribution of the soil texture in the study area.

No quantities measurements were made for soil structures and so reliance must be placed on qualitative field assessment. Generally, there are marked structure differences between the top and second horizon, which is usually a direct reflection of soil texture and moisture, though salinity also has some effect. The soils were almost invariably dry in close proximity to the surface and almost dry or only slightly moist at depth. The soils tended to become harder or firmer down the profile, being soft to slightly hard in the top 15 to 20 cm, and becoming hard or firm below 50 cm. Below 50 to 90 cm, where the sub soils are calcareous matrix, the consistency becomes very hard and usually remains very hard or extremely hard (Younis, 1985).

Land capability classification

Three classes were recognized in the capability order as S1, S2 and S3 and two classes in the unsuitable order as N1 and N2. Figure 9 shows the spatial distribution of the land suitability classes.

Conclusion

The outcomes from this study refer to the effective role

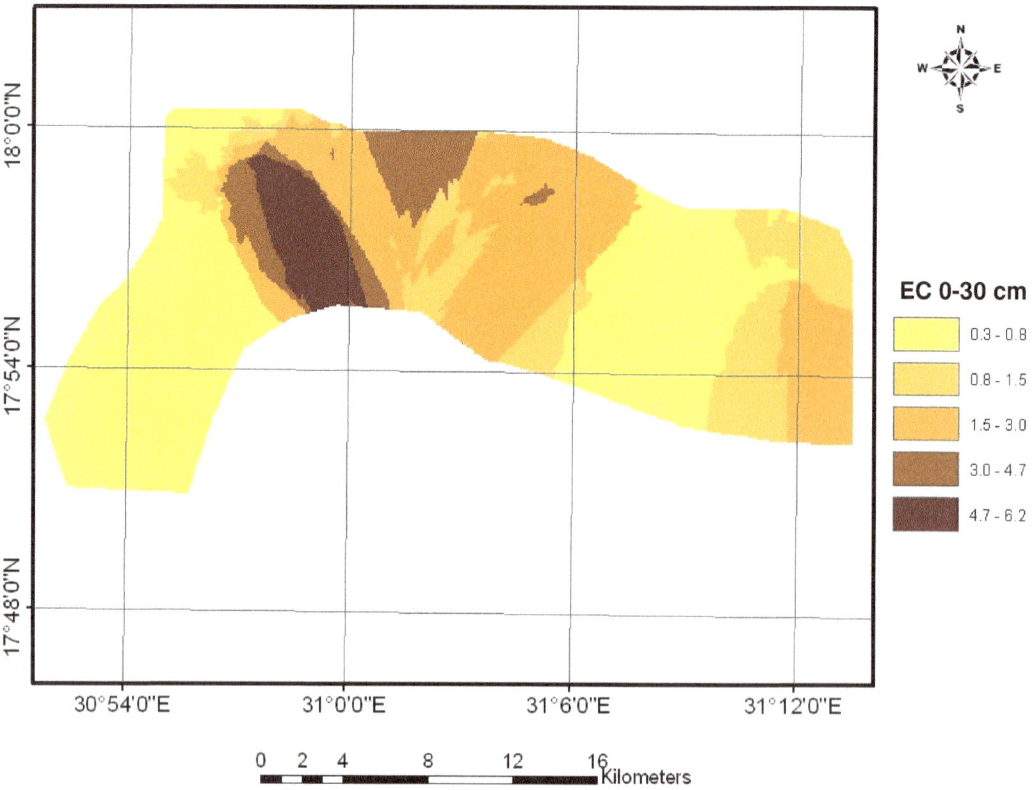

Figure 4. Spatial distribution of EC in 0 to 30 cm depth.

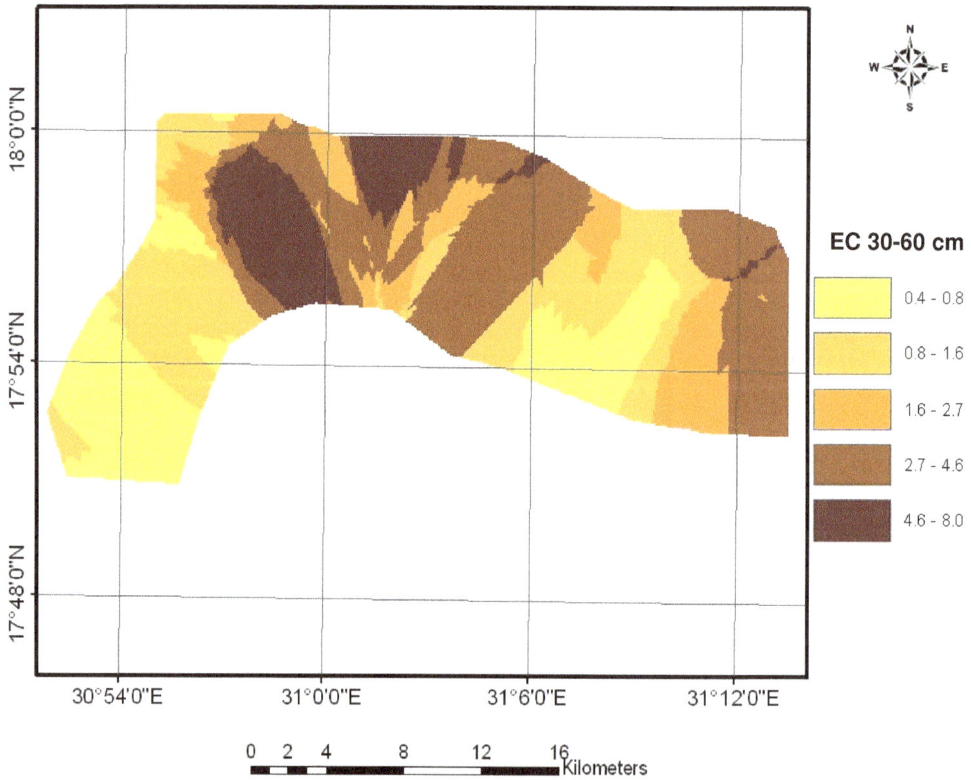

Figure 5. Spatial distribution of EC in 30 to 60 cm depth.

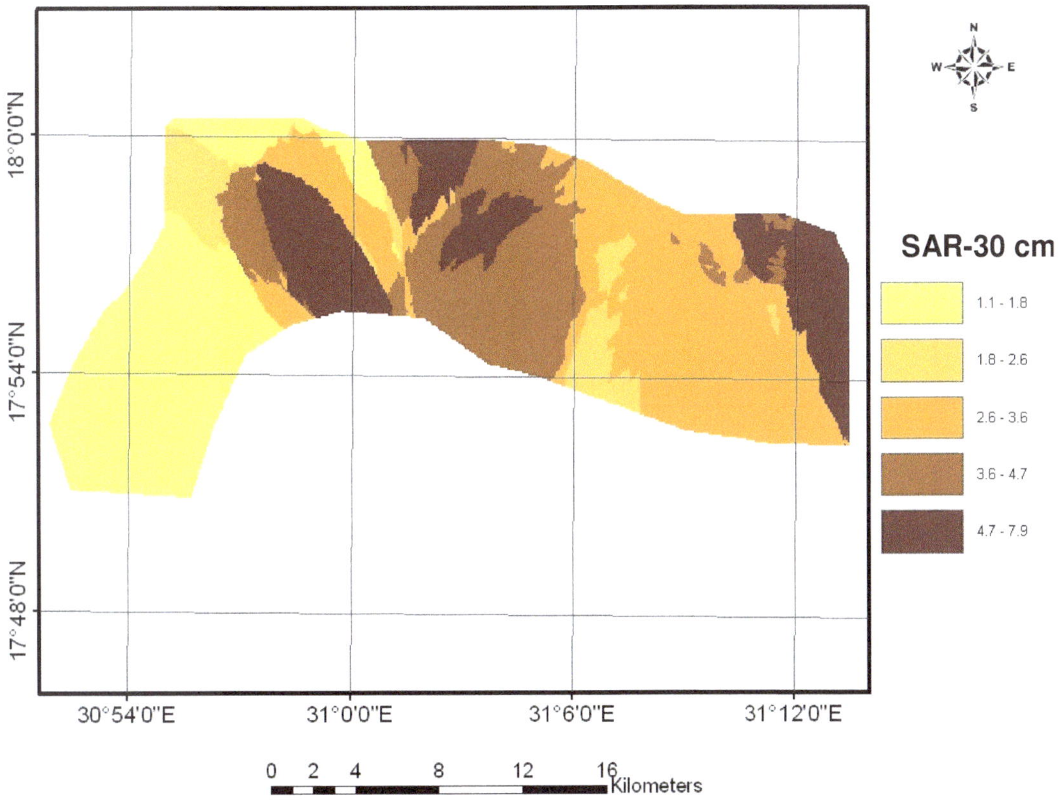

Figure 6. Spatial distribution of SAR in 0 to 30 cm depth.

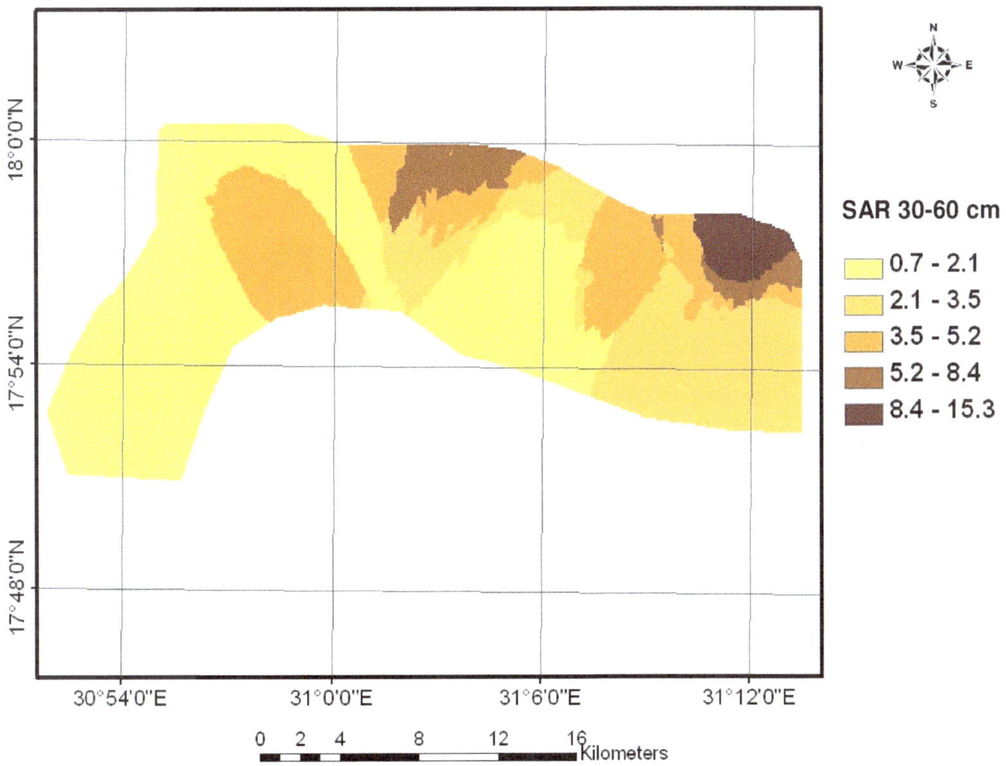

Figure 7. Spatial distribution of SAR in 30 to 60 cm depth.

Table 1. Soil particle size distribution in different depths.

Depth (cm)	Sand %	Silt %	Clay %	Soil texture
0 to 30	48.32	13.07	38.61	Loamy sand
30 to 60	45.05	15.20	39.75	Clay loam

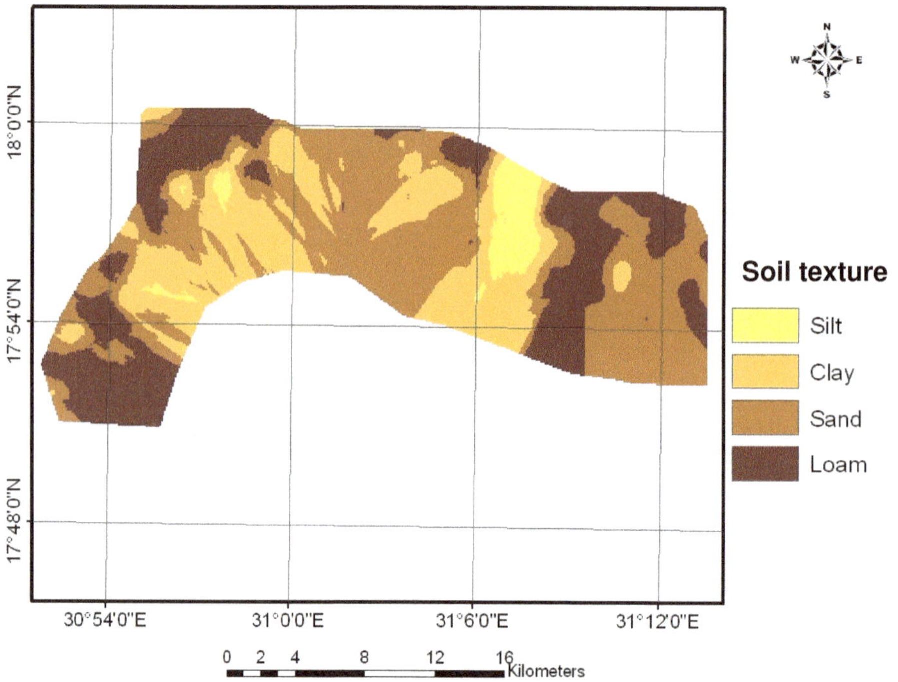

Figure 8. Spatial distribution of soil texture in the study area.

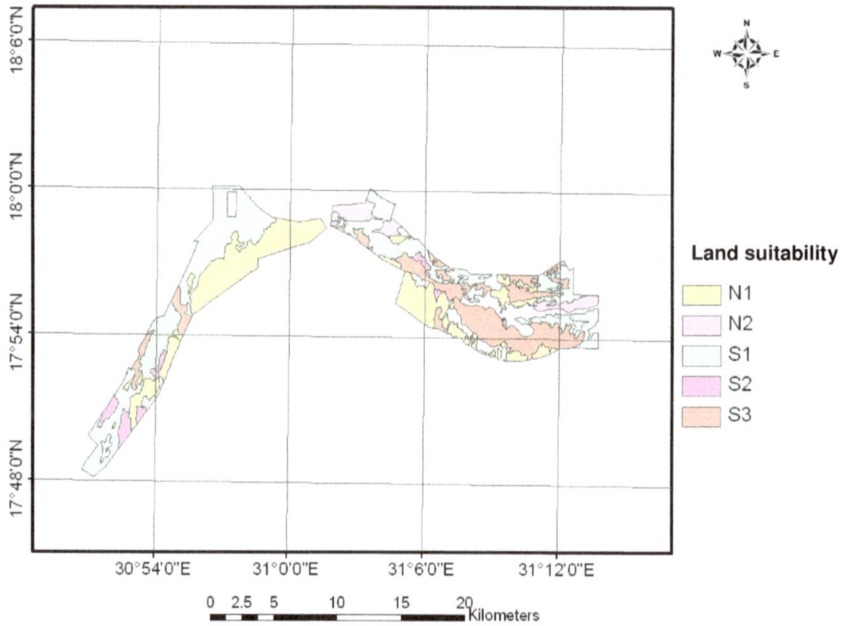

Figure 9. Spatial distribution of the land capability classes.

that can be played by GIS, especially in the stage of interpolation and analysis of soil classification of the soil suitability, in producing thematic maps such as soil physical and chemical properties maps. The use of spatial analysis in particular, shows that there is a difference in most of the moral characteristics of the soil in the study area; as such, the results reflect the soil properties of the study sites. The obtained results showed the alkaline reaction of soil, where the range of pH was between 7.5 and 8.2. The range of EC was found to be 2.34 to 3.30 where the value of 3.38 dS/m was found in the depth of 30 to 60 cm. However, the SAR values fall within the slight, moderate and strong degrees of salinity. According to the results of this study on soil types and impact of land use, it can be concluded that the soil is saline, and its texture varied from clay to clay loam content in the study area. The result showed a decrease of sand fractions in the cultivated area. It can be recommended here that the utilization of remote sensing and geographic information systems in the process of early warring desertification erosion due to wind and water help to supplement the survey and study the soil properties as well. Using the technology of GIS and digital mapping system to implement the variable rate technology is the logic direction to increase the agricultural operations efficiency.

REFERENCES

Brown G, Newman ACD, Rayner JH, Weir AH (1978). The structure and chemistry of soil clay minerals: in The Chemistry of Soil Constituents, D. J. Greenland and M. H. B. Hayes, eds., Wiley, New York, pp. 29-178.

FAO (1988). Salt affected soil and their management, Soils Bul., p. 39.

Ministry of Agriculture and Forestry (MAF) (1999) detailed Soil survey and land suitability classification of Elmultaga Scheme by land and water research center.

Younis MGA (1985). Emirates and Sudan investment CO.LTD Agricultural scheme, Jubel Aulia – Khartoum province, p. 1.

The integration of variable rate technologies for a soil-applied herbicide in leafy green production

Kurt D. Nolte

The University of Arizona, Yuma County Cooperative Extension, Yuma, Arizona (USA) 85364, USA.
E-mail: knolte@ag.arizona.edu

While lettuce is an intensively managed specialty crop for which there is little tolerance for weeds, the herbicide benefin (Balan®) is commonly used in preplant and preemergence applications to control weeds during lettuce stand establishment. In fields with high clay and low organic matter contents, considerably more of the herbicide is required for adequate weed control than in soils with higher sand contents where excessive herbicide rates can injure young lettuce roots. A three year effort involving variable rate application (VRA) of Balan® was conducted on three, 10 ha grower-cooperator leaf lettuce fields in Yuma, AZ (USA) each consisting of three distinct soil textures and application rates; a clay (2.8 kgai/ha), a loam (2.2 kgai/ha) and either a sandy loam or loamy sand (1.7kgai/ha). The effectiveness of VRA was determined by comparing conventional applications at the standard rate of 2.2 kgai/ha and quantifying lettuce injury, yields at harvest, and overall weed control.Site-specific placement of Balan® resulted in over 30% reducedlettuce seedling injury in loamy sand textured soils with 35% lower application amounts than using conventional methods; weed control was similar among all treatments. As over 40% more marketable lettuce was demonstrated in sandy loam textures using VRA, the application shows promise for weed control in leafy greens where non uniform soil textures show a severe challenge to production.

Key words: Benefin, variable rate application, variable rate application, precision agriculture, soil texture, soil uniformity, lettuce, site specific, herbicide, weed control, crop injury.

INTRODUCTION

Since in the mid 1990's, advances in global positioning systems (GPS), microprocessors, actuators, controllers, plant/soil sensors and geographical information systems (GIS) have spawned the development of new, site-specific, variable-rate techniques for chemical applications. At the same time, agricultural production has embraced other new technologies that increase the productivity of mechanized operations such as tractor auto-guidance systems. The combination of automatic tractor steering and variable rate technology is well suited for site-specific application of pre-emergence herbicides. With tractor guidance control and variable rate controllers, growers can increase the efficiency of chemical application by eliminating swath overlap, while increasing the efficacy of herbicide action by applying optimum rates based on soil texture.

These technologies have primarily been adopted by growers of major crops such as corn, wheat and soybeans (Koch and Khosla, 2007). Recently however, vegetable producers have become increasingly interested in using these technologies for variable rate application of soil applied preemergence herbicides. In many fields where vegetables are grown in the United States, there is large spatial variation of soil properties, including soil texture. In fact, Bauer and Schefcik (1994) found that recommended application rates of preemergence, soil applied herbicides can vary as much as 50% in a given field due to varying soil textures.

Soil-applied preemergence herbicides are widely used in vegetable production for controlling weeds during stand establishment. These types of herbicides are used to control weeds after sowing, but prior to crop emergence. In vegetable production, they are typically broadcast applied to the soil surface and then incorporated into the top several inches of soil. After weed seed germination, the herbicide is absorbed by the

Table 1. Physical properties of common preemergence herbicides used in Arizona vegetable production.

Product name	Common name	Water solubility (ppm)[1]	Herbicide soil/solution distribution Index (K_d)	Herbicide soil/organic affinity index (K_{oc})[1]	Soil half-life (days)[1]	Mobility rating[1]
Balan® 2.5-G	Benefin	<1	16.7 – 48.6[2,3,4]	9,000	40	Extremely low
Kerb® 50-WSP	Pronamide	15	6.2 – 15.4[2]	200	60	High
Prefar® 4-E	Bensulide	6	5.4 – 13.5[2,5]	1,000	120	Moderate

[1]Weed Science of America, 2007, [2]Weber et al., 2000, [3]Jacques and Harvey, 1979, [4]Weber, 1990, [5]Carlson et al., 1975.

roots and/or shoot of the seedling. This kills the weed seedling, typically before it emerges from the soil. Contrary to conventional wisdom, preemergence herbicides do not kill plants by preventing seed germination (Ross and Childs, 1996).

After a herbicide has been applied to soil, it has one of three fates. It can either remain dissolved in the soil solution, be adsorbed by smaller soil colloidsor be absorbed by plant roots and shoots (Colquhoun, 2006). Herbicide that is adsorbed by soil is bound to soil and is not readily available for weed absorption. Soil properties that affect soil adsorption can include the presence of mineral oxides, soil organic matter and clay composition, the level of crop residues in the soil, soil pH and soil water content. However, typical of western soils with high clay and low organic matter contents, the amount of herbicide adsorbed by most western desert soils is highly dependent on the soil's clay content - the higher the clay content, the greater the amount of soil adsorption (Moomaw et al., 1992). Because of this, recommended rates for soil-applied herbicides are commonly based on soil texture. Herbicide adsorption to soil can be described using soil/organic matter (K_{oc}) and soil/soil solution indices (K_d) (Table 1). Both parameters are defined as the ratio between the level of herbicide adsorbed by a reference and the amount found in soil solution. Herbicides with high sorption indices are typically prone to soil adsorption when applied in soils high in clay or organic matter contents. This reduces herbicide available for root uptake in the soil solution, and higher herbicide application rates are needed to increase soil solution herbicide concentrations for effective weed control. When the same high herbicide rates are applied to soils with low clay content, less herbicide is adsorbed by soil and crop injury can result from excess herbicide in the soil solution. For herbicides with low sorption indices, herbicide effectiveness is much more consistent and less likely to cause crop injury problems since the herbicide is not readily adsorbed by soil.

The three most common soil incorporated preemergence herbicides used in Arizona vegetable production are listed in Table 1 along with the physical properties that are most important to their fate in soil. Of these, benefin (Balan®; active ingredient, N-butyl-N-ethyl-α,α,α-trifluoro-2,6-dinitro-*p*-toluidin) requires the most critical consideration of soil texture when determining the proper application rate to use. The herbicide has a very high soil sorption index making it strongly adsorbed by soil colloids and highly immobile (Table 1). In soils with high clay content, a greater amount of the herbicide is required for adequate weed control as compared to sandy soils. If rates necessary for good weed control in high clay content soils are used on sandy soils, excessive herbicide concentrations in soil solution can cause injury to lettuce seedling roots (Tickes and Kerns, 1996). This high concentration problem is not easily resolved since Balan®has an extremely low mobility rating and is relatively insoluble in water, and therefore not likely to leach below the depth where crop seedlings develop. The objectives of this study were: (i) to quantify and demonstrate the benefits of variable rate technologies for mediating the damage caused by the over application of the herbicide Balan® across non-uniform soil textures; and (ii) to evaluate crop yield in response to Balan® VRA.

MATERIALS AND METHODS

Study location

The feasibility of site-specific, variable rate application strategies for preemergence herbicides in vegetable production was investigated on three commercial lettuce fieldsnear Yuma, Arizona (USA). Typically, candidate fields for variable rate technologies are chosen based on past experience with herbicide injury and the variable nature of soil textures within a field. Prior aerial imagery of the field may also indicate distinct zones of non-uniformity. The 10 ha fields used in this study were selected because the fields had significant variability in soil type and the grower-cooperators had experienced some form of crop injury when applying uniform rates of soil incorporated herbicides in previous seasons.The ten hectares field locations included those within the Yuma Valley (latitude, 32.726442; longitude, -114.674720), Dome Valley (latitude, 32.724776; longitude, -114.310878) and Gila Valley (latitude, 32.727126; longitude, -114.793710) whose soil characteristics are described in Table 2.

VRA map generation

To generate site specific application maps, a onehectare grid sampling scheme was used to prepare a geo-referenced soil textural map based on soil saturation percentage (SP). Saturation

Table 2. Soil characteristics, leaf lettuce cultivars and dates of various field operations.

Experimental descriptor	Variable	Location: Yuma valley 2006	Yuma valley 2007	Dome valley 2007	Dome valley 2008	Gila valley 2007	Gila valley 2008
Soil description	Name	Indio / Ripley		Glenbar / Indio		Indio / Ripley	
	Texture	SC[1] / SCL[2] / LS[3]		C[4] / SCL / LS		C / SC / SL[5]	
Soil composition	SP[6] (%)	49 / 32 / 18		66 / 38 / 16		69 / 47 / 25	
	Sand (%)	46 / 65 / 80		23 / 51 / 88		17 / 47 / 58	
	Silt (%)	1 / 5 / 15		22 / 15 / 1		8 / 5 / 35	
	Clay (%)	53 / 30 / 5		55 / 34 / 11		65 / 48 / 7	
	OM[7] (%)	0.4 / 0.4/ 0.5		0.4 / 0.3 / 0.4		0.5 / 0.5 / 0.4	
	pH	7.8 / 7.7 / 7.8		7.8 / 7.8 / 7.7		7.6 / 7.7 / 7.7	
Lettuce cultivar	Green leaf /Red leaf	Two star / Vulcan	Two star / Red fire	Green star / Blackhawk	Green star / Firecracker	Green star / Red sails	Tropicana / Vulcan
Balan® application	Date	25 Sept	28 Sept	4 Oct	28 Sept	1 Oct	30 Sept
Lettuce planting	Date	28 Sept	1 Oct	8 Oct	1 Oct	5 Oct	5 Oct

[1]SC: Sandy Clay, [2]SCL: Sandy Clay Loam, [3]LS: Loamy Sand, [4]C: Clay, [5]SL: Sandy Loam, [6]SP: Saturation Percentage, [7]OM: Percent Organic Matter.

percentage, expressed as grams of water required to saturate 100 g of soil, is a variable correlated with soil texture (Peacock, 1998) where the higher the SP, the greater the soil clay content. Each experimental field was composed of three distinct soil textures found within three distinct area locations as summarized in Table 2. A herbicide application rate map was created (Farm Works Site Software, Trimble Navigation, Limited, Sunnyvale, CA, USA) based on soil texture with zones having the highest clay contents receiving greater rates of herbicide (2.8 kgai/ha) than in sandy areas (1.7 kgai/ha), an example of which is shown in Figure 1. The regions within fields containing transitional clay contents were assigned to receive an intermediate level of the herbicide (2.2 kgai/ha). The intermediate rate was also applied within each soil textural zone and served as the standard treatment.

VRA techniques

Balan® was applied either at a standard rate or variably and in response to soil conditions (2.8 – 1.7 kgai/ha) in lengthwise strips (200 × 20 m) across the field. Each strip (six for each variable-rate and standard rate treatments) represented one replication in the experiment's randomized complete block design. Variable rate, site-specific herbicide applications were made using a 2004, model 4640 SpraCoupe® (AGCO Corp., Atlanta, GA, USA) integrated with Trimble® GPS guidance technology and a Raven Viper® flow control module. Two key components in this variable rate technology spraying system are the vehicle ground speed sensor and the CAN-bus (Controller-Area Network) hardware that includes: (a) a boom control node that regulates the operation of the sprayer in sections (booms) through the interface with the operator and a switch box; and (b) a CAN control node which receives information from pressure and flow meter sensors and then sends a signal to the flow control valve.

Application rates were adjusted by a variable rate controller, which incorporated an embedded computer that read sensor inputs, the GIS prescription map, user commands, and ground speed to calculate the correct application rate through an appropriate algorithm. The correct rate was then translated into actual product output through actuators, through opening/closing of solenoid valves and changes in rotational speed of hydraulic motors driving the pumps. Numerous variable rate controllers are available commercially which can interface with different devices through standard connectors. An important consideration when selecting a variable rate control system is the response times of its various components, because a rapidly responding system is needed to make the quick rate changes while moving from different application zones. The system used in this study incorporated an automatic look a head feature capable of providing the predictive speed compensation required to correctly synchronize product rates with the correct zones after subtracting inherent system lag times.

VRA assessment

Following disc incorporation of Balan® to a depth of 10 to 15 cm, rows were prepared on 100 cm centers and shaped to a height of 25 cm. Green and ruby red leaf lettuce (*Lactuca sativa* var. *crispa*) was planted within one week following herbicideapplicationat two seed-lines per bed and germinated using sprinkler irrigation. Flood irrigation (~30 cm) was used to irrigate the crop to harvest. Four weeks after germination, seedling injury assessment was determined by counting plants within each soil textural class which showed stunting and a leathery leaf appearance, similar to those described by Tickes and Kerns (1996). All injured lettuce seedlings and weeds which survived herbicide effects within an area measuring 25 m × 5 min each plot were counted. Herbicide efficacy was determined by counting the number of weeds in each plot at 14 and 28 days after germination, and weed control was similar among all treatments (Table 3). Weed species controlled during the study included pigweed (*Amaranthus* sp.), common purslane (*Portulaca oleracea*), lambsquarters *(Chenopodium album*) and ground cherry (*Physalis wrightii*). Treatment effects were also determined at lettuce maturity by counting unharvested lettuce heads within each experimental plot. All original observations were square root transformed prior to statistical analyses using SAS (Statistical Analysis System Institute, Inc.). Fisher's LSD means separation

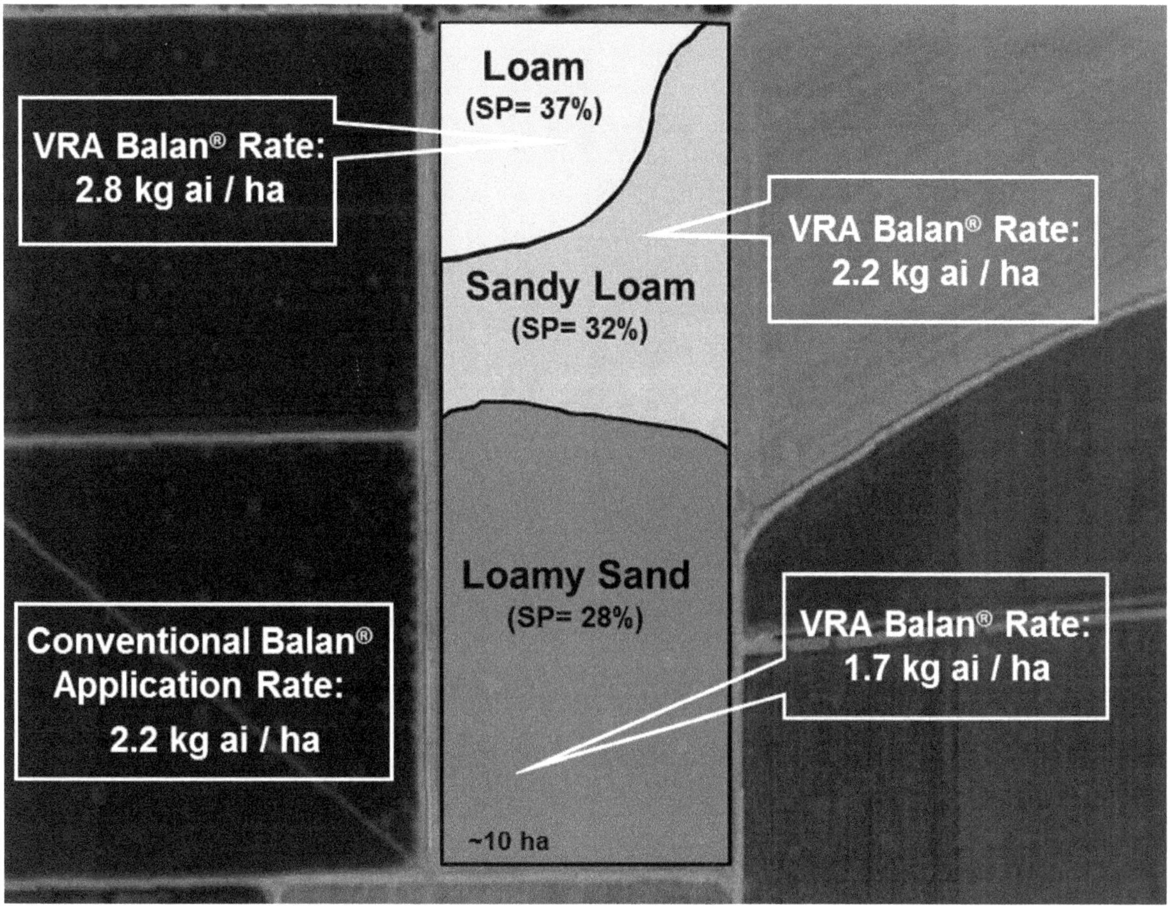

Figure 1. An example of site-specific Balan® application using standard and variable rates made on a 10 hectare commercial lettuce field near Yuma, Arizona. Applications were based on a geo-referenced soil textural map. A standard rate of 2.2 kg ai/ha was used as the control application rate.

tests (P=0.05) were used to determine if weed control efficacy, lettuce injury and crop yield loss was significantly affected by herbicide treatment within each soil textural class.

RESULTS AND DISCUSSION

Site-specific placement of Balan® in a field with non-uniform soil textures resulted in no significant differences in weed control efficacy as compared to the conventional application method (Table 3). Use of variable rate application technology however, significantly reduced levels of seedling injury from 30 to 35% in regions of our experimental fields where soil textures were classified as light or sandy while application amounts were 35% lower than in conventional application approaches (Table 4). Crop injury effects were visually striking in loamy sand conditions four weeks after planting (Figure 2), emphasizing the consequence of enhanced herbicide availability to plant roots in light textured soils. Reduced lettuce leaf injury also resulted in more uniform crop growth and consequently, increased yields of marketable lettuce up to 40% at harvest. In loamy sand and sandy clay loam textured soils, the quantity of unmarketable lettuce was reduced by approximately 3,000 and 5,000 heads per hectare, respectively, when VRA is utilized (Table 4).

The interaction between herbicide chemistry and soil properties greatly affects herbicide weed control efficacy and the potential for crop injury. Because of this, fields with significant variability in soil properties are good candidates for variable-rate application of soil-applied herbicides. A study conducted on ten hectare commercial field in this study showed that, use of variable rate technology with the herbicide Balan® resulted in significantly less crop injury and significantly more marketable yield as compared to uniform application. In the portions of the lettuce field with loamy sand textured soils, 35% less herbicide was applied and up to 40% more heads were harvested, which saved the grower approximately $32.50 (US)/ha and increasing gross revenues by over $3,750 (US)/ha. Furthermore, there were no significant differences in weed control efficacy found between the two application methods examined. In

Table 3. Weed control 4 weeks following variably and conventionally applied preemergence Balan® at three field locations during 2006-2008.[1]

Location	Soil texture	Treatment (kg ai/ha, am)[2]	Weed control (%)											
			Common purslane			Lambs quarters			Pigweed			Total weeds		
			Year			Year			Year			Year		
			2006	2007	2008	2006	2007	2008	2006	2007	2008	2006	2007	2008
YV[3]	Non-treated control		0	0	--	0	0	--	0	0	--	0	0	--
	SC	2.2, Conv.[4]	96^{a}[6]	95^{a}	--	92^{a}	88^{a}	--	90^{a}	94^{a}	--	96^{a}	95^{a}	--
	SC	2.8, VRA[5]	94^{a}	94^{a}	--	93^{a}	94^{a}	--	91^{a}	85^{a}	--	96^{a}	96^{a}	--
	SCL	2.2, Conv.	89^{a}	95^{a}	--	94^{a}	85^{a}	--	95^{a}	96^{a}	--	94^{a}	94^{a}	--
	SCL	2.2, VRA	95^{a}	92^{a}	--	98^{a}	96^{a}	--	94^{a}	89^{a}	--	95^{a}	95^{a}	--
	LS	2.2, Conv.	88^{a}	88^{a}	--	96^{a}	89^{a}	--	95^{a}	94^{a}	--	95^{a}	95^{a}	--
	LS	1.7, VRA	87^{a}	96^{a}	--	94^{a}	94^{a}	--	92^{a}	90^{a}	--	89^{a}	88^{a}	--
DV[7]	Non-treated control		--	0	0	--	0	0	--	0	0	--	0	0
	C	2.2, Conv.	--	97^{a}	96^{a}	--	86^{a}	91^{a}	--	96^{a}	95^{a}	--	94^{a}	94^{a}
	C	2.8, VRA	--	96^{a}	95^{a}	--	95^{a}	94^{a}	--	96^{a}	96^{a}	--	86^{a}	86^{a}
	SCL	2.2, Conv.	--	96^{a}	85^{a}	--	95^{a}	85^{a}	--	91^{a}	94^{a}	--	94^{a}	94^{a}
	SCL	2.2, VRA	--	95^{a}	94^{a}	--	84^{a}	96^{a}	--	91^{a}	95^{a}	--	94^{a}	94^{a}
	LS	2.2, Conv.	--	93^{a}	87^{a}	--	93^{a}	89^{a}	--	95^{a}	95^{a}	--	96^{a}	93^{a}
	LS	1.7, VRA	--	95^{a}	96^{a}	--	91^{a}	94^{a}	--	91^{a}	90^{a}	--	88^{a}	91^{a}
GV[8]	Non-treated control		--	0	0	--	0	0	--	0	0	--	0	0
	C	2.2, Conv.	--	95^{a}	92^{a}	--	95^{a}	95^{a}	--	96^{a}	92^{a}	--	94^{a}	92^{a}
	C	2.8, VRA	--	96^{a}	91^{a}	--	96^{a}	96^{a}	--	96^{a}	93^{a}	--	86^{a}	96^{a}
	SC	2.2, Conv.	--	94^{a}	95^{a}	--	94^{a}	94^{a}	--	91^{a}	94^{a}	--	94^{a}	94^{a}
	SC	2.2, VRA	--	95^{a}	94^{a}	--	95^{a}	95^{a}	--	91^{a}	98^{a}	--	94^{a}	95^{a}
	SL	2.2, Conv.	--	95^{a}	96^{a}	--	95^{a}	95^{a}	--	95^{a}	96^{a}	--	96^{a}	95^{a}
	SL	1.7, VRA	--	84^{a}	89^{a}	--	90^{a}	91^{a}	--	91^{a}	94^{a}	--	88^{a}	90^{a}

[1]Non-treated control not included in statistical analysis; data were square root transformed before analysis, [2]Kilograms of active ingredient per hectare(kg ai/ha), application method (am), [3]YV: Yuma valley, Arizona, 2006 and 2007, [4]Conv.: Conventional application method, [5]VRA: Variable rate application method, [6]Means followed by the same letter in each column are not significantly different at P=0.05 according to analysis of variance and the Fisher's LSD means separation test,[7]DV: Dome valley, Arizona, 2007 and 2008, [8]GV: Gila valley, Arizona, 2007 and 2008.

summary, this project shows the potential for using variable rate technologies in vegetable production to apply preemergence herbicides. Although these technologies have been used commercially to apply fertilizers for over a decade, they are just now being explored for use with preemergence herbicides in vegetables. While precision application approaches have the potential to lower production costs and improve farm profitability, the actual level of savings realized will vary from field to field depending on the degree of spatial variability, costs associated with sampling, and the capital investment for technology and equipment (Roberts et al., 2001). Future studies should address the specific costs of implementing this technology, as well as including more variable inputs. Also, a thorough risk analysis would be beneficial in future

Table 4. Damaged lettuce seedlings[4] and unmarketable lettuce at harvest following conventionally and variably applied preemergence Balan® at three Arizona field locations during 2006-2008.[1]

Location	Soil texture	Treatment (kg ai/ha, am)[2]	Damaged seedlings (seedlings/ha)			Unmarketable lettuce (heads/ha)		
			Year			Year		
			2006	2007	2008	2006	2007	2008
YV[3]	Non-treated control		0	0	--	0	0	--
	SC	2.2, Conv.[4]	3,355[a][6]	3,995[a]	--	1,280[a]	2,130[a]	--
	SC	2.8, VRA[5]	3,668[a]	4,152[a]	--	1,223[a]	2,735[a]	--
	SCL	2.2, Conv.	7,089[a]	6,045[a]	--	3,775[a]	6,715[a]	--
	SCL	2.2, VRA	6,495[a]	5,181[a]	--	3,108[b]	5,490[b]	--
	LS	2.2, Conv.	13,970[b]	12,410[b]	--	11,400[a]	8,220[a]	--
	LS	1.7, VRA	9,087[a]	8,505[a]	--	6,825[b]	4,770[b]	--
DV[7]	Non-treated control		--	0	0	--	0	0
	C	2.2, Conv.	--	5,197[a]	3,547[a]	--	4,256[a]	5,745[a]
	C	2.8, VRA	--	4,884[a]	3,054[a]	--	3,995[a]	6,010[a]
	SCL	2.2, Conv.	--	6,854[a]	4,253[a]	--	6,215[a]	4,236[a]
	SCL	2.2, VRA	--	4,987[a]	2,941[a]	--	4,964[b]	3,096[b]
	LS	2.2, Conv.	--	15,295[a]	14,965[a]	--	11,236[a]	14,863[a]
	LS	1.7, VRA	--	10,524[b]	10,025[b]	--	7,241[b]	9,244b
GV[8]	Non-treated control		--	0	0	--	0	0
	C	2.2, Conv.	--	1,395[a]	2,415[a]	--	6,521[a]	7,523[a]
	C	2.8, VRA	--	1,901[a]	2,248[a]	--	6,354[a]	7,214[a]
	SC	2.2, Conv.	--	2,104[a]	3,687[a]	--	8,698[a]	6,409[a]
	SC	2.2, VRA	--	1,769[a]	2,689[a]	--	6,981[a]	5,687[a]
	SL	2.2, Conv.	--	13,521[a]	11,984[a]	--	9,547[a]	10,950[a]
	SL	1.7, VRA	--	9,523[a]	7,952[a]	--	6,465[a]	7,790[a]

[1]Non-treated control not included in statistical analysis, [2]Kilograms of active ingredient per hectare(kg ai/ha), application method (am), [3]YV: Yuma valley, Arizona, [4]Conv.: Conventional application method, [5]VRA: Variable rate application method, [6]Means followed by the same letter in each column are not significantly different at P=0.05 according to analysis of variance and the Fisher's LSD means separation test, [7]DV: Dome valley, Arizona, [8]GV: Gila valley, Arizona.

explorations. However, as researchers, manufacturers and growers learn more about precision management approaches, the productivity and cost savings gains of these technologies can be expected to only improve.

Precision herbicide management is an effective tool for spatial application of soil-incorporated herbicides which have a tendency for soil adherence. Although field

summary, this project shows the potential for using variable rate technologies in vegetable production to apply preemergence herbicides. Although these technologies have been used commercially to apply fertilizers for over a decade, they are just now being explored for use with

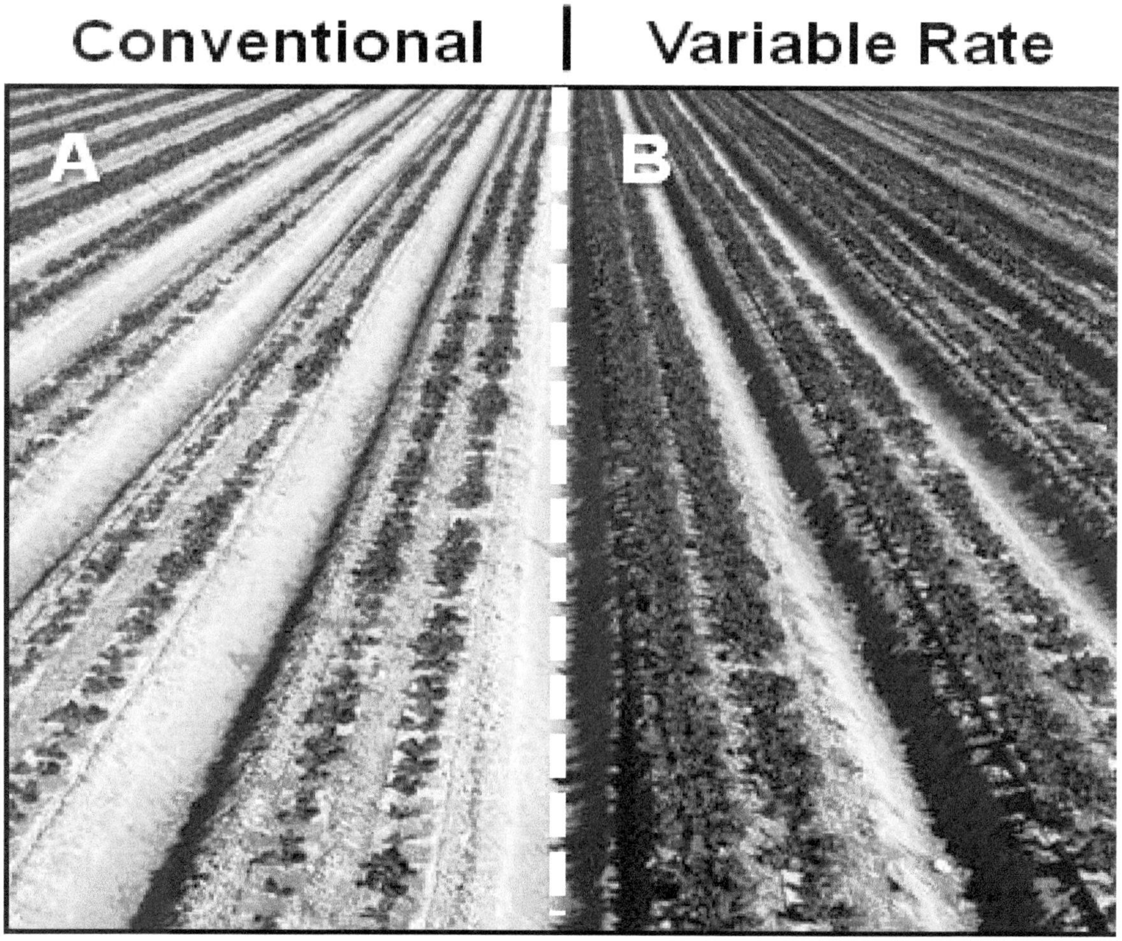

Figure 2. Ruby red leaf lettuce seedling injury (stunting) 4 weeks after planting as a result of applications of Balan® preemergence herbicide using standard (A) and variable rate (B) application schemes in loamy sand textured soil.

preemergence herbicides in vegetables. While precision application approaches have the potential to lower production costs and improve farm profitability, the actual level of savings realized will vary from field to field depending on the degree of spatial variability, costs associated with sampling, and the capital investment for technology and equipment (Roberts et al., 2001). Future studies should address the specific costs of implementing this technology, as well as including more variable inputs. Also, a thorough risk analysis would be beneficial in future explorations. However, as researchers, manufacturers and growers learn more about precision management approaches, the productivity and cost savings gains of these technologies can be expected to only improve.

Precision herbicide management is an effective tool for spatial application of soil-incorporated herbicides which have a tendency for soil adherence. Although field implementation depends on previous knowledge of soil textural variability (soil test and texture evaluations), site-specific technologies show promise for Arizona vegetable fields with non-uniform soils. Regardless of the method used for textural characterization, growers should keep in mind that textural differences do not change in the short/medium term, so the costs associated with defining texture-based management zones can be spread over many years.

REFERENCES

Bauer WD, Schefcik M (1994). Using differential GPS to improve crop yields.GPS World 5 (2): 38-41.

Carlson WC, Lignowski EM, Hopen HJ (1975). Uptake, Translocation, and Adsorption of Pronamide. Weed. Sci., 2(2): 148-154.

Colquhoun J (2006). Herbicide Persistence and Carryover. Publ. No. A3819. University of Wisconsin-Extension, College of Agricultural and Life Sciences, Cooperative Extension. URL: http://corn.agronomy.wisc.edu/Management/pdfs/A3819.pdf.

Roberts RK, English BC, Mahajanashetti SB (2001). Assessing Spatial Break-Even Variability in Fields with Two or More Management Zones. J. Ag. Appl. Econ., 33(3): 551-565.

Tickes BR, Kerns DL (1996). Lettuce Injury from Preplant and Herbicides. IPM Series 9. Publ. No. 196007. University of Arizona, College of Agriculture and Life Sciences, Cooperative Extension, Tucson, Arizona.4 pp.URL: http://valueaddedag.org/crops/vegetables/weeds/lettuceinjury.html.

Jacques GL, Harvey RG (1979). Adsorption and Diffusion of

Dinitroaniline Herbicides in Soils. Weed. Sci., 27 (4): 450-455.
Koch B, Khosla R (2007). The Role of Precision Agriculture in Cropping Systems. J. Crop Prod., 9: 361-381. URL: http://classes.css.wsu.edu/css403/8_PrecisionAg.pdf.
Moomaw RS, Klein RN, Martin AR, Roeth FW, Shea PJ, Wicks GA, Wilson RG (1992). Factors That Affect Soil-Applied Herbicides. University of Nebraska, Institute of Agriculture and Natural Resources, Cooperative Extension Service, Publication No.: G92-1081-A. URL: http://digitalcommons.unl.edu/cgi/viewcontent.cgi?article=2219&context=extensionhist.
Peacock B (1998). The Use of Soil and Water Analysis. University of California Cooperative Extension.Publ. No.IG3-96. URL: http://cetulare.ucdavis.edu/pubgrape/ig396.htm.
Ross MA, Childs DJ (1996). Herbicide Mode-Of-Action Summary. Cooperative Extension Service Publication WS-23-W, Purdue University, West Lafayette, IN. URL: http://www.extension.purdue.edu/extmedia/WS/WS-23-W.html.
Weber JB (1990). Behavior of Dinitroaniline Herbicides in Soils. Weed Tech., 4(2): 394-406.
Weber JB, GWilkerson GG, Linker HM, Wilcut JW, Leidy RS, Senseman S, Witt WW, Barrett M, Vencill WK, Shaw DR, MuellerTC, Miller DK, Brecke BJ, Talbert RE, Peeper TF (2000). A Proposal to Standardize Soil/Solution Herbicide Distribution Coefficients. Weed. Sci., 48: 75-88.
Weed Science Society of America (2007). Herbicide Handbook, 9th edition.Lawrence, KS; Allen Press.

Morphogenetic characterization of soils formed from basement complex rock in the humid tropical rainforest of Nigeria

O. T. Ande

Institute of Agriculture, Research and Training, Obafemi Awolowo University, P. M. B. 5029, Apata, Ibadan, Oyo State, Nigeria. E-mail: funmiande@yahoo.com.

The morphogenetic and physico-chemical properties of four soil units occupying different slope facets on a rolling landscape were studied to understand distribution of soils on a typical landscape. The soil colours range from dark brown to reddish brown (10YR3/3-5YR3/3) in the surface and strong brown (7YR4/6) to yellowish red (5YR5/8) in the sub soil. The texture varies from loam/sandy clay loam in the surface to clay/clay loam in the sub soil. The solum was moderately acidic to neutral (5.5 - 8.0) at the surface and strongly acidic to moderately acidic (4.65 - 5.80) in the sub soils. The soils are characterized by low exchangeable bases and cation exchange capacity. Pedon 1 on the crest was classified as Typic Hapludult (Haplic Acrisols). Pedon 2 at the mid slope was classified as Dystric Hapludult (Dystric Luvisol) while Pedons 3 and 4 were classified as Albic plinthudult (Albic plinthosols) and Dystric Haplaquent (Dystric Gleysols) respectively. Pedogenesis was influenced by physiographic positions resulting in different soil types on the landscape. Where similar parent material exists, different soil types were formed due to different exposure to morphogenesis processes such as eluviation, illuviation, deposition, cementation of pedi-sediments and multiple stratification as determined by the position of the pedon on the landscape.

Key words: Morphogenesis, pedoturbation, quartz vein, plinthite, argillic horizon, base saturation.

INTRODUCTION

Theories had been put forward to explain the genesis of soils formed from rocks of the basement complex in the humid tropical zone of Nigeria. Smyth and Montgomery (1962) observed that weathered rock materials from which the soils develop have a complicated history reflecting the influence of previous climatic variation, cycles of erosion and periods of intense soil forming activities. Distinct erosional phases in South western Nigeria and the superficial deposits of pediment and secondary sediment have been observed. Ojanuga (1978) noted that multiple stratification of soil parent materials of varying Pedi sediments (over saprolite) and existence of inherited pedological features tend to indicate that cycles of erosion must have alternated repeatedly with cycles of pedogenesis. Cyclic changes in climate in West Africa were also reported by Thomas and Thorp (1985).

Smyth and Montgomery (1962) reported that on a landscape shoulder, the soil properties reflect a combination of characteristics inherited during erosional evolution of the landscape. This is as a result of pedologic processes acting on the present surface. The relative importance of these processes was considered to vary within the same landscape segments because different soil types have been noticed to be associated with different parts of the landscape. Ojo-Atere et al. (1988) also reported from the study of upper slope soils that the surface horizons may be developed in transported material while the subsurface horizons may be derived from the underlying rock. The soils of the basement complex vary a lot in soil characteristics, therefore intense understanding of the slope processes and spatial distribution is of great importance for sustainable management of the soil.

This research was conducted with objective of studying the spatial distribution of soils on a rolling topography using morphogenetic and physico-chemical properties.

METHODS

The study area

The area of study lies approximately within longitude 4° 30E and 4° 50E and latitude 7° 21N and 7° 30N.

The area is chracterised by rolling landscape of 3 - 9%. The land use consisted of a mixture of bush regrowth and arable crops. The area had been under cultivation for more than ten years. The area of study has humid tropical climate with distinct dry and wet seasons of about 4 and 8 months, respectively. The mean annual rainfall and air temperature of the area is 1300 mm and 13°C, respectively. This area specifically is underlain by coarse/medium grained granite gneisses (Boesse and Ocan, 1988). The terrain is unique with schist intrusion existing at the mid slope/lower slope facets on the landscape.

Field work

Representative soil units were identified from mini pits during reconnaissance field study after which profile pits were dug. Soil samples were collected from the horizons designated in the profile pits. The soils were described according to Soil Survey Staff (2003). Samples were prepared and routinely analyzed following the guidelines of IITA (1979). The parameters that were determined include: Sand, silt clay, pH, %C and cations from which sum of bases and effective cation exchange capacity were calculated for each horizon. The soils were classified according to USDA Soil Taxonomy (1999) and FAO/UNESCO (1994). At series level Smyth and Montgomery (1962) classification was used.

RESULTS AND DISCUSSION

Morphological description of the study area

The landscape consists of soils that occupy gentle to moderate slope (3.4 - 9.0%) positions. Four pedons were identified on different slope facets (Table 1). The upland soils (that is, on the mid slope to hill crest) have deeply weathered profiles with surface texture being sandy clay loam or sandy loam to loam (Pedons 1 and 2). The colour of the upland soils from surface to subsoil gradually changed from dark brown (10YR 3/3) to reddish brown (5YR 3/3). The darker values of the surface soils were attributed to organic matter content. The structure of Pedons 1 and 2 was strong, gradually changing from medium to coarse sub-angular blocky in the sub soils. Pedon 2 maintains the rock structure up to about 68 cm from the land surface showing the various bands of light and dark minerals. Their descriptions in general are similar to that of partly residual soils described by Ojanuga (1978) in South western Nigeria.

Transition horizons are also well developed in the upland pedons. They have combined features of the corresponding horizons with diffuse way to irregular boundary. Horizon boundary in Pedons 1 and 2 is gradual and wavy to diffuse and irregular. This could be due to gradual transformation, homogenization, various pedoturbation or erosional/depositional processes. Generally, these upland soils have a layer of gravel accumulation. This could be due to any of the previously mentioned processes or their combinations. Eluviation and illuviation are active in all the profiles as evidenced by clay skin. Mass movement is important in pedon 2 because stones and gravels are oriented down slope. The prominent quartz vein diffusing to the surface horizon serves as the source of the gravels. The materials greater than 2 mm vary from 1.80 to 56.4% (Table 1).

The lowland pedons are very different from upland pedons due to formation of frequent Manganese nodules which also is a function of seasonal change in water table. Upland soils are also well drained thus having brown to reddish colour which become yellow down the landscape indicating the influence of soil moisture. Similar colour changes were reported by Okusami and Oyediran (1985)

Soil physical and chemical properties

The textures vary from sandy clay loam to sandy loam for surface horizons. While the B horizons are clay, the C horizons vary from clay loam to loam except in pedon 1 where the C horizon is also clay. The lower silt content in pedon 1 indicates higher degree of weathering than mid slope soils with medium to high silt content while high silt content reflects the deposition process in the lowland pedons (Table 2). These materials increase with increasing depth to a maximum level and decrease in the C horizons.

The chemical properties are also displayed in Table 2. The organic carbon content in surface horizons are relatively high, between 1.56 to 3.98%. The organic C decreases down the profile. Okusami and Oyediran (1985) observed similar distributions in soils of Ife area in Nigeria. The pH of the surface soils varies from slightly acidic to neutral (6.0 – 7.5), while that of sub soils are moderately acidic to strongly acidic. Aluminum saturation is low on the surface and increases down the profile. The dominant exchangeable bases are Ca and Mg in the surface soils. Potassium and Na are low (0.01 – 0.06 Cmol/kg). Exchangeable Mg is high in the surface horizons (1.25 – 10 Cmol/kg) probably as a result of nutrient recycling and in B-horizons because of the increased CEC. associated with higher clay content. Base saturation decreases down the profiles randomly from the surface. Base saturation is highly variable, it ranges from 85.08 to 99.53% in the surface and 26.32 to 97.77% in the sub soil.

Surface horizons have more variable composition being area of active depositional processes. The distribution tends to be more uniform in the saprolite according to the degree of weathering it had undergone. The base saturation decreases concomitantly with increase in

Table 1. Morphologic description of soils from Ile -Ife rolling landscape.

[illegible]	[illegible]	[illegible]	[illegible]	[illegible]	[illegible]	[illegible]	[illegible]	[illegible]	[illegible]	[illegible]	[illegible]	[illegible]
Pedon 1												
A	0 – 12	7.5YR 4/2	-	SCL	sab	Fr	-	I	-	FM	DS	-
AB	12 – 27	7.5YR 4/2	-	SCL	sab	Fr	-	I	-	FM	cw	-
Bt	27 – 67	7.5YR 5/6	-	Slg C	sab	Fr	-	I	Fe-Mn	FM	cw	-
BC	67 – 130	7.5YR5/6	Yellow	C	M	Fr	-	I	-	F	dw	C
C	130-190	7.5YR5/6	Yellow	C	M	Fr	-	I	-	F	-	A
Pedon 2												
A	0 – 12	10YR3/3	-	Sl SL	Cr, Ls	Fr sl P	-	I	-	f,m	ds	-
AB	12 – 28	10YR4/4	-	gSL	W Gr	Fr -	-	I	Fe Mn	f, m	cw	-
Bt	28 – 68	7.5YR5/6	-	C	sab	F p	F M	I	-	F	ds	-
Bw	68 – 115	7.5YR/5/6	Yellow, red	C	sab	Fr SP	-	I	-	-	ds	C
Pedon 3												
A	0 – 6	10YR3/2	-	L	Cr	Fr sl S	-	I	-	F	cw	-
E	6 – 20	10YR4/4	-	slg, SCL	sab	Fr SP	ff	I	f	F	cs	-
Bt	20 – 50	7.5YR5/6	-	C	sab	F SP	m	I	-	F	ds	-
Bw	50-70	5YR5/8	Red, yellow,	Slg,C	sab	F SP	f	I	f	F	gs	-
CBcx	70-105	2.5YR4/6	Yellow	C	sab	Fr SP	f	I	mFe, Mn	F	gs	C
Ccx	105-157	10YR6/8, 5YR4/6, 10YR5/6	Pink, yellow	CL	ab	F SP		I	cFe	-	-	M
Pedon 4												
[illegible]	[illegible]	[illegible]	[illegible]	[illegible]	[illegible]	[illegible]	[illegible]	[illegible]	[illegible]	[illegible]	[illegible]	[illegible]
[illegible]	[illegible]	[illegible]	[illegible]	[illegible]	[illegible]	[illegible]	[illegible]	[illegible]	[illegible]	[illegible]	[illegible]	[illegible]
[illegible]	[illegible]	[illegible]	[illegible]	[illegible]	[illegible]	[illegible]	[illegible]	[illegible]	[illegible]	[illegible]	[illegible]	[illegible]
[illegible]	[illegible]	[illegible]	[illegible]	[illegible]	✓	[illegible]	[illegible]	[illegible]	[illegible]	[illegible]	[illegible]	[illegible]

Texture: SCL- sandy clay loam, slgC-slightly gravelly clay, C-clay, g-gravelly, SL-sandy loam; Structure: sab-sub angular blocky, M-massive w-weak, cr-crumbs, ls-loose; Consistence: Fr-firm, p-plastic, sl p- slightly plastic ; Cutan: f-few, c-common, m-many; Drainage: I- well drained, II-moderately well drained, III-somewhat poorly drained, IV-poorly drained; Concretions: Fe-Mn- iron –manganese; Roots: F-fine, M-mediun. Boundary:ds-diffuse wavy, cw-clear wavy, dw-diffuse wavy, ds-diffuse-smooth; weatherable minerals: f-few, c-common, m-many, a-abundant.

Table 2. Physical and chemical properties of soils formed from rocks of basement complex in the humid tropical rainforest of Nigeria.

Hor	Depth	%Gravel	%Sand	%Silt	%clay	pH H_2O	Al^{3+}	Ex.ac	Ca^{2+}	Mg^{2+}	K^+	Na^+	SB	ECEC	BS	%C	Tex
Pedon 1																	
A	0 - 12	12.8	52	28	28	6	0.2	0.4	6.25	4.29	0.03	0.01	10.58	10.94	96.71	3.98	SCL
AB	27 - Dec	33.9	57.7	11.2	11.2	5.75	0.1	0.4	1.25	1.83	0.02	0.01	3.11	3.51	88.6	1.08	SCL
Bt	27 - 67	47	40	10.2	10.2	5.55	0.01	0.4	1.5	4.29	0.02	0.01	7	7.4	94.46	0.49	C
BC	67 -130	21	33.9	15	15	4.84	0.4	1.6	1.25	1.17	0.02	0.01	2.45	4	60.05	0.43	C
C	130 -190	21.2	32.3	18.4	18	4.75	1.2	1.8	1.2	1.08	0.01	0.01	2.35	4.15	56.63	0.18	C
Pedon 2																	
A	0 - 5	44.8	69.4	13.6	17	6.4	0.1	0.2	5	4.29	0.05	0.01	10.35	10.55	98.1	3.82	SL
AB	28 - Dec	58.5	69.1	14.9	16	6.3	0.1	0.4	1.25	1	0.02	0.01	2.28	2.68	85.08	0.78	SL
Bt	28 - 68	56.4	45.2	7.2	47	4.9	1.6	2.4	0.25	0.92	0.06	0.02	1.25	3.65	34.25	0.74	SC
BW1	68 - 115	29.9	35	14	51	4.7	1.8	2.8	0.25	0.71	0.03	0.01	1	3.8	26.32	0.2	C
BW2	115 - 155	16-6	49.4	19.6	31	4.65	2	3	0.25	0.79	0.01	0.01	3.31	6.31	52.4	0.2	SCL
Pedon 3																	
A	0 – 6	25	64.7	15.1	7.2	20.2	0.1	0.5	10	2.21	0.05	0.02	12.28	12.78	96.09	3.7	SCL
E	20-Jun	21	59.4	18.6	8	22.2	0.1	0.4	5	1.08	0.02	0.01	6.11	6.5	94	1.29	SCL
B	20 - 50	34.5	37.6	17.7	5.7	50.7	0.2	0.6	25	1.25	0.02	0.01	26.88	26.88	97.77	0.66	C
2BWt	50 - 70	33.5	33.2	11.2	5.4	55.6	1.4	2	1.25	1.46	0.04	0.01	2.16	4.16	57.98	0.27	C
2BW2	70 - 105	28.4	36.7	15.9	5	47.4	1.3	2	1.25	0.96	0.03	0.01	2.25	4.25	52.94	0.12	C
2Ccx	105 - 157		43.2	18.5	5.1	38.3	1.3	2	1.25	0.79	0.01	0.01	2.06	4.06	50.74		CL
Pedon 4																	
A	0 – 10	2.2	56	20	24	5.8	0	0.3	4.5	1.76	0.02	0.01	6.29	6.32	99.53	1.56	L
BA	20 – 46	2.7	46.1	27.1	27	6	0.1	0.3	10	1.88	0.03	0.01	11.9	12	99.17	1.17	SCL
Bt	20 – 46	24.2	48.9	15.9	36	5.3	0.8	1.5	2.5	1.08	0.01	0.01	3.6	5.1	70.39	0.59	SCL
BWt	46 – 63	38.8	46.3	15.3	39	5.3	0.5	1.2	1.25	1.13	.0.01	0.1	2.4	3.6	66.67	0.43	SC
BW2	63 – 90	26.6	48.9	18.9	34	5.5	0.7	1.1	2.5	1.33	0.02	1	3.8	4.9o	66.67	0.54	SCL

Ex.ac: exchangeable acidity; SB: sum of bases; ECEC: effective cation exchange capacity; BS: base saturation; %C: percent Carbon; Tex-texture.

exchangeable acidity. On the crest, the soil is formed from coarse granitic rock with deep weathering giving rise to kaolinised sub soils. The position on the landscape seems to be stable thus allowing a lot of residual accumulation of weathered materials. On the crest, moderate cementation of pediments characterizes soil formation as evidence in the medium to coarse sub angular block structure in the surface horizons. This was similar to the observation of Folster and Von (1977) that locally restricted and rather moderate cementation of pediments characterizes soil formation in South western Nigeria. The existence of inherited pedological features such as smooth round nodules also suggested multiple stratification of soil parent materials that is, varying pedisediments over saprolite. These coupled with clear wavy boundary, indicates that cycles of erosion must have alternated repeatedly with cycles of pedogenesis. In pedon 2 located at mid slope, the presence of flecks of white mica, saprolite and quartz veins indicate another parent material different from the crest. The characteristics is similar to soils formed from sericite schist described by Smyth and Montgomery (1962).

Soil pedogenetic characterization

On the crest, the soil is formed from coarse granitic rock with deep weathering giving rise to kaolinised sub soils. Quartz vein was very prominent in pedon 2 and it diffuses to the BA horizon which is attributed to its position and elevation, which made accumulation of hill creep materials possible. This quartz gravel was traceable to the prominent quartz veins which diffuses to the transition horizon (AB) and probably colluvial deposits. Soil residuum in the middle slopes is identified by massive and strong structure. Hence various pedogenic processes have contributed to soil development. Fasina et al. (2007) reported similar result where different hydrolytic and geomorphic processes in various landscape positions cause differences in type and distribution of soil taxi on the landscape.

Pedon 3 at lower slope showed influence of deposition at surface soils with the presence of eluvial (E) horizon. The B horizons consists of argillic horizon formed from eluviation of deposited materials from upper slope and illuvial clay as evident from the underlying saprolitic horizon with coarse mineral nodules (Manganese) which indicate residual formation. Thus, soil profile variation therefore occurs due to landform processes of transformation and deposition. These have led to non-uniformity and discontinuity in parent materials. Therefore the profile seems to be developed from two parent materials: Transported material and the underlying rock (granite gneiss and schist) as indicated from the profile morphology (Table 1). This view was supported by Graham et al. (1990) and Ande (1995) reported that different parent materials are still possible in areas under laid by a single type of rock. They also reported from the study of upper slope soils that the surface horizons may be developed in transported material while the subsurface horizons may be derived from the underlying rock. Pedon 1 on the crest was classified as Typic Hapludult (Haplic Acrisols-Igbajo series) due to its low CEC, increasing clay with depth and low base saturation of < 99%. While Pedon 2 at the mid slope was classified as Dystric Hapludult (Dystric luvisol-Mamu series), and Pedons 3 and 4 at lower and valley bottom were classified as Albic plinthudult (Albic plinthosols-Oba series) and district Haplaquent (Dystric Gleysols-Jago series), respectively.

Conclusion

In conclusion, landscape invariably influences the pedogenetic processes leading to formation of different soil types on the landscape. Pedogenesis was influenced by physiography resulting in different soil types on the landscape. Where similar parent material exists, different soil types were formed due to different exposure to morphogenesis processes such as eluviation, illuviation, deposition, cementation of pedisediments, and multiple stratification as determined by the position of the pedons on the landscape.

REFERENCES

Ande OT (1995). Morphogenetic characterization of upland soils in south western Nigeria.PHD Thesis.Unpublished. Obafemi Awolowo University, Ile Ife, Nigeria, pp. 25-40.

Boesse JM, Ocan OO (1988). Geology and Evolution of the Ife-Ilesa Schist Belt, S.W. Nigeria. Proceedings of Benin – Nigeria Geotraverse International Meeting on Proterozic Geology and Tectonic of High Grade Terrains. Nov. 28th to 30th, Obafemi Awolowo University, Ile – Ife, Nigeria, pp. 200-225.

FAO/UNESCO (1994). Soil map of the world, 1:500,000 Vol-Legend sheet and memoirs. Paris, pp. 1-65.

Fasina AS, Omolayo JO, Falodun AA, Ajayi OS (2007). Grranitic derived soils in humid forest of southwestern Nigeria-Genesis classification and sustainable management. American –Eurasian J. Agric. Environ. Sci., 2(2): 189-195.

Folster H, Von Christen (1977). Quarter Uplift Altitude Zonation. CATENA. 3: 233.

Graham RC, Daniel RB, Boul SW (1990). Soil Geomorphic Relations on the Blue Ridge Front: I Regolith Types and Slope Processes. Soil Sci. Soc. Am. J., 54: 1362-1367.

International Institute of Tropical Agriculture, IITA (1979). Selected methods for soil and plant analysis, manual series No 1, 3rd Edition, IITA, Ibadan, Nigeria, p. 34.

Ojanuga AG (1978). Genesis of Soils in the Metamorphic Forest Region of S.W. Nigeria. Pedologie. Ghent, 28(1): 105-117.

Ojo-Atere JO, Oladimeji MO (1988). Characteristics and Classification of Some Slope Soils in Savanna Ecosystem of South Western Nigeria. Ife J. Agric, Ile-Ife, 5(1-2): 22, 442.

Okusami TA, Oyediran GO (1985). Slope-Soil Relations on an Aberrant Toposequence in Ife Area of South Western Nigeria: Variabilities in Soil Properties. Ife J. Agric., Ile-Ife, 7(1-2): 1-13.

Soil survey staff (2003). keys to Taxonomy SMS Tech. Monograph (9th Ed.). Virginia, Virginia polytechnic Institute state University pp. 41-148.

Soil Survey Staff (2003). Soil Taxonomy. A basic system of soil classification for making and interpreting soil surveys, 2nd edition. Agricultural Handbook 436, Natural Resources Conservation Service, USDA, Washington DC, USA, p. 869.

Smyth AJ, Montgomery RF (1962). Soils and Land use in Central Western Nigeria, Ibadan Government Printer, p. 265.

Thomas MF, Throp MB (1985). Environmental Change and Episodic Ectoplantation in the Humid Tropics of Sierra Leone: The Koidu Etchplain: In Douglas, and Spencer, T. (ed.) Environmental Change and Tropical Geomorphology. George Allen and Unwin, London, pp. 239-267.

12

Yield and nitrogen accumulation in five cassava varieties and their subsequent effects on soil chemical properties in the forest/savanna transitional agro-ecological zone of Ghana

S. Adjei-Nsiah

University of Ghana Agricultural Research Centre, Kade, Institute of Agricultural Research, University of Ghana, Legon, P. O. Box 38, Legon, Ghana. E-mail: y_nsiah@yahoo.co.uk.

We evaluated five varieties of cassava for yield, N accumulation and their effect on soil chemical properties. Soil phosphorus content increased drastically after 14 months of crop growth while soil organic carbon, N and exchangeable K contents reduced during the same period. Fresh root yield ranged from 17 t ha^{-1} to 35.9 t ha^{-1}, while total dry matter production also ranged from about 18 t ha^{-1} to about 25 t ha^{-1}. Total N accumulation in the total plant biomass varied from about 228 kg N ha^{-1} to about 288 kg N ha^{-1}. The amount of total N uptake exported from the soil through crop harvest ranged from about 59 kg N ha^{-1} to about 123 kg N ha^{-1} while the total N uptake recycled into the soil through the return of crop residue ranged from 127 kg N ha^{-1} to about 189 kg N ha^{-1}. The study suggests that cassava varieties with high root dry matter yield have the potential of exporting large quantities of nutrients from the soil while varieties with high leaf litter deposition is likely to recycle substantial amount of the plant nutrient uptake.

Key words: Biomass, leaf litter, nutrients, root dry matter, soil chemical properties.

INTRODUCTION

Cassava (*Manihot esculenta* Crantz.) is the most important staple crop in Ghana in terms of area under cultivation and production output (ISSER, 2005). Besides being a major staple crop in Ghana, cassava serves as a source of income for farmers and processors.

In spite of its role in the national economy, cassava production has not been promoted to any appreciable extent with the belief that cassava degrades soil. This perception which is based on the ability of cassava to grow on depleted soils where other crops would fail (Howeler, 2001) is however, contrary to perception by farmers in some parts of Ghana (Adjei-Nsiah et al., 2004) and elsewhere in Africa, (Saidou et al., 2004; Fermont, 2009), that cassava improves soil productivity. The perception that cassava degrades soils may be correct if the crop extracts large amount of nutrients and/or causes severe erosion if it has slow initial growth rate thus resulting in soil degradation as a result of heavy rains (Puthacharoen et al., 1998). In the forest/savannah transitional agro-ecological zone of Ghana, Adjei-Nsiah et al. (2007) reported that yields of maize grown after cassava were higher than those grown after nitrogen-fixing legumes such as cowpea (*Vigna unguiculata*), pigeonpea (*Cajanus cajan*) or mucuna, due possibly to large amount of nitrogen returned into the soil after cassava harvest through incorporation of leaf litters and green leafy biomass into the soil. Although farmers in the forest/savannah transitional zone assume that cassava varieties preceding maize crop usually provide substantial amount of nutrients to the maize crop through litter falls and leafy biomass incorporated into the soil, the relative contributions of nutrients by different cassava varieties is not fully known.

We evaluated different cassava varieties developed by the National Agricultural Research System (NARS) in Ghana in both farmer-managed and researcher-managed experiments on farmers' field. In the researcher-managed experiment, dry matter production, nitrogen accumulation and the effects of cropping on soil chemical properties were evaluated.

The objectives of the study were to assess (1) the yield of elite cassava varieties developed by the NARS and the local varieties being cultivated by farmers; (2) the residual effect of cropping of the various cassava varieties on soil chemical properties and (3) the total N accumulated in the plant biomass of the various cassava varieties and recycled through leaf litters and other crop residues for uptake by subsequent crop.

MATERIALS AND METHODS

The study was conducted in three communities namely Asuoano (7°41´ N, 2°05´ W), Beposo (7°42´ N, 2°05´ W) and Droboso (7°43´ 2°05´ W) near Wenchi in the Brong-Ahafo Region of Ghana. Wenchi is located in the forest/savannah transitional agro-ecological zone of Ghana. The study site is characterised by a bimodal rainfall pattern with a 30-year average of 1271 mm. The major growing season is from April to July and the minor growing season is from September to November followed by a dry season from December to March. The annual rainfall amounts during the study period were 1396 mm (2003) and 1350 mm (2004). The soils which developed on Voltaian sandstone (coarse-grained type) are mainly Lixisols (Asiamah et al., 2000). The topography is gently flat to undulating and soils are well drained, friable, porous and sandy loams.

Researcher-managed experiment

The experimental plot had been under continuous maize cropping for three years (1999 - 2001) and had been abandoned for one year (2002) due to poor soil fertility and high infestation of spear grass (*Imperata cylindrica*). The initial land preparation consisted of harrowing after first and second ploughing. Composite soil samples were taken from 0 - 20 cm and 20 - 40 cm soil depth before the cassava was planted.

The trial consisted of six treatments replicated four times in four blocks. The treatments consisted of five cassava varieties namely Abasafitaa, Tekbankye and Afisiafi (which are improved varieties) developed by the NARS, Bensre and Tuakentemma (which are local varieties usually cultivated by farmers in the study area). To assess the effects of the cassava cropping on soil fertility a fallowed plot which was under spear grass was included as a control.

Plot size was 10 x 8 m. There were 3 m alleys between plots and replicates. Distance between rows was 100 cm and distance within rows was 50 cm. The cassava was planted during the last week of May 2003.

Six months after planting, 5 litter traps each measuring 50 by 50 cm and raised about 30 cm from the ground were placed under the cassava canopies to trap fallen leaf litters. The litters were collected every four weeks, oven dried at 70°C for 2 days for dry matter determination.

In September 2004 which was approximately 64 weeks after planting the cassava, an area of 12 m^2 (2 rows of 6 m) of the cassava plot was harvested. The plants were separated into roots, stems and foliage and weighed. Sub-samples of all harvested components were taken and oven dried at 70°C for 2 days for dry matter determination.

Sub-samples of plant parts of all cassava varieties were analysed for nitrogen (N). These plant parts were then grouped into two categories: (1) plant parts that would normally be removed from the field in the harvested products; and (2) plant parts that would remain in the field and be incorporated into the soil. The amounts of DM and N that would either be removed from the field or reincorporated into the soil were also calculated.

After harvesting the cassava in September 2004, soil samples were collected from 0 - 20 cm and 20 - 40 cm depth layers of the soil and analysed for soil chemical properties.

Farmer-managed experiments

Selection of farmers was based on interest and preparedness to spend one day in a week on the experimental field. The experiments were conducted on collective plots in each of the communities.

Land preparation on farmers' fields consisted of ploughing by tractor followed by harrowing using hoe. On the farmer-managed plots, one other additional variety namely Gblemo duade was included by the farmers in addition to the five varieties used in the researcher-managed experiments because this variety is high yielding but has high water content. Planting, field management and harvesting decisions were made by farmers. All participating farmers in each of the communities met with the researcher and decided when to plant, weed and harvest. The idea of having the experiments on collective plots was moved by the farmers. There were no replications within each experiment. Instead, each community represented a replicate.

Litter traps measuring 50 by 50 cm and raised 30 cm above the ground were placed under the canopies of the cassava to trap falling leaf litter. The litters were weighed together with the farmers every four weeks to determine the amount of litter produced by each of the varieties. The yield of the different cassava varieties was assessed with the farmers through ranking and scoring as well as by harvesting and weighing the roots.

Plant and soil analysis

Percent nitrogen in various plant parts of all the cassava varieties was determined by kjeldahl digestion and calorimetric analysis of NH_4-N in the digest.

Soil pH was determined in water suspension at 1: 1 ratio; organic C by Walkley-Black procedure; total N by kjeidahl method; available P by Bray-1 method and exchangeable bases (K, Ca and Mg) by 1 M NH_4 OAC method (ASA - SSSA, 1982).

Statistical analysis

Data were subjected to analysis of variance (ANOVA) using the general linear model (GLM) procedure (SAS, 1996). Data on soil chemical properties, cassava yield and DM and N contents of component parts of cassava varieties were analysed using randomized complete block design.

RESULTS AND DISCUSSION

The organic carbon at the start of the experiment were 8.0 and 5.4 g kg^{-1} for the (0 - 20) and (20 - 40) cm depth respectively in 2003 (Table 1) compared to an average of 6.7 and 4.7 g kg^{-1} for the (0 - 20) and (20 - 40) cm depth respectively in 2004 (Table 2). The reduction in organic carbon (-16% in the uppermost layer and -13% for the lower layer) could be attributed to mineralization of organic matter to supply nutrients for plant uptake (Ayanaba et al, 1976). Observations by Ayanaba et al. (1976) and Zingore et al. (2005) that organic matter content of tropical soils decline rapidly confirms the observed

Table 1. Initial soil chemical and physical characteristics of the 0 - 20 cm and 20 - 40 cm layers of the soil before planting the cassava in 2003.

Soil layer (cm)	pH	OC	Total N	P-Bray	Avail. K	Ca	Mg	Sand	Silt	Clay
	(1:1 H_2O)	(g kg^{-1})		($mgkg^{-1}$)	(me 100 g^{-1})				%	
				Researcher-managed experiment						
0 - 20	6.0	8.0	0.96	4.6	0.21	2.6	1.9	73	12	15
20 - 40	6.0	5.4	0.60	2.5	0.20	2.1	1.5	72	12	16
				Farmer-managed experiment						
0-20	5.6	7.4	0.80	6.1	0.22	2.0	1.9	76	7	17
20-40	5.8	4.3	0.70	3.3	0.12	1.4	1.5	72	6	22

Table 2. Effect of preceding cassava variety on chemical characteristics of 0 - 20 cm and 20 - 40 cm soil layers in September, 2004.

Soil chemical properties	Abasafitaa	Tekbankye	Afisiafi	Bensre	Boakentemma	Bush fallow	Mean	SED	Significance
					0 - 20 cm				
pH (1:1 H_2O)	6.2	6.2	6.3	6.2	6.2	6.2	6.2	0.06	NS
Organic C (g kg^{-1})	6.8	6.3	5.2	7.1	7.5	7.3	6.7	0.86	NS
Total N (g kg^{-1})	0.6	0.60	0.60	0.57	0.60	0.57	0.59	0.04	NS
P-Bray (mg kg^{-1})	11	7.7	8.3	9.5	8.4	8.6	8.9	1.8	NS
Avail. K (me 100 g^{-1})	0.09	0.09	0.09	0.09	0.09	0.09	0.09	0.016	NS
Ca (me 100 g^{-1})	2.7	2.5	2.4	2.8	2.4	2.8	2.6	0.16	NS
Mg (me 100g^{-1})	2.3	2.2	2.3	1.9	2.0	2.5	2.2	0.18	NS
					20 - 40 cm				
pH (1:1 H_2O)	6.2	6.2	6.2	6.1	6.1	6.2	6.2	0.06	NS
Organic C (g kg^{-1})	5.3	4.2	4.6	4.6	4.8	4.6	4.7	0.28	0.05
Total N (g kg^{-1})	0.4	0.3	0.4	0.4	0.4	0.4	0.4	0.05	NS
P-Bray (mg kg^{-1})	7.2	4.9	4.9	4.8	5.0	4.8	5.3	1.21	NS
Avail. K (me 100 g^{-1})	0.08	0.07	0.07	0.07	0.06	0.07	0.07	0.010	NS
Ca (me 100 g^{-1})	2.0	2.1	2.0	2.2	2.0	2.3	2.1	0.13	NS
Mg (me 100 g^{-1})	1.7	1.8	1.8	1.7	1.7	2.1	1.8	0.23	NS

SED = Standard error of difference.

Table 3. Root yield and N removal by five cassava varieties grown at Wenchi expressed in both kg ha^{-1} and kg t^{-1} harvested product.

Variety	Root yield (t ha^{-1})		N		
	Fresh	Dry	Kg ha^{-1}	Kg t^{-1}DM produced	
	Researcher-managed experiment				Farmer-managed experiment
Abasafitaa	35.9	11.4	101.4	9.1	28.0
Tekbankye	26.5	10.4	96.0	9.1	22.8
Afisiafi	30.5	9.4	87.1	9.1	23.0
Bensre	31.2	14.6	122.9	8.4	30.7
Boakentemma	17.4	6.4	59.3	9.1	21.0
SED	4.9	0.25	19.4	0.92	40.4
Significance	0.05	0.01	NS	NS	7.2

SED = Standard error of difference.

reduction in the organic carbon content of the soil.

Total soil N was significantly reduced from an initial amount of 0.96 g kg-1 in 2003 (Table 1) to an average of about 0.60 g kg-1 in 2004 (Table 4) in the uppermost soil layer and from 0.6 g kg^{-1} to an average of 0.40 g kg^{-1} in the 20 - 40 cm depth. The rapid decline in the soil N content in this study, confirms CIAT (1982) report that cassava depletes a large amount of N for root yield formation. Soil available P was generally increased from 4.6 (Table 1) to an average of 8.9 mg kg^{-1}(Table 2). Although mycorrhizal associations were not studied in this experiment, studies have shown that cassava roots form an association with mycorrhiza which allows for an increased surface area through which diffusion of P into the roots could take place (Howeler, 2001; Howeler and Sieverding, 1983). Exchangeable K was generally significantly reduced (-60%) in the uppermost layer of the soil after fourteen months of crop growth (Tables 1 and 2). Howeler and Cadavid (1983) reported of similar significant reduction in soil exchangeable K after 12 months of crop growth and attributed it to plant uptake as well as to leaching and soil erosion. While Ca content was not affected by cassava cultivation, Mg content was increased from 1.9 to 2.1 me 100 g^{-1} after fourteen months of cropping (Tables 1 and 2). The observed general increase in the Mg status of the soil could be attributed to release of Mg^{2+} from the cassava leave litter that was deposited on the soil surface (Pellet and El-Sharkawy, 1997). Cassava varieties however did not have any significant effect on any of the soil chemical properties.

On the researcher-managed plots, statistically significant ($P < 0.05$) differences in fresh cassava root yield were found between different cassava varieties (Table 3). Fresh root yield ranged from 17.4 t ha^{-1} with 'Boakentemma' to 35.9 t ha^{-1} with 'Abasafitaa'. Root dry matter yield also ranged from 6.4 t ha^{-1} with 'Boakentemma' to 14.6 t ha^{-1} with 'Bensre' with statistically significant ($P < 0.01$) differences between the varieties. On the farmer-managed plots, yields were generally low, with no statistically significant differences among the different varieties. The differences in fresh root yield in the farmer-managed and the researcher-managed fields could be attributed to differences in management practices.

The fresh root yields of all the cassava varieties studied in both the researcher and the farmer managed experiments were generally high compared with the current national average of 12 t ha^{-1} (ISSER, 2005). The yields reported in both the researcher and farmer managed experiment in this study are also comparable to those reported in Ghana under experimental conditions which range between 16 and 32 t ha^{-1} (MOFA, 2004) and elsewhere in other parts of the world (Puttthacharoen et al., 1998; Howeler and Cadavid, 1983). The amount of total dry matter (dry matter of various plant parts of cassava removed in the crop harvest plus those returned to the soil) produced by the various cassava varieties ranged from 17.9 t ha^{-1} with 'Boakentemma' to 24.9 t ha^{-1} with Bensre (Table 4). Except for the variety Boakentemma, in which the largest contribution was made by the stem, the plant part that made the largest contribution to the total dry matter was the root while the leaf made the smallest contribution to the total dry matter. The percent dry matter removed from the field as storage roots by the various cassava varieties varied from 35.7% with Boakentemma to 58.7% with Bensre. In Thailand, Putthachareon (1998) reported that with total dry matter production of about 33.5 t ha^{-1} in four crops of cassava, the percentage DM that was removed from the field was 45%. In Columbia, Howeler (1985) estimated that percentage dry matter removed from the soil was about 60% when total dry matter production was about 23 t ha^{-1}.

The total amount of N accumulated by the different cassava varieties (made up of total N in plant parts of harvested crop removed from the field and total N in plant parts of harvested crop returned to the field) within the fourteen months of crop growth ranged between 228.4 kg ha^{-1} with 'Abasafitaa' and 287.6 kg ha^{-1} with 'Bensre' (Table 4). The amounts of N removed from the system through crop harvest were 59.3, 87.3, 96, 101.4 and 122.9 kg ha^{-1} for Boakentemma, Afisiafi, Tekbankye, Abasafitaa and Bensre respectively.

Table 4. Dry matter (t ha^{-1}) and N content (kg ha^{-1}) of various plant parts of five cassava varieties removed in the crop harvest and those returned to the soil.

Cassava variety	Roots		Leaves		Litters		Stems		Total	
	DM	N	DM	N	DM	N	DM	N	DM	N
			Plant parts of harvested crop removed from the field							
Abasafitaa	11.4	101.4							11.4	101.4
Tekbankye	10.4	96.0							10.4	96.0
Afisiafi	9.4	87.2							9.4	87.2
Bensre	14.6	122.9							14.6	122.9
Boakentemma	6.4	59.3							6.4	59.3
SED	2.0	19.4							2.0	19.4
Significance	0.05	NS							0.05	NS
			Plant parts of the harvested crop returned to the field							
Abasafitaa			0.93	21.6	4.8	71.9	2.8	33.5	8.5	127.0
Tekbankye			0.83	29.5	5.0	80.6	5.3	68.4	11.1	178.4
Afisiafi			0.92	26.7	4.6	69.1	4.0	47.6	9.5	143.3
Bensre			0.75	32.2	3.9	65.5	5.6	67.0	10.3	164.7
Boakentemma			1.06	41.2	3.9	63.8	6.6	84.5	11.6	189.5
SED			0.13	4.2	0.31	7.04	1.02	15.6	1.3	41.0
Significance			NS	0.01	0.01	NS	0.05	0.05	NS	NS

SED = Standard error of difference.

These values are higher than those reported by Putthachareon et al. (1998) but lower than that reported by Howeler and Cadavid (1983). Howeler (2002) reported that the amount of nutrients removed in the root harvest, is highly dependent on crop growth rate and yield, which, in turn depends on climate, soil fertility and variety.

The amount of N that were recycled through the return of crop residues into the soil were 127.0, 143.3, 164.7, 178.4 and 189.5 kg N ha^{-1} for Abasafitaa, Afisiafi, Bensre, Tekbankye and Boakentemma respectively. These represents 56, 62, 57, 65 and 76% of the total N accumulated by Abasafitaa, Afisiafi, Bensre, Tekbankye and Boakentemma respectively. Thus incorporation of the stems, leaves and leaf litter into the soil would greatly diminish total nutrient export and thus the requirement for fertilization.

Table 4 shows that on average, the largest contribution to the total biomass nitrogen returned to the soil was made by the leaf litter. While there were significant differences among the different cassava varieties with respect to leaf litter production with Tekbankye and Abasafitaa having the highest litter production and the Boakentemma, the least, the amount of N returned into the soil through litter production did not differ very much. The amount of leaf litter reported in this experiment (3.9 to 5.0 t ha^{-1}) is however, comparable to that reported by Carsky and Toukourou (2003) in Benin who collected a total of between 3.4 and 4.1 during 12 months of growth. Our estimate is however higher than that reported by Pellet and El-Shakawy (1997) who reported of values of up to 3 t ha^{-1} for four varieties grown for 10 months. The higher cumulative total of leaf litter reported in this study was partly due to the fact that in our study, measurement was done over a long period of crop growth. Leaf deposition offers a mulch soil-cover and may reduce soil erosion and also contribute to the recycling of substantial amount of nutrients removed from the soil. Pellet and El-Shakawy (1997) reported that leaf litter constituted about 32% of total Ca, 19% of total Mg and 16% of total N uptake in the total cassava biomass. In the present study, leaf litter constituted between 23 and 32% of total N uptake in the total cassava biomass. Our N values were higher, possibly because of differences in methods of litter collection. It is probable that our monthly litter collection reduced losses of nutrients compared with collection of nutrients from the soil surface at the time of harvesting cassava roots.

Although, less than 50% of the total N uptake is lost from the soil through crop harvest, it is estimated that about two thirds of the total amount of K uptake accumulates in the roots and would thus be removed through crop harvest (Howeler and Cadavid, 1983). Thus in this study where root dry matter yield ranged between 6.4 and 14.6 t ha^{-1}, large amount of K could be removed through crop harvest and would thus have to be replaced through fertilization especially as in the case of these soils that contained 0.21 me 100 g^{-1}.

Conclusion

The study suggests that cassava mobilizes and extracts

large amounts of N and even larger amount of K (whereas the soil is already suffering from K-limitation) from the soil but recycle substantial amount of the N into the soil through leaf litter deposition. Reduction in the nutrient content of the soil cannot exclude the possibility of losses through erosion and leaching. The study also shows that cassava varietal choice seems unimportant from soil fertility enhancement point of view but from a nutritional point of view since all the cassava varieties studied mobilize similar amount of N. However, as some varieties remove more N through better crop harvest, some varieties recycle more of its total N uptake for subsequent crops (if it is not leached).

ACKNOWLEDGEMENTS

The author gratefully acknowledges the financial contribution made by the Interdisciplinary Research and Education Fund (INREF) of the Wageningen University and Research Centre, the Netherlands and the Netherlands Ministry of International Co-operation (DGIS).

REFERENCES

Adjei-Nsiah S, Leeuwis C, Giller KE, Sakyi-Dawson O, Cobbina J, Kuyper TW, Abekoe MK, Van Der Werf W (2004). Land tenure and differential soil fertility management practices among native and migrant farmers in Wenchi, Ghana: implication for interdisciplinary action research. NJAS-Wageningen J. Life Sci. 52: 331-348.

Adjei-Nsiah S, Kuyper TW, Leeuwis C, Abekoe MK, Giller KE (2007). Evaluating sustainable cropping sequences with cassava and four other legume crops: Effects on soil fertility and maize yields in the forest savannah transitional agro-ecological zone of Ghana. Field Crops Res. 103: 87-97.

Agronomy Society of America-Soil Science Society of America (1982). Methods of Soil Analysis, Part 2. Chemical and Microbiological Properties-Agronomy Monograph no. 9, 2nd ed., ASA-SSA, Madison, USA.

Asiamah RD, Adjei-Gyapong T, Yeboah E, Fening JO, Ampontuah EO, Gaise E (2000). Report on Soil Characterization and Evaluation at four Primary Cassava Multiplication Sites (Mampong, Wenchi, Asuansi and Kpeve) in Ghana. Technical Report No. 200. Soil Research Institute, Kumasi.

Ayanaba AJ, Tuckwell SB, Jenkinson DS (1976). The effects of clearing and cropping on the organic reserves and biomass of tropical forest soils. Soil Biol. Biochemist. 176(8): 519-525.

Carsky RJ, Toukourou MA, (2003). Cassava leaf litter estimation in on-farm trials. Expl. Agric. 40: 315-326.

Centro International Agricultura Tropicale (CIAT) (1982). Report of Cassava Programme: Annual Report for 1981; CIAT: Cali, Columbia.

El-Sharkawy MA (1993). Drought-tolerant cassava for Africa, Asia and Latin America: breeding projects work to stabilize productivity without increasing pressures on limited natural resources. Bioscience 43: 441-451.

Fermont AM (2009). Cassava and soil fertility in intensifying smallholder farming systems of East Africa. Published doctoral dissertation. Wagenigen University and Research Centre, Wageningen

Howeler RH, (1985). Mineral nutrition and fertilization of cassava. In: Cassava: Research, Production and Utilization. UNDP-CIAT Cassava Programme, Cali, Columbia: pp.249-320.

Howeler RH (2001). Long-term effect of cassava cultivation on soil productivity. Field Crops Res. 26: 1-18.

Howeler RH (2002). Cassava mineral nutrition and Fertilization. In Cassava: Biology, Production and Utilization, Hilocks, R.J. Thresh J.M. and Bellotti A.C. (eds): 115-147. CAB International, Wallingford, UK.

Howeler RH, Cadavid LF 1983. Accumulation and distribution of dry matter and nutrients during a 12-month growth cycle of cassava. Field Crops Res. 7: 12-139.

Howeler RH, Sieverding E (1983). Potentials and limitations of mycorrhizal inoculation illustrated by experiments with field-grown cassava. Plant Soil 75: 245-261.

ISSER (2005). The state of the Ghanaian Economy in 2004. Institute of Social, Statistical and Economic Research, University of Ghana, Legon.

MOFA (2004). Cassava Production Guide. A resource and Reference Manual. Root and Tuber Inprovement Programme (RTIP), Ministry of Food and Agriculture, Kumasi, Ghana.

Pellet DM, El-Sharkawy MA, (1997). Cassava varietal response to fertilization: growth dynamics and implications for cropping sustainability. Experimental Agriculture 33: 353-365.

Putthacharoen S, Howeler RH, Jantawat S, Vichukit V (1998). Nutrient uptake and soil erosion losses in a Psamment in eastern Thailand. Field Crops Res. 57: 113-126

Saidou A, Kuyper TW, Kossou DK, Tossou R, Richards P, (2004). Sustainable soil fertility management in Benin: Learning from farmers. NJAS-Wageningen J. Life Sci. 52: 349-369.

SAS (1996). SAS User's Guide: Statistics, SAS Institute, Cary, NC.

Zingore S, Manyame C, Nyamugafata P, Giller KE (2005). Long-term changes in organic matter of woodland soils cleared for arable cropping in Zimbabwe. Eur. J. Soil Sci. 56: 727-736.

Effects of farm yard manure on cadmium and lead accumulation in Amaranth (*Amaranthus oleracea* L.)

M. Alamgir[1,2], M. G. Kibria[1] and M. Islam[1]

[1]Department of Soil Science, University of Chittagong, Chittagong- 4331, Bangladesh.
[2]School of Agriculture, Food and Wine, University of Adelaide, SA 5005, Australia.

Two pot experiments were carried out to study the effect of farm yard manure (FYM) on Cd and Pb accumulation by Amaranth. Six levels of FYM (0, 2.5, 5, 10, 15 and 20 t ha^{-1}) were applied to Cd and Pb treated soil separately in the pot experiments. The weight of the shoot and root of Amaranth significantly increased following the application of FYM to both Cd and Pb treated soil. The application of FYM in soil significantly decreased Cd and Pb content in Amaranth. Cadmium content in the shoot and root gradually decreased with the increase in level of FYM up to 20 t ha^{-1}. On the other hand, FYM at 10 t ha^{-1} and above drastically reduced the Pb content in both parts of *Amaranth*. The contents of both Cd and Pb in the shoot and root of Amaranth showed a significantly negative correlation (r = -0.84 to -0.87) with the rates of FYM applied to the soil.

Key words: Farm yard manure (FYM), heavy metal, phytoremediation, cadmium (Cd) uptake, lead (Pb) uptake.

INTRODUCTION

Cadmium (Cd) and lead (Pb) are heavy metals that have adverse effect on the soil quality. Wide areas of earth are polluted by heavy metals which may not be degraded by biotic process and may enter the surface and/or underground water resources and may be absorbed by plants and thus enter the food chain and consequently become hazardous to humans and animals. Major sources of heavy metal inputs to ecosystems are mining, smelting and metallurgical industries, sludge disposal and agricultural practices (Horak and Friesl, 2007). Cadmium and Pb accumulation in plants has a positive correlation with their availability and plant growth inhibition (Greger and Bertell, 1992; Kibria et al., 2006; Kibria et al., 2007). Amelioration and rehabilitation of polluted soils can be achieved by *in situ* treatment, such as phytoremediation or immobilization of metals in soil. Immobilization of metals in soil is aimed to significant reduction of their bioavailability by liming and addition of adsorptive materials such as iron oxides, clay minerals, zeolites and organic matter. The efficiency of the metal immobilization in the soil on reduction of absorption of these metals depends strongly on type of the crop (Hornburg and Brümmer, 1990). Plants that hyperaccumulate metals have attracted attention for many years as geobotanical indicators of mineral deposits. More recently, their ability to extract metals from the soil and to concentrate them in the shoot, has given rise to the idea of practical applications such as phytomining and phytoremediation (Lombie et al., 2000). (Yassen et al., 2007) found that application of Farm yard manure (FYM) significantly decreased Cd concentration (41to 31%) in spinach plant grown on soil without heavy metal application and under contaminated soil conditions. Another study conducted by (Kos et al., 2003) showed that Zn and Cd uptake was 3.3 times higher in Amaranthus spp. compared to the control. Amaranth is a popular leafy vegetable crop in Bangladesh and Cd and Pb pollution of the soil of Bangladesh has been reported by Khan (2001) and Kibria (2008). Of late, integrated nutrient management in Bangladesh involving organic manure and chemical fertilizer had received considerable attention. Farm yard manure (FYM) positively influence crop production (Kaihura et al., 1999), improve soil physical properties (Chen et al., 1996) and can be used to reduce heavy metal hazards in plants (Yassen et al., 2007). With this

*Corresponding author. E-mail: alamgircu@hotmail.com.

view in mind, the aim of this study was to investigate the effect of FYM on Cd and Pb absorption by Amaranth.

MATERIALS AND METHODS

Pot experiments

Two pot experiments were carried out with Amaranth on sandy loam soil in the net house of the Department of Soil Science, University of Chittagong. The soil was collected from the top 15 cm of an uncontaminated agricultural field of Syedpara at Hathazari Upazilla under Chittagong district of Bangladesh. Soil was air dried and ground to pass through a 2 mm sieve to separate dry roots, grasses and gravels, etc. A portion of the soil passed through 2 mm sieve was retained for laboratory analyses.

In first pot experiment, each earthen pot was filled with 8 kg of soil. Cadmium was applied at 10 mg kg^{-1} soil in 18 pots in solution form as 3Cd $(SO_4)_2.8H_2O$. Then FYM was mixed separately with the soil at the rate of 0, 2.5, 5.0, 10.0, 15.0 and 20.0 ton ha^{-1} on air dry weight basis. Each treatment was replicated thrice. The pots were arranged in randomized block design. Second pot experiment was conducted similar to first experiment, however, instead of Cd, Pb was applied at 50 mg kg^{-1} soil in 18 pots in solution form as Pb $(NO_3)_2$. Potted soils mixed with FYM, Cd and Pb were allowed to equilibrate for two weeks in moist condition prior to sowing of seeds. Nitrogen (as Urea), P (as TSP) and K (as MP) were applied in the pots (99 kg N, 18 kg P, 60 kg K ha^{-1} soil). According to Bangladesh Agricultural Research Council (2005) recommendation, half of N and K, and whole of P were applied during soil preparation. The remaining N and K were applied in two equal installments after 2 and 4 weeks of seedling emergence. Healthy and uniform seeds of Amaranth were sown in each pot. After two weeks of seedling emergence, five seedlings were kept in each pot and care was taken to keep uniform seedlings in the pots. Water was applied regularly to the pots to maintain the field capacity of the soils. Plants were harvested after five weeks of sowing seeds. Shoots and roots were separately collected. Roots were washed thoroughly with tap water first to remove adhering soil particles and then with distilled water. Oven dry (at 65°C to constant weight) weights of shoots and roots were recorded.

Soil analysis

The particle size distribution was determined by hydrometer method of Day (1965). Soil pH was measured in a 1:2.5 soil/water suspension with glass electrode pH meter. The organic carbon, cation exchange capacity (CEC), and total elemental analyses were done following the methods described by Jackson (1973). Briefly organic carbon was determined by the wet-oxidation method using potassium dichromate and organic matter content was calculated by multiplying the values with 1.724. The cation exchange capacity (CEC) was determined by saturation with 1 N NH_4OAc at pH 7.0. The soil samples were digested with aqua regia on a sand bath for the determination of total Cd, Pb, Zn, Fe, Mn, P and K. Total nitrogen was determined by micro - Kjeldahl method. Chemicals of analytical reagent grade were used in all the analyses. Phosphorus was determined by vanadomolybdo phosphoric yellow color method in nitric acid system. Potassium was measured by flame photometer and Fe, Mn, Zn, Cd and Pb were determined by flame atomic absorption spectrophotometer (Varian spectra AA-220).

Plant analysis

Oven dried and ground plant samples were digested with a tri-acid mixture (HNO_3, H_2SO_4 and $HClO_4$ mixture at the ratio of 5:1:2) for the determination of Cd and Pb content in the plant tissues (Huq and Alam, 2005). The concentrations of Cd and Pb in the plant digests were measured by atomic absorption spectrophotometer.

Data analysis

Significance of differences among the treatments was evaluated by one way analysis of variance followed by Duncan's multiple range test at the significance level of $p< 0.05$. Pearson's correlation of both Cd and Pb in the shoot and root of Amaranth was determined with the rates of FYM applied to the soil. The statistical software SPSS version 12 (SPSS Inc. 2003) was used in the analysis.

RESULTS AND DISCUSSION

Soil properties

The soil contained 63% sand, 20% silt, 17% clay; and had pH 5.1, organic carbon 0.61%, cation exchange capacity (CEC) 6.43 cmol kg^{-1} soils; total N 0.04%, total P 0.02%, total K 0.26 % and aqua regia extractable Pb 9.6 mg kg^{-1}, Zn 52.02 mg kg^{-1}, Fe 1.472% and Mn 86.90 mg kg^{-1}. While aqua regia extractable Cd was below detection limit (< 0.002 mg kg^{-1}).

Biomass production

The dry matter yields of shoot and root of Amaranth were significantly increased by the addition of FYM to both Cd and Pb treated soils (Table 1). However, there was no significant increase in the dry weight of shoot up to 5 and 2.5 t ha^{-1} FYM application in Cd and Pb treated soil, respectively. An increase of 152 and 231% in the dry weight of shoot and root was observed respectively in Cd treated soil at the highest rate (20 t ha^{-1}) of FYM application in comparison to control. In Pb treated soil, application of FYM at the highest rate (20 t ha^{-1}) increased the shoot and root dry weight by 181 and 209% of the control, respectively.

Cd and Pb accumulation in Amaranth

Cadmium and Pb concentration in Amaranth decreased with increasing rates of FYM added to the soils, with the decrease being more pronounced in shoot for Cd and root for Pb (Table 2). Increasing the rate of FYM addition up to 20 t ha^{-1} gradually reduced the Cd concentration in shoot and root while the addition of FYM at 10 t ha^{-1} and above reduced the Pb concentration to zero in both parts of Amaranth. This decrease may result from the increase of soil sorption capacity in the presence of FYM (He and Singh, 1993a; McBride, 1995; Blum, 1999). The little absorption of Pb by Amaranth may be attributed to its high retention in the soil complex system (Hooda and Alloway, 1994; Berti and Jacobs, 1996).

Table 1. Effects of farm yard manure (FYM) on biomass production (g pot^{-1}) of Amaranth grown in Cd and Pb treated soil.

Treatment (FYM, t ha^{-1})	Biomass production (g pot^{-1})			
	Cd treated soil		Pb treated soil	
	Shoot	Root	Shoot	Root
0	4.25^{d}	0.48^{d}	3.75 ^{e}	0.51^{d}
2.5	4.59^{d}	0.86^{c}	4.41 ^{e}	0.87^{c}
5	5.53^{d}	0.91^{c}	5.16 ^{d}	0.87^{c}
10	7.21^{c}	1.16^{bc}	6.98 ^{c}	1.14^{bc}
15	8.87^{b}	1.47^{ab}	8.82 ^{b}	1.39^{ab}
20	10.73^{a}	1.59^{a}	10.54 ^{a}	1.58^{a}
Significance of F value (P)	0.001	0.05	0.01	0.05

Mean values in the column followed by the same letter(s) are not significantly different according to Duncan's multiple range test (DMRT) ($P \leq 0.05$). Each value is the mean of 3 replicates.

Table 2. Effects of farm yard manure (FYM) on Cd and Pb accumulation in Amaranth grown in Cd and Pb treated soil respectively.

Treatment (FYM, t ha^{-1})	Cd concentration (mg kg^{-1}plant tissue)		Pb concentration (mg kg^{-1} plant tissue)	
	Shoot	Root	Shoot	Root
0	30.36^{a}	49.98^{a}	2.59^{a}	15.42^{a}
2.5	20.57^{b}	40.91^{b}	1.47^{b}	5.58^{b}
5	18.00^{bc}	39.48^{bc}	0.89^{c}	3.75^{c}
10	15.83^{bc}	35.39^{bc}	0.00	0.00
15	13.57^{cd}	30.41^{cd}	0.00	0.00
20	9.26 ^{d}	24.52^{d}	0.00	0.00
Significance of F value (P)	0.001	0.05	0.01	0.05

Mean values in the column followed by the same letter(s) are not significantly different according to DMRT ($P \leq 0.05$). Each value is the mean of 3 replicates.

The typical behaviour of Pb in contaminated soils includes high retention (McBride, 1995), low mobility (Sheppard and Thibault, 1992) and low bioavailability (Kádár, 1994; Alloway, 1995). This behaviour is associated with its high affinity to soil organic matter (McBride, 1995). Increased adsorption of Cd by soil components with increasing amounts of organic matter has also been reported by others (Christensen, 1989; Eriksson, 1998). Addition of 320 g kg^{-1} organic matter decreased the soil pH by around one pH-unit, but it increased the soil CEC by six fold in the sand and 1.5 fold in the sandy loam and clay loam soil (He and Singh, 1993b). He and Singh (1993a) attributed the decreased Cd concentration in plant with higher levels of organic matter addition was predominantly due to the effect of increasing CEC in soil. In the present study, addition of FYM might cause an increase of soil CEC that have increased the ability of soils to adsorb Cd and Pb ions. He and Singh (1993a) reported that addition of 20 g kg^{-1} organic matter (sphagnum peat) to the sand and sandy loam soil significantly reduced the Cd concentration in ryegrass. Eriksson (1988) also reported that Cd uptake by ryegrass and winter rape was considerably reduced in both sand and clay soils when peat was added, and the reduction in uptake of Cd by plant was more pronounced in the sand than in the clay soil even though the exchangeable Cd did not differ between the two soils. In another study by Kibria et al. (2011), the grain Cd content of rice was significantly reduced by 27% by FYM and 62% by FYM mixed with lime.

A significant negative Pearson's correlation was found between Cd concentration in *Amaranth* and rate of application FYM ($r = -0.841$ and -0.869 for shoot and root, respectively); and between Pb concentration in Amaranth and FYM ($r = -0.859$ and -0.868 for shoot and root, respectively) applied to the soil. It indicates a low mobility as well as low availability of Cd and Pb in soil caused by FYM. In both pot and field investigations, Eriksson (1988) also found a negative relationship between the Cd concentration in plants and the organic matter content in soils. Organic matters with suitable reactive groups, such as hydroxyl, phenoxyl, and carboxyl effectively control the adsorption and complex of heavy metals with soil, and the activity of metals in soils (Alloway, 1995). The rise in pH caused by FYM applied in acid soil (Kaihura et al., 1999) would also decrease soil Cd and Pb availability through increasing Cd and Pb adsorption. High pH increases adsorption sites by increasing negative charges on the soil surface and reducing competing cations.

CONCLUSION

Farm yard manure can be a very effective amendment to reduce the accumulation of Cd and Pb in Amaranth and other crops growing on Cd and Pb contaminated soils. Further research in field condition is needed to fully understand the effects of FYM on crop Cd and Pb accumulation.

ACKNOWLEDGEMENT

The authors would like to thanks the anonymous reviewers whose suggestions have improved the quality of the manuscript.

REFERENCES

Alloway BJ (Ed.) (1995). Heavy metals in Soils. 2nd Ed. Blackie Academic and Professional. London.

Bangladesh Agricultural Research Council (2005). Fertilizer Recommendation Guide. BARC Soils Publications No 42.

Berti WR, Jacobs LW (1996). Chemistry and phytotoxicity of soil trace elements from repeated sewage sludge applications. J. Environ. Qual., 25:1025-1032.

Blum WH (1999). Cadmium uptake by higher plants. p. 109-110. Proceedings of the 5th International Conference on the Biogeochemistry of Trace Elements, Vienna.

Chen L, Dick WA, Streeter JG, Hoitink HAJ (1996). Ryegrass utilization of nutrients released from composted biosolids and cow manure, Compost Science Utilization 4: 73–83.

Christensen TH (1989). Cadmium soil sorption at low concentrations. VIII. Correlation with soil parameters. Water, Air, Soil Pollut., 44: 71-82.

Day PR (1965). Particle fractionation and particle size analysis. 1965. p. 545-567. In: Methods of Soil Analysis (Black CA Ed.). Part I. Agronomy Monograph, Academic Press, New York.

Eriksson JE (1988). The effect of clay, organic matter and time on adsorption and uptake by plants of cadmium added to the soils. Water, Air, Soil Pollut., 40: 359-373.

Greger M, Bertell G (1992). Effects of Ca^{2+} and Cd^{2+} on the carbohydrate metabolism in sugar beet (*Beta vulgaris*). J. Exp. Bot., 43 (2): 167-173.

He QB, Singh BR (1993a). Effect of organic matter on the distribution, extractability and uptake of cadmium in soils. J. Soil Sci., 44: 641-650.

He QB, Singh BR (1993b). Plant availability of cadmium in soils. I. Extractable cadmium in newly and long term cultivated soils. Acta Agric. Scand. Sect. B. Soil Plant Sci., 43: 134 -141.

Hooda PS, Alloway BJ (1994). Changes in operational fractions of trace metals in two soils during two-years of reaction time following sewage sludge treatment. Int. J. Environ. Ana. Chem., 57: 289-311.

Horak O, Friesl W (2007). Soil additives immobilizing heavy metals in contaminated soils. Nova Biotechnol., 7(1): 5-9.

Hornburg V, Brümmer GW (1990). Cadmium availability in soils and content of wheat grain. In '5th Spurenelement Symposium on Iodine and other Trace elements. F. Schiller University, Jena' (Anke M, H Brümmer, C Brümmer, B Groppel Eds) p. 916.

Huq SMI, Alam MD (Eds.) (2005). A handbook on Analyses of Soil, Plant and Water. BACER-DU, University of Dhaka, Bangladesh.

Jackson ML (1973). Soil Chemical Analysis. Prentice Hall of India Private Limited, New Delhi, India.

Kádár I (1994). Effect of heavy metal load on soil and crop. Acta Agron. Hung. 43: 3-9.

Kaihura BS, Kullaya IK, Kilasara M, Aune JB, Singh BR, Lal R (1999). Soil quality effects of accelerated erosion and management systems in three eco-regions of Tanzania. Soil Tillage Res., 53: 59-70.

Khan AH (2001). Heavy metals profile in Bangladesh Environment: Soil and irrigation Water Components. Final Completion Report. Bangladesh Agricultural Research Council, Dhaka.

Kibria MG, Osman KT, Ahmed MJ (2006). Cadmium and lead uptake by rice (*Oryza sativa* L.) grown in three different textured soils. Soil. Environ., 25(2): 70-77.

Kibria MG, Osman KT, Ahmed MJ (2007). Cadmium and lead uptake by radish (*Raphanus sativus* L.) grown in three different textured soils. Soil Environ., 26(2): 106-114.

Kibria MG (2008). Dynamics of cadmium and lead in some soils of Chittagong and their influx in some edible crops. Unpublished Ph. D. Thesis, University of Chittagong, Bangladesh. p. 292.

Kibria MG, Osman KT, Ahammad MJ, Alamgir M (2011). Effects of farm yard manure and lime on cadmium uptake by rice grown in two contaminated soils of Chittagong. J. Agric. Sci. Technol., 5(3): 352-358.

Kos B, Grcman H, Lestan D (2003) Phytoextraction of lead, zinc and cadmium from soil by selected plants. Plant Soil Environ., pp. 49, 548–553.

Lombie E, Zhao FJ, Dunhams J, McGrath SP (2000). New Phytol., 145, 11-20.

McBride MB (1995). Toxic metal accumulation from agricultural use of sewage sludge. Are USEPA regulations protective? J. Environ. Qual. 24: 5-18.

Sheppard J, Thibault A (1992). Desorption and extraction of selected heavy metals from soils. Soil Sci. Soc. Am. J., 56: 415- 423.

SPSS Inc. (2003). SPSS Inc. Chicago, USA.

Yassen AA, Nadia BM, Zaghloul MS (2007). Role of some organic residues as tools for reducing heavy metals hazards in plant. W. J. Agri. Sci., 3(2): 204-207.

Comparative influence of benomyl on rhizosphere and non-rhizosphere bacteria of cowpea and their ability to solubilise phosphate

Ekundayo F. O.

Department of Microbiology, Federal University of Technology, P. M. B. 704, Akure, Ondo State, Nigeria.
E-mail: foekundayo2002@yahoo.com.

Six concentrations of benomyl (g/ml): 0.00 (control), 0.05, 0.10, 0.20, 0.40 and 0.80 were assayed by using 1.5 kg of top soil collected from nursery of Federal University of Technology, Akure, Ondo State, Nigeria. Cowpea (IT - 93K - 452 - 1) seeds were planted in the soil sample in different polyethylene bags. Harvesting of cowpea seedlings was done on 5, 10, 15, 20, 25 and 30 days of germination. Nodule and soil bacterial counts were done; nodulating as well as rhizosphere and non-rhizosphere bacteria were cultured and identified. The activity of the isolated bacteria was evaluated by rock phosphate solubilisation. Cowpea seedlings were healthy without much loss of root nodules and beneficial rhizosphere bacteria at 0.10 and 0.20 g/ml of benomyl. Nodule loss of 94.3% occurred at 0.80 g/ml of benomyl when compared to control. *Bradyrhizobium japonicum* was responsible for nodulation of cowpea. *Bacillus subtilis*, *Staphylococcus aureus* and *Micrococcus luteus* were identified from rhizosphere while *Arthrobacter simplex*, Azotobacter *chroococum*, *Pseudomonas aeruginosa* and *B. subtilis* also were identified from non-rhizosphere of cowpea. *Bacillus subtilis* significantly ($p < 0.05$) increased solubilisation of rock phosphates in the presence of benomyl. As benomyl concentration increased from 0.00 to 0.80 g/ml, there was a decrease in bacterial population in the rhizosphere. The optimum concentration of benomyl on beneficial soil bacteria was 0.10 to 0.20 g/ml. *Bacillus subtilis* could be inoculated to rhizosphere soil of cowpea to increase the amount of soluble phosphate available for the growth of cowpea seedlings in benomyl treated soil.

Key words: Benomyl, rock phosphate solubilisation, rhizosphere and non-rhizosphere bacteria.

INTRODUCTION

Cowpea cultivation has been intensified in developing countries to provide protein (Omokaro and Ajakaiye, 1989). Cowpea (*Vigna unguiculata* (L.) Walp) is widely grown in the tropics and subtropics of African, Asian, and Latin American. The origin of the cultivated species was from Africa. Cowpea is a widely adapted annual legume, fodder and cash crop in the dry savannas of West and Central Africa. Although the fresh seeds and pods can be frozen or canned, most of the crop is grown for home consumption of the dried seed. At present, this species is the second most important pulse crop in Africa where Nigeria is the biggest producer (Omokaro and Ajakaiye, 1989). Benomyl is a systemic benzimidazole foliar fungicide that is selectively toxic to microorganisms. It is effective in controlling fungal diseases such as *Fusarium*, *Cercospora*, *Sclerotinia*, *Rhizoctonia*, *Pythium* and *Colletotrichum* of cowpea and other crops (Newsham et al., 1994; Ekundayo and Oladunmoye, 2007). Although benomyl is a chemical designed for the control of plant pathogenic fungi (Tomlin, 1994), it may nonetheless prove detrimental to non-target organisms (Akobundu, 1987) such as phosphate solubilising and root nodule bacteria.

In soil fertility, phosphorus is a very limiting factor in the growth and yield of plants. Most of the phosphate in soil occurs in the insoluble organic form, thus total phosphate reaches the plants in soluble form (Saha and Biwas, 2009). The soil bacteria are therefore responsible for the continuous availability of the element as they solubilize the inorganic phosphate compounds (Etok et al., 2010). Phosphate solubilizing bacteria belong to diverse taxonomic groups and differ in structure and response to environmental changes (Behbahani and Behbahani, 2009). There is a need to know the concentration at which benomyl may be detrimental to useful soil microorganisms especially the symbiotic and non symbiotic nitrogen fixers like Rhizobium and Azotobacter in the soil. This is necessary so that these beneficial microorganisms may be preserved to play their roles in soil fertility, phosphorus solubilisation and biotransformation.

MATERIALS AND METHODS

Collection of samples and sterilization of materials

The soil sample (top soil) used for this project was collected with hand trowel behind greenhouse section of the Department of Biology, Federal University of Technology, Akure Ondo State, Nigeria. This soil sample was used for planting cowpea during the experimental set up. The cowpea - (*Vigna unguiculata* IT93K-452-1) was obtained from the International Institute of Tropical Agriculture, Ibadan, Oyo State, Nigeria. Powdered rock phosphate was obtained from Ogun State, Nigeria.

Soil sample (that is, 738 kg) used for planting were sterilized in the oven (Gallenkamp BS 250) at 180℃ for three hours. This was then used for planting after cooling. Hutchinson agar, yeast mannitol agar and ammonium sulphate yeast extract glucose medium used in this project were sterilized by autoclaving at 121℃ for 15 min by using (Amsco 2022 Isothermal autoclave - Eagle series).

Physical and chemical analysis of soil samples

These were performed according to the methods of A.O.A.C (1990) and Carter (1992). Soil sample used for this experiment was analyzed to determine it's physicochemical according to the methods of (Carter, 1992) and Association of Analytical Chemist, A.O.A.C., (1990). Ten millilitres of sodium hexametaphosphate was added to 40 g of soil sample in a beaker. Ninety millilitres of water was added to soil sample and left over night and made up to 1000 ml mark of the cylinder. A hydrometer was then used to take the reading and the values of sand, clay and silt were calculated.

pH determination

The pH of the soil sample was determined by using the method of Halder et al. (1990). A sample of soil and water was prepared in ratio 1:1 and the pH value determined with Jenway 3015 pH meter.

Percentage carbon determination

The Walkey-Blakey wet oxidation method of A.O.A.C (1990), Carter (1992) and Ibitoye (2006) were used. Ten millilitres of potassium dichromate (1 M $K_2Cr_2O_7$) was added to a pre-weighed 5 g well ground soil sample. Twenty millilitres of concentrated H_2SO_4 was added, shaken gently and left to cool. The suspension was made up to 150 ml mark with distilled water in a conical flask. Ten drops of orthophenanthroline ferrous complex indicator were introduced. It was titrated against 0.51 M ferrous ammonium sulphate until end point (solution changed from blue to red colour). The percentage of carbon was determined from the value of $K_2Cr_2O_7$ reduced using the fol1owing formula:

$$\frac{(10.0-0.4y)\times 0.004\times 100}{W}$$

y = volume in millilitres of 0.51 M ferrous ammonium sulphate used.
W = weight of soil taken

Determination of calcium, magnesium, potassium and sodium

Atomic absorption spectrophotometer method of A.O.A.C (1990) and Carter (1992) was used to determine the soil calcium, magnesium, potassium and sodium content. One gramme each of the soil was transferred into 100 ml conical flask. Three millilitres of concentrated hydrogen fluoride was added to the conical flasks and shaken vigorously for 30 min. This was followed by the addition of 2 ml of aqua regia. The conical flasks were left to stay for 3 days before they were made up to the 50 ml mark with distilled water that is, 4 ml of distilled water was added. An atomic spectrophotometer (Buck scientific model 200A) was used to determine the presence of the elements.

Determination of phosphorus

Five grammes of air-dried soil samples were transferred into 250 ml flask. Bray one solution was added and the mixture was left to stand for one minute and filtered. Eight millilitres of sample of standard solution or blank was pipetted into glass vials. Five drops of phosphorus B reagent (ammonium molybdate solution and 5 drops of phosphorus C reagent ($FeSO_4$ solution) were added and carefully mixed. This was allowed to stand for 15 min. The samples were read in the colorimeter using a green filter (600 milimicrons wavelength) against a blank. The standard curve was then used to convert to appropriate units. The colorimeter readings for the standards and phosphorus were determined from the graph (A.O.A.C., 1990; Carter 1992).

Determination of total nitrogen

The titrimetric method of A.O.A.C. (1990) and Carter (1992) was used. Two grammes of the soil were weighed into a kjeldahl flask and 10 ml of concentrated sulphuric acid was introduced into the sample. A tablet of catalysts (copper sulphate) was added. Heat was applied on digestion rack and the sample left to boil for 3 h until a clear solution was obtained. This indicates complete digestion, after digestion, the solution was left to cool; it was made to the mark with distilled water in a 100 ml volumetric flask. This solution was then titrated against 0.1 M HCl until end point was reached, that is, colour changed from blue to black.

Experimental design

The experimental design for planting of cowpea was 2 x 6 factorial experiments in a complete randomized design with 3 replications. The factors were soil type (non rhizosphere and rhizosphere) and addition of benomyl (0.00, 0.05, 0.10, 0.20, 0.40 and 0.80 g/ml). The average weight of soil in each polyethylene bag was 1.50 kg. For serial harvesting at 5, 10, 15, 20, 25 and 30 days after planting of cowpea, the total experimental unit was repeated six times.

Planting of cowpea

Different concentrations of benomyl were added to the soil samples in all the polyethylene bags and mixed properly with the soil. Seeds of *V. unguiculata* were surface sterilized for 2 min in 70% alcohol, rinsed twice in water and planted on day 6 (after the inoculation of soil samples) at a depth of 3 cm, at two seeds per pot. After growth, all seedlings were thinned to one per pot. The plants were grown at 25 ± 2°C and the pots were watered regularly to maintain a good soil moisture condition. Plants were harvested on the 5, 10, 15, 20, 25, and 30th days after planting. Harvesting of cowpea seedling in each polyethylene pot was done by using hand trowel after 30 days of planting. Each seedling was gently shaken to collect the rhizosphere soil around its root.

Identification of Rhizobium

Nodules were collected from harvested plants after 30 days of planting and Rhizobium from nodules was identified according to the method of Somasegaran and Hoben (1994). The roots were washed carefully with distilled water and the nodules detached from it. They were immersed in 95% ethanol for 10 s to soften the seed coats and then transferred to another beaker containing 3% hydrogen peroxide which acted as sterilant for 4 min aseptically. They were later removed and rinsed in six changes of sterile distilled water with the aid of a sterile forceps. Each sterilized nodule was placed in a sterile Petri dish containing few drops of sterile water and crushed with a sterile rod to make slurry. One loopful of the slurry was then streaked on the prepared yeast-mannitol agar plate containing bromothymol-blue indicator. The streaking was repeated twice for each nodule slurry to ensure accurate result. The inoculated plates were incubated at 25 - 28°C for 10 days. Rhizobium growth was indicated by change in indicator colour (Somasegaran and Hoben, 1994).

Isolation, enumeration and characterization of non-rhizosphere and rhizosphere soil bacteria

Eight-fold serial dilution was performed on the soil sample used and 0.1 ml of the eight- fold diluent was pipetted on Petri dish and flooded with molten Hutchinson agar (40°C). It was allowed to cool, solidify before incubating at 37°C for 24 h. Characterization of the isolate was done according to the methods of Olutiola et al. (1991) and Holt (1989).

Determination of plasmid DNA from bacteria isolated from the soil using gel electrophoresis

Extraction of plasmid from Gram-negative and positive bacteria as well as preparation of agarose gels was done according to the method of Meghrous et al. (1997). The bacteria were sub-cultured from the slant culture on MacConkey agar and incubated at 37°C for 24 h. A loopful of the 24 h culture of all isolates was each transferred into 1.5 ml Eppendorf tubes containing 100 µl Tris EDTA buffer and mixed vigorously (using auto vortex mixing machine). A 400 µl of 0.2 N NaOH and 1% sodium duodecyl sulphate (SDS) was also added to the solution. The resulting suspension was mixed gently by inverting the tubes rapidly 5 times and the tubes were stored on ice for 5 min. To this mixture, 150 µl of ice-cold 3 M-sodium acetate, glacial acetic acid and distilled water was added to the suspension in the tubes. This was then vortexed for 10 s, stored on ice for another 5 min; then centrifuged at 12,000 × g for 5 min at 4°C and the supernatants were transferred into another set of 1.5 ml Eppendorf tubes. An equal volume of phenol was added into the tubes and vortexed again.

The Eppendorf tubes were centrifuged at 12,000 × g for 5 min at 4°C. The supernatants were transferred into fresh set of 1.5 ml Eppendorf tube using a micropipette. A 2 volume of 100% ethanol was added into each test tube and these tubes were allowed to stand for 2 min at room temperature. These tubes were again centrifuged at 12,000 × g for 5 min at 4°C after which the supernatants were discarded and the pellets were air-dried. These pellets were then washed with 1 ml 70% ethanol at 4°C and centrifuged at 12,000 × g for 5 min at 4°C. The supernatants were discarded and the pellets air dried for 5 min. The nucleic acids redissolved in 5 µl of Tris EDTA (TE) buffer (pH 8.0, 10 mMTris, 1 M EDTA and distilled water) containing DNASE free RNASE (Meghrous et al., 1997).

The bacteria were twice sub-cultured on MacConkey agar and incubated at 37°C for 24 h, a loopful of each organism was transferred into Eppendorf tubes containing 0.04 M Tris-acetate pH 8.0 and 0.02 M EDTA and 200 µl lysis buffer was added. The tubes were put on a SB - 16 Techne shaking bath for 45 min at 60°C. Afterwards, 6 µl of phenol was added to the sample and mixed gently to complete emulsification. The tubes were centrifuged at 10,000 × g for 20 min at room temperature to separate phases. The aqueous phase, which contained the nucleic acids, was used directly for gel electrophoresis (Meghrous et al., 1997). Approximately 0.89% agarose gel containing 5 ml Tris-boric acid EDTA buffer was prepared in 250 ml flask. The content in the flask was heated to boiling; it was removed and allowed to cool to 50°C. A 3 µl of ethidium bromide was added into the heated agarose after cooling. It was then poured into the gel electrophoresis tray with an already inserted comb. It was left for some minutes to gel. Before loading the gel, 20 µl of the DNA sample was mixed with 2 µl of loading buffer (that is, 0.05 M EDTA), 20% Fiscoll, 0.25% bromophenol blue in water was placed on paraffin paper and 0.5 µl of plasmid DNA was mixed with it.

Afterwards, a fourteen welled comb was removed and 1 µl of DNA markers was introduced into the first well using micropipettes. Twenty microlitres (20 µl) of DNA samples were introduced into subsequent wells. Tris boric acid EDTA buffer was used as the electrophoresis buffer. The electrophoresis tray was carefully placed into the electrophoresis compartment containing the buffer. It was ensured that the buffer covered the gel completely. The electrophoresis compartment has two poles (that is, negative and positive poles). This compartment was then connected to a power pack which supplied high voltage (power pack supplies 64 volts at 0.328 Amps). The compartment lid was tightly fitted on the tray. An ice bag was placed on the lid to reduce the temperature generated during the process. Plasmid DNA was positively charged. After running for 1 h, plasmid DNA was seen to move from negative pole to the positive pole. At the end of the process, DNA bands were visible under short wavelength UV light (Meghrous et al., 1997).

Rock phosphate solubilization experiment

A loopful of all rhizosphere bacterial isolates from slants was cultured in nutrient broth at 37°C for 24 h. The broth was centrifuged at 5000 rpm with Beckman Coulter Allegra 6R centrifuge at 28°C for 5 min. The supernatants were decanted and the sediment was resuspended with sterile distilled water. This was centrifuged again at 5000 rpm at 28°C for 5 min to wash the cells. Harvested cells were obtained after decanting the supernatant. This was used for rock phosphate solubilization experiment. Harvested cells are prepared in order to get a pure culture of fungal isolates. Fungi was cultured in potato dextrose broth and centrifuged at 3, 000 rpm with Beckman Coulter Allegra 6R centrifuge at 28°C for 5 min.

The supernatants were decanted and sediment was resuspended with sterile distilled water. The sediment represented harvested cells and was used for rock phosphate solubilization

Table 1. Influence of benomyl concentration on nodule count and rhizosphere bacteria of 30 day old cowpea seedlings.

Concentration (g/ml)	Bacteria (cfu × 108) /g	Nodule count
0.00	12.6c	3.1c
0.05	12.2c	3.0c
0.10	12.1c	2.5c
0.20	11.2c	2.4c
0.40	10.4b	1.7b
0.80	9.9a	0.5a

Values followed by similar along the same column are not significant at $P \leq 0.05$.

experiment. Powdered rock phosphate at 2 g/litre was added as a source of insoluble phosphate into already prepared ammonium sulphate yeast extract glucose (AYG) medium. The medium was maintained at pH 6-7 (so that pH of the medium will correspond with that of soil) using Corning pH meter 450 with nouva stirrer II. Fifty millilitres of AYG medium in conical flasks were inoculated with 2 ml of the harvested cells and a loopful of bacterial colony separately (for control experiment) from slants in MaCartney bottles and incubated at 28°C on rotary shaker (New Brunswick Scientific incubator shaker and Lab-line orbit shake mahose park, ILL) at 150 rpm for 6 days. Contents of the culture flask at the end of incubation period were centrifuged using Beckman Coulter Allegra 6R centrifuge at 15000 rpm for 15 min to remove biomass and unsolubilised materials. The soluble phosphate expressed as equivalent phosphorus was determined as a measure of extent of solubilisation of rock phosphate using autoanalyser (Technicon II analyser) (Halder, 1990).

RESULTS

Physical and chemical analysis of soil sample

The result of the mechanical analysis of the soil sample used was 80% sand, 17% silt and 3% clay. This indicated that the soil was loamy sand. The mean value of pH of this soil was 6.62%, organic carbon was 2.18%, organic matter was 3.75% nitrogen was 1.09, phosphorus content was 9.48 mg/kg, potassium content was 5.33 cmol/Kg, sodium was 4.70 cmol/Kg, calcium was 1.5 cmol/Kg and magnesium was 1.27 cmol/Kg.

Rhizosphere and non-rhizosphere bacterial isolates of cowpea

B. subtilis, *S. aureus* and *M. luteus* and *P. damnosus* were isolated from rhizosphere. *B. japonicum* was responsible for nodulation of cowpea. *A. simplex*, *A. chroococum*, *P. aeruginosa* and *B. subtilis* also were isolated from non-rhizosphere of cowpea.

Influence of benomyl on nodule count and rhizosphere bacteria of cowpea

Cowpea seedlings were healthy without much loss of root nodules and beneficial rhizosphere bacteria at 0.10 and 0.20 g/ml of benomyl. Nodule loss of 94.3% occurred at 0.80 g/ml of benomyl when compared to control. Also, there was decrease in the mean value of nodule count (3.1 to 0.5) with increase in benomyl concentration as shown on Table 1. As benomyl concentration increased from 0.00 to 0.80 g/ml, there was a decrease in bacterial population (12.6 to 9.9 cfu × 108 /g) in the rhizosphere.

Influence of increase in number of days on nodule count and rhizosphere bacteria of cowpea

From 5 to 30 days, there was an increase in the mean value of nodule count (0.0 to 9.4) and mean count for bacteria (1.1 to 24.8 cfu × 108 /g) rhizosphere of cowpea as indicated on Table 2.

Influence of benomyl on ability of non rhizosphere and rhizosphere bacteria to solubilise phosphate

A. simplex released 0.233 mg/kg of phosphorus in benomyl treated non rhizosphere soil above the amount released by the same organism in untreated non rhizosphere soil. The phosphorus released by *B. subtilis* in non rhizosphere soil was slightly higher than that released in rhizosphere soil at 0.00 g/ml of benomyl. However, the same amount of phosphorus was released at 0.80 g/ml of benomyl as shown in Table 3. Benomyl (0.8 g) led to decrease in phosphorus released by *S. aureus* (2.481 to 1.564 mg/kg) and *M. luteus* (from 2.440 to 1.362 mg/kg) rhizosphere soil.

Relationship between plasmids of non rhizosphere and rhizosphere bacteria and their ability to solubilise phosphate after addition of benomyl

Plates 1 to 2 show the plasmid sizes of rhizosphere and non rhizosphere bacteria in the presence and absence of benomyl while Plate 3 shows the plasmid sizes of only rhizosphere bacteria in the presence and absence of benomyl. Benomyl led to decrease in molecular weight of

Table 2. Influence of increase in number of days on nodule count and rhizosphere bacteria of cowpea at all concentrations.

Time (Days)	Bacteria (cfu × 108)/g	Nodule count
5	1.1a	0.0a
10	5.2b	0.0a
15	12.7c	0.0a
20	9.3d	0.0a
25	15.8e	4.1b
30	24.8f	9.4c

Values followed by similar along the same column are not significant at $P \leq 0.05$.

Table 3. Amount of phosphorus released in (mg/kg) into AYG medium by loopful of non-rhizosphere and rhizosphere soil bacteria in presence and absence of benomyl (mean ± SD).

Organisms	Rhizosphere		Non – rhizosphere	
	At 0.00 g/ml of benomyl	At 0.80 g/ml of benomyl	At 0.00 g/ml of benomyl	At 0.80 g/ml of benomyl
Bacillus subtilis	1.317±0.0107	1.5955±0.0089	1.3615±0.0185	1.5955±0.0155
Staphylococcus aureus	2.4810±0.0107	1.5645±0.0197	-	-
Micrococcus luteus	3.5750±0.0312	2.4395±0.0061	-	-
Arthrobacter simplex	-	-	0.7570±0.0000	0.9835±0.0015

plasmid from 1100 to 100 base pairs in *A. simplex* and *P. aeruginosa* as shown on Table 3. The size of the plasmids remained the same after addition of benomyl in *A. chroococcum* (100 base pairs) and *B. subtilis* (1100 base pairs). Plasmid size of *M. luteus* and *S. aureus* were 1100 and 100 base pairs respectively (Table 4).

DISCUSSION

The presence of *B. subtilis*, *S. aureus*, *M. luteus* and *P. damnosus* in the rhizosphere of cowpea may be due to their affinity to the benomyl residues. *A. simplex*, *A. chroococcum* and *P. aeruginosa* were isolated from the non - rhizosphere soil. Among the three later organisms, *A. chroococcum* and *P. aeruginosa* are asymbiotic nitrogen fixers. The use of benomyl at higher concentration is not encouraging because it reduced nodulation. Also, there was a decrease in the bacterial count in rhizosphere soil of cowpea as the concentration of benomyl increased.

Bacteria from untreated rhizosphere soil of cowpea produced higher soluble phosphorus than benomyl treated rhizosphere soil except *B. subtilis*. However, *A. simplex* and *B. subtilis* from untreated non rhizosphere soil of cowpea produced low soluble phosphorus than benomyl treated non rhizosphere soil. The variation in the degree of phosphate solubilisation by different bacterial isolates may be due to difference in their genetic make-up as reported by Moritz and Hergenrother (2007). The increase in the distance moved by the DNA marker was the only reason that can be adduced for increased phosphorus released by *B. subtilis* into the soil. For *A. simplex* isolated from non rhizosphere, the increase in phosphorus released may be due to the decrease in molecular weight of the plasmid as a result of treatment with benomyl. This is because physiological or genetical changes of an organism may increase their metabolic activities concerning biodegradation of xenobiotics, organic chemicals and pesticides (Zhang et al., 2005). On the other hand, *S. aureus* and *M. luteus* reduced their level of phosphate solubilization in the presence of benomyl. Due to this, it is not advisable to use benomyl at high concentration.

Conclusion

The optimum concentration of benomyl on beneficial soil bacteria was 0.10 to 0.20 g/ml. *B. subtilis* could be inoculated to rhizosphere soil of cowpea to increase the amount of soluble phosphate available for the growth of cowpea seedlings in benomyl treated soil.

ACKNOWLEDGEMENTS

The author wish to thank Dr. (Mrs.) Stella and Mr. Omonigbehin of Nigerian Institute of Medical Research, Yaba, Lagos for allowing me to use their facilities and equipment for the plasmid analysis of my bacterial isolates.

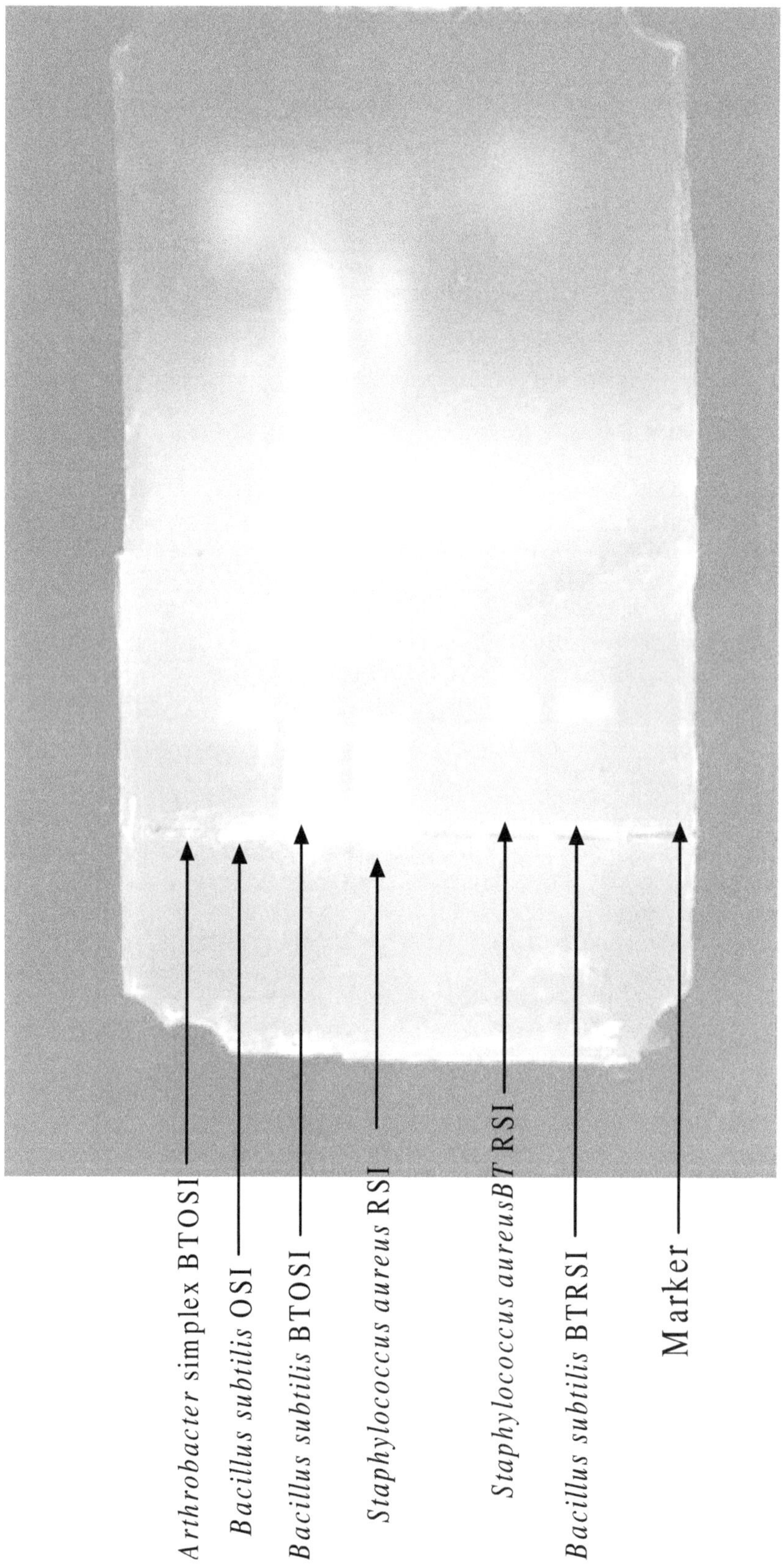

Plate 1. Plasmid sizes of rhizosphere and non rhizosphere bacteria in the presence and absence of benomyl. OSI = Ordinary (non rhizosphere) soil isolate. RSI = Rhizosphere soil isolate. BTOSI = Benomyl treated ordinary (non rhizosphere) soil isolate (at concentration 0.80 g/ml). BTRSI = Benomyl treated rhizosphere soil isolate (at concentration 0.80 g/ml).

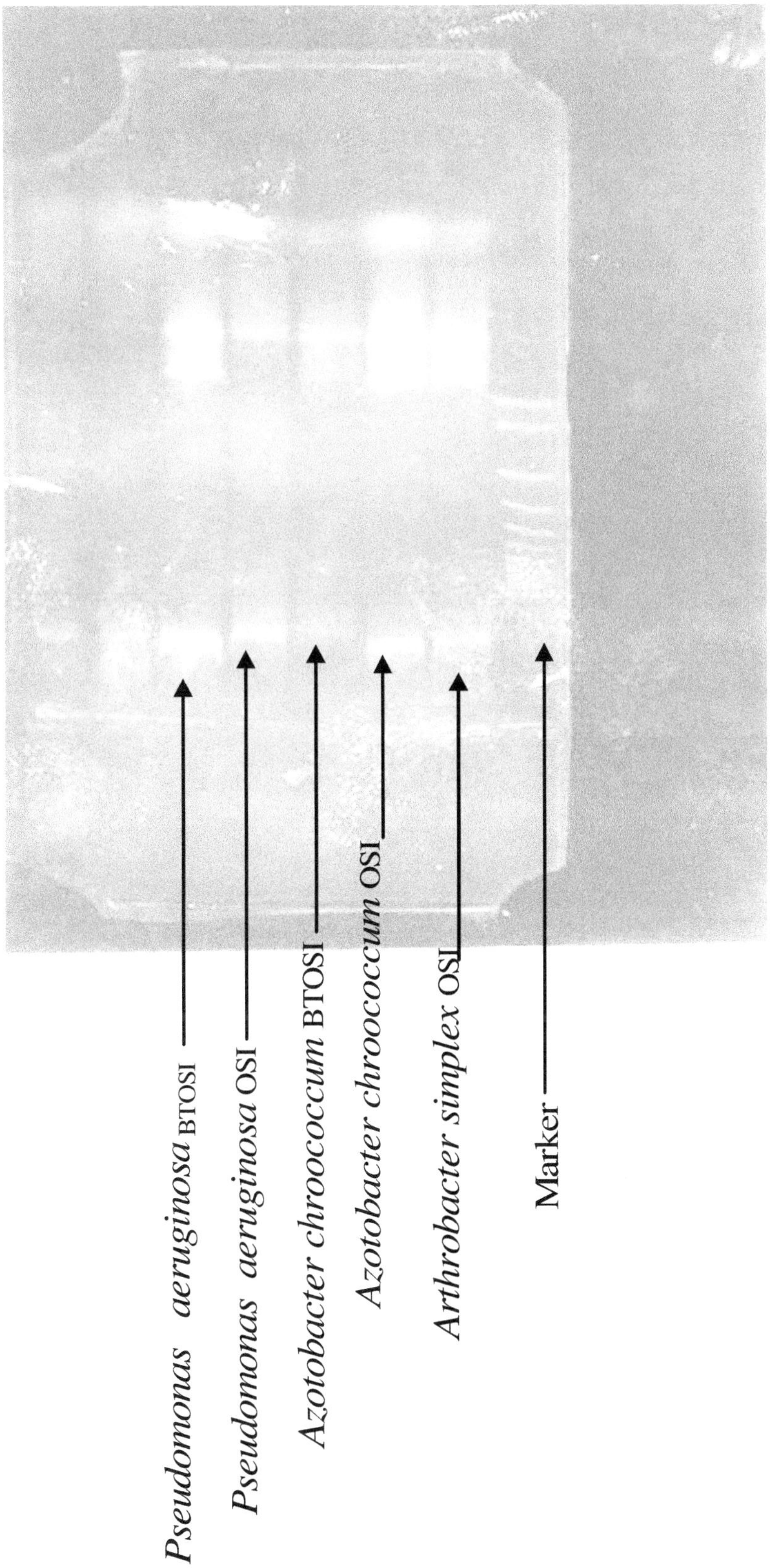

Plate 2. Plasmid sizes of rhizosphere and non rhizosphere bacteria in the presence and absence of benomyl. OSI = Ordinary (non rhizosphere) soil isolate. RSI = Rhizosphere soil isolate. BTOSI = Benomyl treated ordinary (non rhizosphere) soil isolate (at concentration 0.80 g/ml). BTRSI = Benomyl treated rhizosphere soil isolate (at concentration 0.80 g/ml).

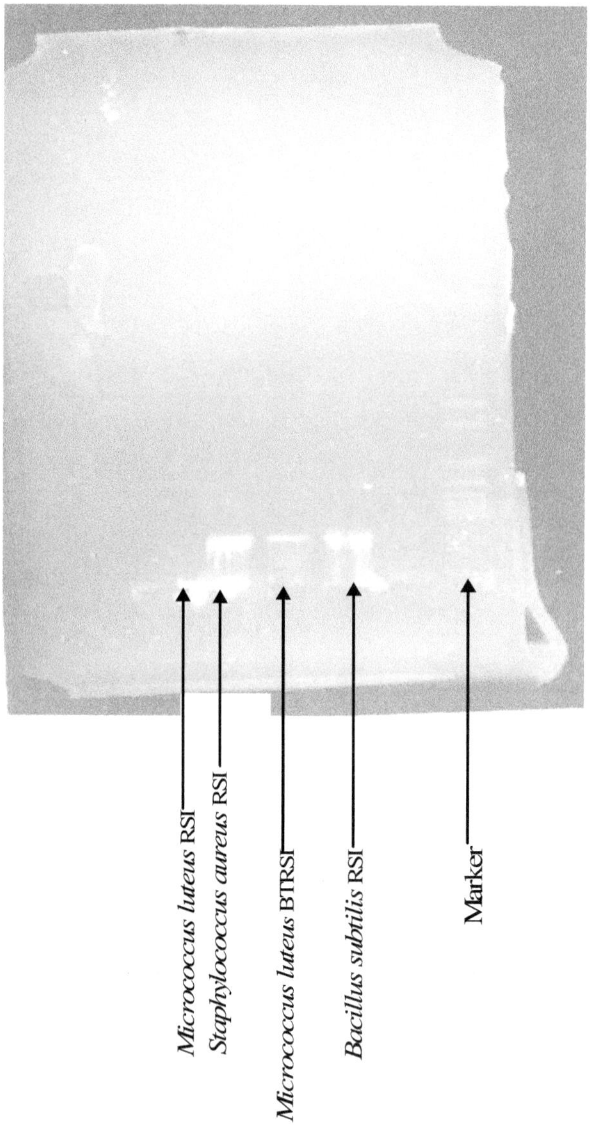

Plate 3. Plasmid sizes of rhizosphere bacteria in the presence and absence of benomyl. RSI = Rhizosphere soil isolate. BTRSI = Benomyl treated rhizosphere soil isolate (at concentration 0.80 g/ml).

Table 4. Influence of benomyl on plasmid weight of some bacteria and the level of phosphorus released into the soils.

Organisms	Molecular weight of plasmid (in bases)	Distance moved by DNA marker (cm)	Phosphorus released (mg/kg)
Non rhizosphere bacteria			
Arthrobacter simplex OSI	1100 (plate2)	1.3	0.757
Arthrobacter simplex BTOSI	100 (plate 1)	0	0.984
Azotobacter chroococcum OSI	100 (plate 2)	1.3	N.D
Azotobacter chroococcum BTOSI	100 (plate2)	1.2	N.D
Pseudomonas aeruginosa OSI	1100 (plate2)	1.2	N.D
Pseudomonas aeruginosa BTOSI	100 (plate2)	1.2	N.D
Bacillus subtilis OSI	>1100 (plate 1)	1.2	1.362
Bacillus subtilis BTOSI	>1100 (plate 1)	1.2	1.596

Table 4. Contd.

Rhizosphere bacteria			
Bacillus subtilis RSI	100 (plate 3)	1.2	1.362
Bacillus subtilis BTRSI	100 (plate 1)	1.2	1.596
Staphylococcus aureus RSI	100 (plate 3)	1.9	2.481
Staphylococcus aureus BTRSI	100 (plate 3)	1.0	1.564
Micrococcus luteus RSI	>1100 (plate3)	1.9	2.440
Micrococcus luteus BTRSI	> 1100 (plate3)	1.9	1.362

OSI = Ordinary (non rhizosphere) soil isolate; RSI = Rhizosphere soil isolate. BTOSI = Benomyl treated ordinary (non rhizosphere) soil isolate (at concentration 0.80 g/ml). BTRSI = Benomyl treated rhizosphere soil isolate (at concentration 0.80 g/ml). N. D. = Not determined.

REFERENCES

Akobundu IO (1987). Herbicides in the environment. Weed Science in the tropics - principles and practices. A. Wiley -Interscience publication, pp. 179-206.

A. O. A. C. (1990). Fertilizers: water, phosphorus, nitrogen, potassium and other elements. In: Official methods of analysis of the Association of official analytical chemists, Inc. 15th edition, Suite 400, 2200 Wilson Boulevard Arlington, Virginia 22201 USA, pp. 9-36.

Behbahani M, Behbahani M (2009). Investigation of biological behavior of Iranian indigenous phosphate solibilizing bacteria and determinant of colonization ability of Potato roots by the bacterial isolation. J. Med. Plants Res., 3(12): 1126-1133.

Carter MR (1992). Soil sampling and methods of analysis. Canadian Soil Society, Lewis Publishers, London, p. 823.

Ekundayo FO, Oladunmoye MK (2007). Influence of benomyl on ability of Fusarium oxysporum and Fusarium solani to produce beauvericin and rhizosphere organisms of cowpea. Int. J. Soil Sc., 2(2): 135-141.

Etok CA, Asamudo NU, Onwuchekwa IS, Nwaugo VO (2010). Effects of quarry plant rock dusts on soil phosphate solubilizing bacteria and enzyme activities in Ishiagu, Ebonyi State, Nigeria. Niger. J. Microbiol., 24(1): 2207-2213.

Halder AK, Mishra AK, Brattacharyya P, Chaknabartlty PK (1990). Solubilisation of rock phosphate. J. Gen. Appl. Microbiol., 36: 81-92.

Holt GJ (1989). The shorter Bergeys Manual of Determinative Bacteriology.Waverly Press Inc. Eight Edition. ISBN 0-683-41051-2.

Meghrous J, Lacroix C, Bouksaim M, Lapointe G, Simard RE (1997). Genetic and biochemical characterization of nisin Z produced by *Lactococcus lactis* sp. lactis biovar diacetyllactis UL 719. J. Appl. Microbiol., 83: 133-138.

Moritz EM, Hergenrother PJ (2007). The prevalence of plasmids and other mobile genetic elements in clinically important drug resistant bacteria. In antimicrobial resistant in bacteria (Carlos F. Amabile-Cuevas ed.). Horizon Bioscience 32 Hewitts lane Wymondham Norfolk N R. 18 OJA, UK. www.horizonbioscience.com.

Newsham KK, Fitter AH, Watkinson AR (1994). Root pathogenic and arbuscular mycorrhizal fungi determine fecundity of a symptomatic plant in the field. J. Ecol., 82: 805-814.

Olutiola PO, Famurewa O, Sontag HG (1991). An introduction to general microbiology – a practical approach. Biochemical reactions of microorganisms. Higiene-Institue der Universitat Heide; berg. pp. 157- 180.

Omokaro DN, Ajakaiye CO (1989). Effects of soil applied herbicides on leaf nitrate reductase and crude protein in the leaf and seed of two cowpea cultivars. Plant and soil, 116.2: 141-146.

Saha N, Biwas S (2009). Mineral phosphate solubilizing bacterial community in agro-ecosystem. Afr. J. Biotechnol., 8(24): 6863-6870.

Somasegaran P, Hoben HJ (1994). Handbook for Rhizobia: Methods in Legume – Rhizobium Technology. Springer-Verlag, New York, Inc., 175 Fifth Avenue, New York NY 10010, USA, pp. 7-23.

Tomlin C (1994). The pesticide manual, 10th ed. British Crop protection,

Zhang G, Jia X, Cheng T, Ma X, Zhao Y (2005). Isolation and characterization of a new carbendazim-degrading Ralstonia sp strain. World J. Microbiol. Biotechnol., 21: 265 -269.

Determinants of adoption of conservation tillage practices in maize-cowpea cropping systems: The case of Makueni District, Kenya

Jane Mutune[1]*, John Mburu[1], Rose Nyikal[1] and Geoffrey Kironchi[2]

[1]Department of Agricultural Economics, University of Nairobi, Kenya.
[2]Land Resource Management and Agricultural Technology, University of Nairobi, Kenya.

The low soil moisture cannot support productive agriculture to meet the increasing population in the low rainfall tropical areas. Ripping and tied-ridging are some of the recent technologies introduced by the Food and Agricultural Organization of the United Nations, and is used to conserve moisture in the semi-arid areas. Although farmers are aware of the technical gains of these technologies, the adoption rates have remained below the expectations of researchers and policy makers. The objective of this study was to analyze household and technology characteristics that influence the adoption of ripping and tied-ridging techniques of conservation tillage. Semi-structured questionnaires were used to interview a random sample of 177 farmers. Using a logit model, different factors that influenced farmers' use of ripping and tied-ridging were identified. The significant variables include availability of off-farm employment, closeness to local markets, group membership, availability of family labour, contact with extension services and conservation tillage promoters, and farmers' farming experience. The paper recommends that future demonstrations of ripping and tied-ridging should target farmers who cannot easily access markets for farm inputs and outputs. Moreover, non-adopters should be encouraged to join or form new groups to establish contacts with extension services, and organizations promoting the tillage practices if adoption rates of these technologies are to be improved in the study area.

Key words: Adoption, adoption rate, conservation tillage, ripping, tied-ridging, semiarid areas, Kenya.

INTRODUCTION

The utilization of appropriate agricultural technologies is an important prerequisite for economic development in developing countries as they raise the production possibility frontier of agricultural commodities. Without such technologies, the scope for increasing agricultural production and rural incomes is very limited. However, while technology advancement is necessary for agricultural development, it is not a sufficient step towards increased agricultural production *per se*, unless critical mass adoption is achieved (Feder and Slade, 1984). Farmers' technology adoption decisions are shaped by personal, economic, socio-cultural and environmental influences under which they operate (Wale et al., 2005). Consequently, for any technology to have an impact on farmers' productivity and thus welfare, the factors that influence its adoption should be well understood. Such an understanding forms an important basis for any attempt aimed at introducing change and improving production levels in smallholder farming systems.

The semi-arid areas of Eastern Kenya, including Makueni District, are characterized by low, erratic and poorly distributed bimodal rainfall that makes crop production difficult under rain fed conditions (Biamah et al., 2000). The four major constraints to arable agriculture in the semi-arid areas are severe soil erosion, low soil fertility, soil crusting and low soil moisture. Of the four, soil moisture is the most critical as it directly affects crop production in these areas (Biamah, 2001). Hagmann et al. (1996) indicated that the concern for improvement of

*Corresponding author. Email: j_mutune@yahoo.com.

crop productivity in the semi-arid areas calls for a radical shift in relying on erosion control practices to adoption of appropriate techniques which can conserve both soil and water with reduced farming costs.

In the Kenyan semi-arid areas, public extension service providers, researchers and non-governmental organizations have, for the last century, been engaged in various endeavors to counter the effects of low soil moisture in agriculture. Some of these efforts have included advising farmers to plant crops such as millets, sorghum, cowpeas, and specially selected maize varieties which are adapted to water stress. Yet, with the conventional tillage the yields of these crops, and particular those of maize and cowpeas, have remained low compared to those realized from areas with higher rainfall or under irrigation (Mellis et al., 1997). Other sustainable techniques advanced in the semi-arid areas for water harvesting and conservation have included terracing, contour ridging, cover cropping, mulching and sub-soiling (Okwach et al., 2004). While considerable success has been achieved in this regard, there is still decline in land productivity even on the best conserved land suggesting that more needs to be done to improve soil moisture.

In the turn of the decade, the Food and Agricultural Organization of the United Nations (FAO) introduced ripping and tied- ridging in Makueni and other arid and semi-arid districts of Kenya (Biamah, 2001; FAO, 2004). Ripping is a form of minimum tillage whereby only parallel furrows are cut using a ripper without disturbing the soil between planting rows (Gachene and Kimaru, 2003). Its function is to restore micro-pores for maximum water infiltration and increase root depth. Tied-ridges are usually discontinuous furrows made by cross-ties that interrupt water flow in the furrow thus creating pools to retain water for a while and to promote slow seepage (Gachene and Kimaru, 2003).

There is sufficient evidence that conservation tillage is superior to conventional tillage in terms of crop yields. Ripping and tied-ridging assure farmers of good timing of farming operations and less labour and energy requirements (Chuma et al., 1998). A 50 to 100% increase in grain yields has been possible (Pretty, 1999). Field experiments carried out by Miriti (2005) in the Makueni district to compare the yields of conventional tillage and tied-ridging that that crop yield of maize was higher, by 55%, in tied-ridging. Although the tied-ridges require high labour during preparation, they are permanent and only need minimal maintenance in the subsequent years (Harper, 2002).

Despite the fact that technical gains of conservation tillage practices are known by farmers in Makueni and other semi-arid districts, the adoption rates have remained low ((Bradshaw and Knowler, 2006). This is in spite of the continued promotional efforts of the conservation tillage techniques by various government and non-government organizations since 1998 (Kaumbutho and Josef, 2004). In Makueni District, these organizations include the Kenya Agricultural Research Institute (KARI), KENDAT, the FAO, Ministry of Agriculture (MoA) and the Agricultural Technology Development Center (ATDC).

One of the reasons given for the low adoption rates is that proper targeting of farmers is not done since farm and farmers' characteristics that influence technology uptake are not precisely known, though availability of draught power and extension services have been noted to play an important role (Kaumbutho and Mutua, 2002). Thus, this study takes the case of ripping and tie-ridging in Makueni District to determine which farmers' characteristics and technology aspects influence adoption of conservation tillage practices in arid and semi-arid areas of Kenya. Understanding these factors will facilitate a targeted approach in promoting use of conservation tillage in order to enhance crop production in the study area. Acquiring this knowledge is important since the Kenyan rural population continues to increase, and more people are moving to arid and semi arid areas where they continue to rely on farming for their livelihoods (Central Bureau of Statistics, 2001).

METHODOLOGY

The conceptual background

This study was conceptualized as a technology adoption investigation. Feder and Slade (1984) define adoption as the degree to which a new technology is used in the long run equilibrium when farmers have complete information about the technology. Since farmers are rational consumers of agricultural technologies, they are conceptualized to choose technology packages that give maximum utility. Adoption is assumed to be a function of the farm and household characteristics, and technology specific attributes (Adesina and Zinnah, 1993).

The decision to adopt an innovation is a behavioral response arising from a set of opportunities and constraints facing the decision maker (Leagans, 1979). These opportunities and constraints can be grouped into incentive and disincentives respectively. Adoption proceeds only when the incentives outweigh the disincentives. Economically, incentives are the returns while the disincentives are the costs. If the benefits are more than the costs, the farmers are motivated to take up a new innovation due to the expected high returns on investment.

Economists usually treat adoption of agricultural innovations as the dependent variable while the independent variables are all interdisciplinary factors comprising a primary set of socio-economics, physical and institutional factors. It is therefore the interaction of these factors, including incentives versus disincentives that create tension that motivates action resulting in change and thus adoption or non- adoption (Leagans, 1979; Adesina and Zinnah, 1993).

Study area

Makueni district has a population of over 245, 768 people (CBS, 2001) who are mainly subsistence farmers. The types of crops mainly grown are maize, cowpeas, beans, pigeon peas, cassava, sweet potatoes, local mangoes, sorghum and kales. The district receives bimodal rainfall with the most reliable rains occurring in October to December (short rains). The long rains occurring in April to May are unreliable in both amounts and distribution (Okwach et al., 2004). A lot of rainfall is lost through water surface runoff, and this, together with poor tillage practices, has resulted in low soil moisture thus low crop yields. To address this problem, various water conservation measures and appropriate conservation tillage practices (Gachene and Kimaru, 2003) have been implemented in the district.

Sampling procedure

This paper used both primary and secondary data. Primary data was collected by using semi-structured questionnaires in Kalawa and Kathozweni divisions in Makueni district from December 2006 to June 2007. These data included household characteristics and the technologies attributes. Secondary data was collected from institutional libraries, that is, World Agro-forestry Centre, University of Nairobi and the Ministry of Agriculture.

In this study, an adopter was identified as a farmer who had been using either ripping or tied ridging or both for the last four years up to the time the study was conducted. A non-adopter was a farmer who had never used the conservation tillage practices and thus purely practiced conventional tillage. A sampling frame of 90 farmers who have adopted conservation tillage practices within maize-cowpea crop systems was gathered from the local non-governmental organizations (mainly Agricultural Technology Development Centre (ATDC) in Machakos) and public extension personnel operating in the study area. All the farmers in this list were selected. However, only 87 of them were interviewed since the rest were unavailable. The sampling frame for the non-adopters consisted of adopters' neighbours who were growing maize and cowpeas but had not adopted the conservation tillage practices. For every adopter, the closest neighbouring household was selected. Thus, a total of 90 non-adopters were sampled and interviewed. The final sample therefore consisted of 177 farm households. All of them were interviewed using a semi structured questionnaires. The data gathered during these interviews included socio-economic characteristics, reasons for adopting/not adopting the conservation tillage practices, agronomic characteristics, market access characteristics and perceptions towards conservation tillage practices.

Econometric model

Since the study was conceptualized as analyzing adoption behaviour, the phenomenon to model was taken as discrete rather than continuous. In this case, the dependent variable took the value of 1 if technology had been adopted and 0 if not adopted.

A form of a qualitative response model was used to analyse this phenomenon. Binary choice models such as logit and probit models are often applied in modelling adoption decisions (Greene, 2004). These models are used for estimating the probability of an event (such as adoption) that can take one of two values (adopt, do not adopt). The basic difference between the two models is that logit assumes a cumulative logistic distribution with a higher variance hence flatter, while probit model assumes cumulative normal distribution with a unit variance hence more bell-shaped. Generally, the interpretation of the two models is similar.

In a logit model, the expectation of the adoption Y is a value P, which is related to the independent variables (X) as follows (Pindyck and Rubenfield, 1991; Maddala, 2001; Greene, 2004):

$$E(Y/X) = P = F(Z) = \alpha + \beta X + \mu$$
$$= 1/(1+e^{-z}) \qquad 1$$
$$= 1/\{1+e^{-(\alpha+\beta x+\mu)}\}$$

Where $P(Y=1/X)$ = conditional probability of being adopter given the values of independent variables (X); Z is the standard normal variable, $Z \sim N(0,\sigma^2)$; α = Constant; β = Regression coefficients; μ = stochastic error term; σ = standard deviation.

In the afore expression, P is referred to as the logistic probability function. When the logistic function is expressed in terms of odds, it is called the logit and takes the following form:

$$probabilit\ y(event)/(no\ event) = \{P/(1-P)\} = e^z = e^{(\alpha+\beta X+\mu)} \qquad 2$$

In order to estimate the logit model, the dependent variable is transformed by taking natural logarithms of both sides to yield "log odds" as follows:

$$In\{P/(1-P) = Z = \alpha + \beta X + \mu \qquad 3$$

Independent variables and their hypotheses

The adoption models were specified using several factors, derived from the adoption literature (Asambu, 1993; CIMMYT, 1993). These were:

i. Education (EDTN): This was the number of years of formal schooling for the household head. Education was hypothesized to positively influence the decision to adopt (Gould *et al.*, 1990). This is because more years of schooling tend to make farmers less risk averse thus, enabling them to try out new innovations (Chuma et al., 1998). Well-educated farmers easily acquire and comprehend new information hence, demand and utilize complex agricultural technologies.

ii. Gender: This variable was coded as a dummy variable, representing the sex of the household head, where male =1, and female = 0. Female-headed households were expected to adopt technologies more as they tend to have less income compared to their male counterparts (Blackden and Bhanu, 1998). In addition the conservation tillage practices are labour intensive in the initial years and women provide most of the labour for food production.

iii. Off farm employment (OFFEMP): This variable was coded as a dummy variable, 1 = household heads with off farm employment and 0 = for those without off farm employment. The financial resources obtained through off farm employments enables a family to acquire the implements, herbicide, sprayers and other inputs needed for conservation tillage practices (CIMMTY, 1997). The presence of a business is also likely to raise the opportunity cost of family labour. This is related once again to cash flow on the farm, given that someone with a side business is likely to have more cash on hand. Off farm employment is therefore expected to have a positive influence on the adoption of conservation tillage.

iv. Family labor (FAMLB): This was the quantity of labour in man-days per year available to the household for farming. The amount of family labour available on-farm was estimated based on the composition of the family and the participation of family members.

Smallholder farmers have a low propensity to hire labour outside their farms, because they are resource constrained. They mainly use family labour to do most of the farm work (CIMMYT, 1997). Family labour is increasingly scarce in the study area. The quantity of family labour available was hypothesized to have a positive influence on the adoption of the conservation tillage practices since they are labour-intensive technologies.

v. Farm size: This was the size of land, in acres, cultivated for maize and cowpea production. Farmers with bigger farm sizes may have extra land to try the new practices before they decide switching from the conventional practices. The larger the acreage under crop production, the greater the likelihood a farmer will be motivated to invest in equipment for conservation tillage (Kaumbutho and Kienzle, 2007). Larger farm sizes are associated with wealth and therefore farmers may plant perennial crops allowing the practice of conservation tillage on nurse crops on the same piece of land. Therefore, farmers with bigger farm sizes were hypothesized to adopt ripping and tied ridging than those with smaller sizes of land.

vi. Experience (EXP): This was the number of years of farming experience of the household head. Frank (1997) observed that individuals assess the utility of new practices by relating their perception of the practice to their experience. Consequently, years of farming experience and exposure are likely to aid adoption.

vii. Extension services and technology promoters (EXTORG): This was the number of times a farmer has had contact with the researchers and extension officers in the last four years. The respondents were asked to recall or refer to farm records for the number of times the extension officers or technology promoters had visited them for the purpose of helping them with conservation tillage. It was hypothesized that the degree of contact would highly enhance adoption.

viii. Membership in farmer organizations (GRPMB): This variable was coded as a dummy variable, where members to a group = 1, and non-members = 0. Members of an organization for instance, farmer groups, and non-governmental organizations are more privileged compared to other farmers in terms of access to information on agricultural innovations (de Harrera and Sain, 1999). Groups in Kenya have taken an active role in mitigating drought impacts on their members and the scope of drought mitigation appeared to expand as groups mature over time (Coopock et al., 2005). Membership in a farmer organization is therefore expected to relate positively to the adoption of conservation tillage practices.

ix. Distance to the nearest market (DTM): This variable was measured as the number of kilometers (km) from the farm to the nearest local market. Farmers who are closer to input and output markets have better access to production inputs and can easily sell their farm produce. Long distance to the market disconnects farmers from the supply chain (Jonas et al., 2008). Low use of farm inputs by smallholder farmers in Sub-Saharan Africa is responsible for the gap between potential farmers yield and actual crop yields at farm level (Jonas et al., 2008). It was hypothesized that the distance to the nearest market (DTM) is negatively related to the adoption process.

RESULTS

Descriptive statistics of explanatory variables used in the model

Table 1 presents the descriptive analysis of some discrete variable for the adopters of ripping and/or tied-ridging. The adopters were more in off farm employment (Table 1). Off farm employment provided more cash flow to the household. The adopters reported to have used off farm income to purchase farm inputs for ripping and/or tied ridging. The inputs included rippers, ridgers, herbicides and wages for hired labour. Personal interview with the adopters found out that most of the rippers (at 4%) and ridgers (at 5%) were owned by farmers in off farm employment. Hence farmers in off farm employment were more likely to adopt ripping and/or tied ridging.

The study found out that public extension officers and non-government organisations promoting ripping and tied ridging preferred working with farmers who showed common interests and hence in groups. Group membership is crucial in the study area since most of the demonstrations and farm trials on ripping and tied ridging were carried out among farmers in groups. The groups are also important in the procurement and ownership of implements for conservation tillage practices. Eighty two percent of the adopters were member to farmers' group (Table 1). Majority of the non-adopters reported that the status was due to lack of information which is accessible from groups. This illustrates that non-adopters joining groups or forming new ones could lead to more adoption of the technologies. The groups mainly included self-help groups, common interest groups and farmer co-operatives. The main objective of these groups was to improve the farm incomes by procuring the farm inputs and marketing the products, collectively.

More women were found to have adopted ripping and tied ridging than their male counterpart 63 % of the adopters were female compared to 37% males (Table 1). This is attributed to the fact that women provide most of the family labour for crop production. In addition, the practices contribute to increased crop yields as a result of soil moisture conservation hence more food to the household. Table 2 compares the crop yields of both maize and cowpea under conservation and conventional tillage.

Descriptive analysis of the sample socio-economic characteristics

The sample socio-economic characteristics for adopters and non adopters were analyzed and presented in Table 3. The mean, standard deviation, maximum and minimum analyses of each continuous variable were considered. Contact with technology promoters and public extension agents were found to increase the probability of using the technologies. Backstopping and information provision for ripping and tied ridging was done by both the governmental and nongovernmental organizations. The major technology promoters in the area of study were reported as KENDAT, MoA and the FAO and promoted the technologies at 36, 21 and 9% respectively in the last four years. Table 3 shows that farmers had been contacted at least once by these organizations for the last four years.

On the average, the distance walked by farmers to the nearest market was 3 km (Table 3). The study found that 45% of the adopters were at least one kilometer away

Table 1. Descriptive analysis of discrete variables for the adopters.

Variable	Category	Number reporting	Percent
Gender	Female	55	63
	Male	32	37
Group membership	No	16	18.4
	Yes	71	81.6
Off-farm employment	No	21	24.4
	Yes	66	75.8

Table 2. Yields of maize and cowpea under conservation tillage and conventional tillage in 90 kg bag per hectare.

Variable	Minimum	Maximum	Mean	Std. dev.	Case (n)
Yield under Conventional tillage	3.000 (0.747)	20.000 (4.000)	8.538 (5.9)	5.0102 (5.234)	90
Yield under Conservation tillage	7.000 (2.780)	30.000 (7.000)	7.384 (10.4)	6.752 (8.750)	87

The figures in brackets are for cowpea yields.

Table 3. Descriptive analysis of some continuous variables for the adopters and non adopters.

Variable	Definition	Min.	Max.	Mean N=177	S.D
Extension services	Number of contact by technology promoters for last four years	1.451	3.042	1.122	0.74
Farming experience	Farming experience of the household head in years	1.046	49.01	18.769	1.876
Farm size	Total farm size in hectares	1.046	15.43	4.676	0.856
Education	Level of education in years in formal school	7.99	15.78	9.231	0.961
Family labour	Family size as per number of family members providing farm labour	1.00	13.00	1.413	2.469
Distance to market	Number of kilometers from farm to nearest local market	0.20	18.09	2.970	1.623
Farm size under crops	Farm size under maize and cowpea in hectares	0.150	3.772	1.109	0.171
Age	Age of farmer in years	16.00	75.00	43.39	9.75

from the market, hence could easily access farm inputs like herbicides, rippers and effectively market their farm produce. The study found out that farmers far away from the market used other conservation measures to improve the soil moisture. Hence accessibility to the market is likely to contribute to adoption of ripping and tied ridging.

Most of the respondents had acquired up to primary education level (Table 3). The adopters and non-adopters were almost equally educated implying that the technologies were simple enough to be understood by all farmers at any educational level. The mean farming experience of the households was 18.8 years while mean landholding size was 4.7 acres. This landholding size can be considered large when compared with that of high rainfall areas (Jaetzold and Schmidt, 1983). Further, the technologies do not require lot of farm labour especially when the crop is in the field. This may be one of the reasons why the household had an average of one person proving farm labour (Table 2)

The mean distance of the household from the nearest market center was about 3 km. This was a considerable distance if farmers have to walk to buy inputs and sell

Table 4. Factors influencing adoption of conservation tillage practices.

Variable	Coefficient	Standard error	P-values
GENDER	0.2448	0.6288	0.9919
EDTN	-0.2479	0.0936	0.3364
FMSZ	-0.3092	0.1539	0.7478
OFFEMP	0.7774*	0.4532	0.0074
FARMEXP	0.03642**	0.0196	0.0406
FAMLOB	0.2532**	0.1790	0.0307
EXT	1.3171*	0.6295	0.0001
GRPMEBR	0.3445**	0.7383	0.0015
DISMRKT	-3.440*	0.6315	0.0524

Model N= 177, Pseudo R^2=0.358; Percent correctly predicted =89.231045, model chi-square = 69.941***; Log likelihood function (LnL) = -101.4686, Log likelihood function (LnLo) = -157.9848; ***, **, * significant at 1%, 5% and 10% level of error probability respectively.

their farm outputs. With this poor accessibility to market, market visits are confined to a single market day in a week in the study area.

Factors influencing adoption of ripping and tied-ridging conservation practices

The results of marginal effects for the factors influencing the adoption of the conservation tillage practices were analyzed by use of the logit model as shown in Table 4. The table also includes the list of independent variables used in the model and the expected effect.

Off farm employment was significant at $P < 0.1$. At the mean score, a unit increase in off farm employment increased the log-odds of adopting ripping and tied ridging by 0.778 when the other variables are held constant.This meant that farmers with off farm employment were more likely to adopt the conservation tillage practices than those without. This agrees with the qualitative information collected in the study. It was found that the presence of off farm employment is important in raising the opportunity cost of family labour.

The coefficient of farming experience was positive as hypothesized and significant at $P < 0.05$. This implies that farmers with more farming experience were more likely to adopt ripping and tied ridging. This relationship shows that as farmers gain experience, their decisions on the choice of conservation tillage practices are influenced positively (Adesina and Zinnah, 1993). At the mean score, a unit increase in farming experience increased the log-odds of adopting ripping and tied ridging by 0.036 when the other variables are held constant (Table 4). This result agrees with the human capital theory, which holds that farmers become less risk averse as they gain more experience (Welch, 1979). Thus, farmers with more farming experience make decisions faster and are able to adopt new technologies at higher rates.

The coefficient of family labour was positive and significant at $P < 0.05$. Hence, households with more members were better placed to adopt ripping and/or tied ridging than those with fewer members. This might be so because of the explanation provided earlier that the conservation tillage practices are labour intensive and particularly when furrows and depressions are being prepared. At the mean score, a unit increase in family labour increased the log-odds of adopting ripping and tied ridging by 0.253 when the other variables are held constant (Table 4).

The coefficient of contact with extension service and technology promoters was positive and significant at 10% level. Direct contact with extension services and organizations promoting the technologies provided technical backstopping in terms of information, rippers, farm demonstrations and other inputs to the farmers. This implies that there is need for frequent farmer contacts with promoters of the technology in order to enhance adoption. At the mean value, a unit increase in public extension contact and technology promoters increased the log-odds of adopting ripping and tied ridging by 1.317 when the other variables are held constant. The results are congruent to those of Kaumbutho and Mutua (2002) who found that provision of information on the existing draught animal resources could boost the use of conservation tillage practices in Makueni District.

At $P < 0.05$, the coefficient of group membership was positive and significant (Table 4). This implies that farmers are more likely to adopt conservation tillage when they are in groups. From personal interviews with farmer groups, it was revealed that non-governmental and governmental organisations preferred disseminating information to farmers in groups mainly due to scarcity of personnel and other resources. This implies that group membership is important in the dissemination of technologies at the grass root level. At the mean score, a unit increase in group membership increased the log-odds of adopting ripping and tied ridging by 0.344 when the other variables are held constant.These findings confirm work done by Grootaert (2001) who described group formation as one way of generating social capital which positively contributes to farm and non-farm economic activities in the rural communities.

As hypothesized, distance to the nearest market negatively influenced the adoption of ripping and tied ridging. The coefficient of this factor was significant at $P < 0.1$. As explained earlier, these results were realized because farmers far from the markets do not have access to farm inputs and information on ripping and tied ridging. Also, farmers far from the market may not be able to market their additional farm yield effectively. This result supports Jonas et al. (2008) argument that long distance disconnects villagers from input supply chain, making them lack market information and miss chances of adopting new technologies. At the mean score, a unit increase in distance to the nearest market decreased the log-odds of adopting ripping and tied ridging by 3.44 when the other variables are held constant (Table 4).

CONCLUSIONS AND POLICY IMPLICATION

This study examined factors that influence the adoption of the conservation tillage practices and used a logistic regression to identify the significant variables. Understanding these factors will facilitate a targeted approach in promoting use of conservation tillage in order to enhance maize and cowpea production in the semi-arid areas. The factors that significantly influenced the adoption of ripping and tied ridging were group membership, off farm employment, farming experience, extension contacts and distance to the market. Therefore, these factors should be incorporated in the design of policies and strategies developed to promote the use of conservation tillage practices.

Some farmers in the district under study are using the tillage practices, an indication that they have acquired the necessary technological knowledge and skills. This already acquired knowledge could be exploited to benefit other farmers and especially the non-adopters. This can be done by encouraging non-adopters to join existing groups or form new ones in order to learn practically from each other and access benefits that adopters in farmer groups have been enjoying.

Extension providers and promoters were found to enhance the confidence of farmers in adopting ripping and tied ridging through demonstrations and farm trials. This is because direct contact with extension services and organizations promoting the technologies provided information and technical backstopping on these technologies. Therefore, extension services and technology demonstration could be intensified among farmers to enhance practice of conservation tillage. It is important to note that this should not be limited to farmers far from the road or market places if the current adoption rate is to be enhanced.

REFERENCES

Adesina AA, Zinnah MM (1993). Technology characteristics, farmer perception and adoption decision. A Tobit model application in Sierra Leone. Agric. Econ., 9: 298-310.

Asambu FK (1993). Farmers' perception and economic analysis of factors affecting farm level maize enterprise performance and adoption of agricultural technological innovations in Semi-Arid South Eastern Kenya. MSc. thesis, University of Guelph.

Blackden CM, Bhanu C (1998). Gender growth and poverty reduction: Special program of assistance for Africa, status reports on poverty in the sub-Saharan Africa. In collaboration with the poverty and social policy working group of the special programme of assistance for Africa. World Bank Technical Paper No. 428. World Bank Washington, D.C.

Biamah EK (2001). Seasonal conservation tillage reports (1999) for liuni Conservation Tillage Trials Site, Machakos, Kenya. RELMA/Sida Report, Nairobi, Kenya.

Biamah EK, Rockstrom J, Okwach GE (2000). Conservation tillage for dry land Farming. Technical option and experience in Eastern and Southern Africa. Workshop Report 8-11 April, Bloemfontein, South Africa. Regional Land and Management Unit, Kenya.

Central Bureau of Statistics (CBS), Kenya (2001). Population projection. CBS, Kenya.

CIMMYT (1993). The Adoption of agricultural technology survey for survey design. Mexico.

CIMMYT (1997). The adoption of conservation tillage in a hillside maize production system in Motozintla, Chiapas. Natural Resource Group. Mexico.

Chuma R, Benites J, Fowler J, Kienzle Nyangumbo K, Steiner R, Molopong J (1998). Conservation tillage for sustainable agriculture. International workshop, Harare Zimbabwe. Deutsche Gesellschaft, Eschborn Germany.

Coopock DL, Desta S, Wako A, Aden I (2005). Origins, Governance, And Roles in Poverty Reduction Women's Groups in Arid Northern Kenya.

de Harrera AP, Sain G (1999). Adoption of maize conservation tillage in Azuero, Panama. Economics Working Paper 99-01. CIMMYT.

Feder G, Slade R (1984). The acquisition of information and the adoption of new technology. Am. J. Agric. Econ., 66: 312-320.

Food and Agricultural Organization (FAO) of the United Nations (2004). Soil and water

conservation in the semi arid lands. Natural resource management and environment department. FAO, Rome.

Frank BR (1997). Constrained limiting innovation adoption in northern Queensland beef industry. A socio economic means of maintaining a balanced lifestyle. Agric. Syst., p. 49.

Gachene CK, Kimaru G (2003). Soil fertility and land productivity. A guide for extension workers in Eastern Africa region. Regional Land and Management Unit, Kenya.

Greene WH (2004). Econometric Analysis. Fourth edition. Prentice Hall International Upper Saddle River, USA.

Grootaert C (2001). Does social capital help the poor? A synthesis of findings from the local level institutions studies in Bolivia, Burkina Faso, and Indonesia. World Bank Local Level Institutions Working Paper No. 10. Washington, DC. World Bank.

Gould BW, Saupe WE, Klemme RM (1990). Conservation tillage: the role of farm and operator characteristics and the perception of soil erosion. Land Econ., 65(2): 167-182.

Hagmann J, Murwira K, Chuma E (1996). Learning together: Development and extension of soil and water conservation in Zimbabwe. Q. J. Int. Agric., 35(2): 142-162.

Harper JK (2002). Economics of Conservation Tillage. Conservation Tillage series no.6, College of Agricultural Sciences, Penn state University.

Jaetzold R, Schmidt H (1983). Farm Management Handbook of Kenya. Vol. II/B. Central Kenya. Ministry of Agriculture, Nairobi, Kenya

Jonas NC, Mairura F, Ekise I, Chianu JN (2008). Farm input marketing in western Kenya: Challenges and opportunities Tropical Soil Biology and Fertility institute of the International Centre for Tropical Agriculture, Afr. J. Agric. Res., 3(3): 167-173.

Kaumbutho P, Kienzle J (2007). Conservation agriculture as practiced in Kenya: two case studies. Nairobi. African Conservation Tillage Network and Food and Agriculture Organization of the United Nations.

Kaumbutho PG, Mutua JM (2002). Experience from on farm conservation tillage trial in Kenya. Proceedings of a regional workshop on conservation tillage held in Lusaka, Zambia. Regional Land and Management Unit, Kenya.
Leagans JP (1979). Adoption of modern agricultural technology by small farm operators. Cornell international agricultural mimeograph no. 69, Cornell University, New York.
Maddala GS (2001). Introduction to Econometrics. 3rd edition. Cambridge University Press.
Mellis D, Matsaert H, Micheni A (1997). Rainwater harvesting for crops in semi-arid areas. In: Kang'ara, Participatory dryland agricultural research east of Mount Kenya, DAREP, RRC, Embu, Kenya. pp. 142-161.
Miriti JM (2005). Tied-riding and integrated nutrient management options for sustainable crop production in the semi-arid eastern Kenya. PhD. Thesis, University of Nairobi.
Okwach EG, Shepherd KD, Kwenah KM, Owiti AO, Omuto CT (2004). Assessment of land degradation in Machakos, Makueni and Kitui districts. Collaboration on Agricultural/Resource modeling and applications in Semi-Arid Kenya (CARMASAK) Technical report presented to Rockefeller Foundation.
Pretty (1999). "Can sustainable agriculture feed Africa? Evidence on progress, processes and impacts." Environ. Dev. Sustain., 1: 253-274.
Pindyck RS, Rubenfield DL (1991). Econometric models and economic forecast. McGraw-Hill New York.
Wale E, Holm-Müller K, Mburu J, Zeller M (2005). "Economic Analysis of Farmers' Preferences for Coffee Variety Attributes: Lessons for On-farm Conservation and Variety Adoption in Ethiopia". Q. J. Int. Agric., 44(2): 121-139.
Waswa F, Mburu J (2006). Potential of Dryland Farming in Kenya and Environmental Implications. In: Waswa F, Otor S, Mugendi D, Eds. Environment and Sustainable Development: A Guide for Tertiary Education in Kenya, Vol. 1. School of Environmental Studies and Human Science, Kenyatta University, Downtown Printing Works Ltd, Nairobi, pp. 70-90.
Welch F (1979). Education in publication. J. Polit. Econ., 78: 32-59.

Extractable micronutrients status in relation to other soil properties in Billiri Local Government Area

Ibrahim, A. K.*, Usman, A., Abubakar, B., and Aminu, U. H

Federal College of Horticulture, Dadin Kowa, P. M. B 108 Gombe, Gombe State, Nigeria.

A study of the micronutrient status of soils of Billiri was carried out at 55 different locations. The objective of the study was to evaluate the status of micronutrients and their relationship with soil properties. Soil sample were collected at 0 to 30 cm depth and analyzed for Diethylenetriamine penta acetic acid DPTA extractable micronutrients (iron, Fe, manganese, Mn, zinc, Zn, and copper, Cu). The DPTA extractable Fe, Cu, Zn ranged from 10.31 to 20.17, 4.04 to 8.43, 2.63 to 7.02 and 10.8 to 26.74 mgkg^{-1} soil respectively. All soil samples had adequate amount of micronutrients. DPTA extractable Fe gave negative non significant correlation with soil properties; Mn gave negative significant correlation with pH, clay, CEC and organic carbon while Cu and Zn were positively significantly correlated with pH, clay, cation exchange capacity (CEC) and organic carbon. Since all values were above the critical limits, there is no threat of Fe, Cu, Zn and Mn deficiency in the soils. Deficiency is however likely to develop with continuous cropping and poor fertilizer management.

Key words: Micronutrients, soil properties, pedogenic, geologic.

INTRODUCTION

Micronutrients are chemical elements required in very small amounts for growth and development of plants, this include iron, zinc, copper, boron, manganese and chlorine. Through their involvement in various plant enzymic systems and other physiologically active molecules, these micronutrients play a vital role in gene expression, biosynthesis of protein, nucleic acids, growth substances, metabolism of carbohydrates and lipids (Rangel, 2003). Goa et al. (2008) and Jiang et al. (2009) opined that original geologic substrate; subsequent geochemical and pedogenic processes determine the total levels of micronutrients in soils. Several authors have indicated that the availability of micronutrients in soils depends on soil pH, organic matter content, adsorptive surfaces and other physical, chemical and biological conditions in the rhizosphere (Kabata-Pendias, 2001; Jiang et al., 2009). Large hectares of arable land in Nigeria have been reported to be deficient in micronutrients and many of these deficiencies were brought about by the continuous use of inorganic fertilizers particularly nitrogen, phosphorus, and potassium by farmers, limited use of organic manures as well as non recycling of crop residues are some of the other factors contributing towards rapids exhaustion of micronutrients in soils.

Realizing the seriousness of the problems and considering the decreasing productivity, the present investigation was undertaken to evaluate the status and distribution of micronutrients and their relationship with soil properties.

*Corresponding author. E-mail: ibrahima764@gmail.com.

MATERIALS AND METHODS

The study was conducted in Billiri Local Government Area, Gombe State, located in the northern guinea savanna of Nigeria and lies in latitude 09° 52"N and longitude 11°31" E. The soils parent materials were derived from the older cretaceous sand stones (Bima sandstones and Yolde formation) and the crystalline basement complex rocks (Kaltungo formation, BSADP (1981). The study area has an average rainfall about 897 mm, while the main daily temperatures range from 28 to 32°C (GSADP, 1999).

Soil sampling

Fifty five (55) composites samples were randomly collected at 0 to 30 cm depth using soil auger to avoid any contamination of the soils. The samples were air dried, ground, and passed through

Table 1. Range and average values of physico-chemical properties of soils of Billiri.

Parameter	Range	Mean
Sand (%)	19 - 33	24.5
Silt (%)	26 - 36	31.3
Clay (%)	38 - 48	44.1
pH	6.2 - 7.6	6.90
OC (gkg^{-1})	10.7 - 13.9	12.39
CEC [cmol(+)kg^{-1}]	15.4 - 27.1	22.4

Cation Exchange Capacity (CEC) is expressed as centimoles of positive charge per kilogram (cmol kg-1)

Table 2. Range and average values of micronutrient of soils of Billiri L.G.A.

Parameter ($mgkg^{-1}$)	Range	Mean
Fe	10.31 - 20.17	14.48
Cu	4.04 - 8.43	6.19
Zn	2.62 - 7.02	4.65
Mn	10.88 - 26.74	17.59

2 mm sieve before analysis.

Analytical methods

Particle size analysis, soil pH, organic carbon and cation exchange capacity were determined following the methods described by Baruah and Barthaker (1997). The available fractions of iron (Fe), manganese (Mn), zinc (Zn) and copper (Cu) were extracted by DPTA according to Lindsay and Norvell (1978) and the metals in the extract were determined using an atomic absorption spectrophotometer.

The relationship between various soil properties and micronutrient content was established using simple correlation (Harry and Steven, 1995).

RESULTS AND DISCUSSION

Physico-chemical properties

The range and average values of the physico-chemical properties of the soil samples are shown in Table 1. The results showed that majority of the soil were slightly acidic to neutral reaction with medium amount of organic matter content while most of the soils were clay in texture. The soil pH ranged from 6.2 to 7.6 (mean 6.9); organic carbon contents ranged from 10.7 to 13.9 (mean 12.39 gkg). The soil pH influences the oxidation reaction and the solubility of several constituents and the ionic form of several elements. The cation exchange capacity ranged from 15.4 to 27.1 (mean 22.4). The relatively higher values of CEC indicate that some 2:1 clay mineral, probably mortmorillonite and/or illite, were present in the clay fractions.

The range and average values of micronutrients are presented in Table 2. By comparing the extractable micronutrients (Fe, Cu, Zn and Mn) contents with the established criteria set by Lindsay and Norvell (1978) and Esu (1991), in Table 3, all the soils were found to be high in iron, copper, zinc and manganese contents.

The highest mean values of micronutrient obtained from the site as supported by significant correlation value (Table 4) might be attributed to higher organic carbon content which might have likely increased the content of micronutrient. This corroborates similar observations made by Sangwan and Singh (1993), Samndi et al. (2007) and Sharma et al. (2000). Similarly, Muneshwar and Sekhon (1993) also stated that organic matter accounted between 40% DPTA extractable Cu and about 60% DPTA extractable Zn in some alfisols and inceptisols of India.

Correlation studies

Simple correlation studies between extractable micronutrients such as iron, copper, zinc and manganese were made with some physico-chemical properties as shown in Table 4.

Relationship between iron and some soil properties

The r-value between extractable Fe and soil pH was -0.298. It showed that there was negative non significant correlation between Fe and soil pH; silt, clay, CEC and organic carbon contents r-value were - 0.077, - 0.279, -0.235 and - 0.285 respectively, their availability in the

Table 3. Critical limits for Interpreting levels of soil nutrients.

Parameter	Rating		
	Low	Medium	High
OC (gkg^{-1})	<10	10 - 15	>15
CEC [$cmol(+)kg^{-1}$]	<6	6 - 12	>12
Fe ($mgkg^{-1}$)	<4.5	4.5 - 10.0	>10.0
Cu ($mgkg^{-1}$)	<0.2	0.2 - 1.0	>1.0
Zn ($mgkg^{-1}$)	<0.8	0.8 - 2.0	>2.0
Mn ($mgkg^{-1}$)	<5	5 - 10	>10

Adapted from Lindsay and Norvell (1978) and Esu (1991).

Table 4. Correlation matrix between micronutrients and some soil properties.

Parameter	Soil properties				
	pH	Silt	Clay	CEC	OC
Fe	- 0.298	- 0.077	- 0.279	- 0.235	- 0.285
Cu	0.464*	0.409	0.791**	0.756**	0.808**
Zn	0.699**	0.299	0.539*	0.843**	0.847**
Mn	0.740**	- 0.065	- 0.505*	- 0.630*	- 0.636*

**1% significant * 5% significant.

area is not dependent on any significant extent on any of these soil properties. These results are in agreement with findings of Kparmwang and Malgwi (1997) while analyzing available micronutrient in profiles of ultisols and entisols developed from sandstone in north western Nigeria.

Relationship between copper and some soil properties

The r-value between Cu and soil pH was 0.464. It showed that there was positive significant correlation between Cu and soil pH. These results were supported by Saddiq et al. (2008) and Samndi et al. (2007) who reported that Cu was positively correlated with soil pH. The correlation value (r) between Cu and Silt was 0.409. It showed that silt was positive non significantly correlated with Cu. Similar results were obtained by Sharma et al. (1996), and Nazif et al. (2006) who obtained positive correlation between Cu and silt contents. The data given in Table 4 shows that Cu was positively correlated with clay, CEC and organic carbon with (r) value 0.791, 0.756 and 0.808 respectively. Similar results were reported by Verma et al. (2005), Samndi et al. (2007), and Elbordiny et al. (2008) who also reported positive significant correlation between Cu and clay and organic carbon and CEC

Relationship between Zinc and some soil properties

The r-value obtained between zinc and soil pH was 0.699. It showed that there was positive significant correlation between zinc and soil pH. Similar results were obtained by Saddiq et al. (2008). A significant positive correlation between zinc and organic carbon (r = 0.854) indicate that complexing agents generated by organic matter promote zinc availability in soils. These results were similar to the findings of Chinchmalatpure et al. (2000), Verma et al. (2005) and Elbordiny et al. (2008). The r-value obtained between Zn and CEC was 0.843. It showed positive significant correlation between zinc and CEC contents. Similar results were obtained by Elbordiny et al. (2008). The r-value obtained between Zn and clay was 0.539. The results showed that there was positive significant correlation between Zn and clay. Similar results were derived by Mustafa and Fagam (2007). The correlation r-value between Zn and Silt was 0.299. It shows that Zn had positive non significant correlated with silt. Similar results were obtained by Nazif et al. (2006) who reported positive correlation between Zn and Silt content.

Relationship between manganese and some soil properties

The correlation value (r) between Mn and soil pH was -0.740. It shows that Mn had negative significant correlation with soil pH. This indicates that as pH increases, availability of Mn decreases. This result corroborates the reports of Patiram et al. (2000). The correlation value (r) between Mn and Clay, CEC and organic carbon was

- 0.505, - 0.630 and - 0.636 respectively; it means that Mn was negatively and significantly correlated with clay, CEC and organic carbon. This result was dissimilar to the findings of Elbordiny et al. (2008) who reported positive correlation between Mn and Clay, CEC and organic carbon content. The correlation value (r) between Mn and silt was - 0.065. The results indicates negative non significant. This result was dissimilar to the findings of Nasiz et al. (2006) and Verma et al. (2005) who reported positive correlation between Mn and silt content.

Conclusion

The study has shown that micronutrients were well supplied in the soils of the area and therefore deficiencies of the elements are unlikely. However, there is strong influence of organic carbon content on the micronutrient distribution and availability.

RECOMMENDATION

To maximize micronutrients solubility and hence availability to plants, it is best to maintain soil pH at favourable range.

REFERENCES

Baruah TC, Barthakur HP (1997). A text book of soil analysis. Vikas publishing, PVT Ltd, New Delhi.

BSADP (1981). Geological and Hydrological Map of Gombe. Bauchi State atlas. BSADP Planning Division and water Survey.

Chinchmalatpure A, Brij-lal R, Challas O, Sehgal (2000). Available micronutrient status of soils on different parent materials and landforms in micro watershed of Wunna catchment near Nagpur (Maharashtra). Agropedology, 10(1): 53 - 58.

Elbordiny MM, El-Dewiny YC (2008). Effect of some saltarfected soil properties on micronutrients availability. J. Appl. Sci. Res., 4(11): 1569 - 1573.

Esu IE (1991). Detailed soil survey of NIHORT farm at Bankure, Kano state, Nigeria. Inst. Agric Res, ABU Zaria.

Gao S, Yan R, Coa M, Yang W, Wang S, Chen F (2008). Effect of Cu on growth, antioxidant enzymes and phenylalanine ammonialyase activities in *Jatropha curcas* L. Seedling Plant. Soil Environ., 54: 111 - 122.

GSADP (1999). Climate and soil types in Gombe State. Reports from Department of Technical Service, pp. 1 - 2.

Harry F, Steven CA (1995). Statistics: Concepts and Applications. Cambridge University Press, Great Britain, p. 853.

Jiang Y, Zhang G, Zhou D, Qin Y, Liang WJ (2009). Profile Distribution of Micronutrients in an aquic brown soil as affected by land use. Plant. Soil. Environ., 155(11): 468 - 476.

Kabata - Pendias A (2001). Trace elements in soils and plants 3rd eds. CRC Press Boca Raton floods.

Kparmwang T, Malgwi WB (1997). Some available micronutrients in profile of ultisols and entisols developed from sandstone in North - Western Nigeria. In Singh BR (eds). Management of Marginal Lands in Nigeria. Proceeding of 23rd Annual Conference of Soil Science Society of Nigeria, pp. 245 - 52.

Lindsay WL, Norvell WA (1978). Development of DPTA soil test for Zinc, Iron, Manganese and Copper, Soil. Sci. Soc. Am. J., 42: 421 - 428.

Mustapha S, Fagam AS (2007). Influence of Parent Marterial on the contents and distribution of B and Zn in upland soils of Bauchi State, Nigeria. Int. J. Environ. Sci. Technol., 4(3): 359 - 362.

Muneshwar S, Sekhon GS (1993). DPTA extractable micronutrients in some benchmark soil series of India. J. Ind. Soil. Sci. Soc., 41(3): 566 - 568.

Nazif W, Sajida P, Saleem I (2006). Status of micronutrients in soils of Districts Bhimber (Azad Jammu and Kashmir). J. Agric. Biol. Sci., 1(2): 35 - 40.

Patiram R, Upadhyaya C, Singh CS, Munna R, Ram M (2000). Micronutrient Cation Status of Mandarin (*Citrus reticulate* Blanco) Orchard of Sikkim. J. Ind. Soc. Sci., 48(2): 246 - 249.

Rangel Z (2003). Heavy metals as essentials nutrients. In: Prasad MNV, Hagemeyer J (eds) ; Heavy metal stress in plants : Molecules to Ecosystems. Springer-Verlag, Berlin, Heidelberg, pp. 271-294.

Saddiq AM, Gungula DT, Mustapha S, Chiroma AM (2008). Micronutrient Status in some soils of selected Local Government Areas of Adamawa State Nigeria. Proceeding of the 32nd annual conference of soil science society of Nigeria held at Futy Yola, pp. 196-214.

Samndi MA, Raji BA, Kparmwang T (2007). Long-term effects of exotic tree species (*Tectona grandis* Linn. F) on the status of extractable micronutrients in the Southern Guinea Savanna of Nigeria. Niger. J. Soil. Environ. Res. 7: 15 - 22.

Sangwan BS, Singh K (1993). Vertical distribution of Zn, Mn, Cu and Fe in some alfisols of Haryana and their relationship with soil properties. J. Ind. Soc. Soil. Sci. 41: 439 - 437.

Sharma BD, Sidhu PS, Gurmel S, Mukhopadhyay SS, Singh G (1996). Elemental distribution and mineralogy of arid zone soils of Punjab. J. Ind. Soc. Soil. Sci. 44(4): 746 - 752.

Sharma BD, Mukhopadhyay SS, Sidhu SP, Katyal J (2000). Pedospheric attributes in distribution of total and DPTA extractable, Zn, Cu, Mn and Fe in Indo - Gangetic Plains. Geoderma, 96: 131 - 151.

Verma VK, Setia RK, Sharma PL, Charanjit S, Kumar A (2005). Pedospheric Variations in distribution of DPTA - extractable micronutrients in soils developed on different physiographic units in central parts of Punjab, India. Int. J. Agric. Biol., 7(2): 243 - 246.

Soil management strategies for rubber cultivation in an undulating topography of Northern Cross River State

J. R. Orimoloye*, I. K. Ugwa and S. O. Idoko

Rubber Research Institute of Nigeria, P. M. B. 1049, Benin City, Edo State, Nigeria.

This study was conducted to evaluate the soil fertility status of an undulating topography of a young rubber plantation located at Ikot-Ukpora, Biase Local Government Area of Cross Rivers State and its implication for rubber cultivation. The result showed that the soils of the area ranged from loamy sand to sandy loam on the surface overlying a slightly heavier textured sub-soil ranging from sandy loam to sandy clay loam. Most areas of the soils studied have gravely layers less than 100 cm from the surface leading to a matured plantation in the area to adopt a survival mechanism of curly tap root geometry with preponderance of extensive lateral root system that compensates for the tap root. The soils were slightly acidic and low in most essential nutrients. Total N, P and ECEC were generally low. Fertilizer recommendations were made for the different stages of rubber growth while other management strategies were suggested aimed at reducing organic matter losses and minimizing soil erosion as well as control of termite infestation.

Key words: Rubber, soil fertility, soil properties, Nigeria.

INTRODUCTION

Rubber (*Hevea brasiliensis* Wild ex A. de Juss. Muell. Arg.) is an economic crop and one of the most important cash crops in the rainforest agro ecological zone of Nigeria. The crop has attracted a lot of attention recently as a result of growing uncertainties in the National income as a result of over dependency on oil and the current attractive prices of the crop in the international market. It is a major source of employment opportunity especially for the restive youths of the Niger Delta region (Anschel, 1965; Aigbekaen and Nwagbo, 1999).

Rubber belt of Nigeria covers land area of about 7.6 million ha of land that lies in the southern part of Nigeria extending from Ogun State to Cross Rivers State. Apart from rubber cultivation, the land is also utilized for the cultivation of other crops, mining and urban development (RRIN, 1998). The soils of the area are predominantly coarse textured ultisols and oxisols derived from cretaceous sandstones and Pleistocene unconsolidated coastal sediments (FDALR, 1990) characterized by high acidity, low nutrient status, low activity clay and are subjected to rapid fertility decline on land clearing for cultivation (Asawalam and Ugwa, 1993; Eshett, 1991; Juo 1981; Kang and Juo, 1986). Therefore, economical and sustainable rubber plantation enterprises depend largely on high rate of fertilizer supplementation because rubber is a high nutrient demanding crop especially during the immaturity phase of their growth and development (1 - 6 years). Optimum rubber growth and high quality latex output therefore depend largely on the ability of the farmer to determine controlling factors and properly adjust them to suit rubber production. A major factor in any plant growth is the soil which determines the availability of nutrients required by the plants. A proper understanding of the physical and chemical properties of the soil is necessary to make proper amendments early enough that would ensure good return on investment at later productive stages of rubber.

This study therefore, evaluates the fertility potential of the undulating topography of soils supporting a 1000 ha of rubber plantation located at Ikot Ukpora, Biase Local Government Area of Cross Rivers State and its implication for rubber cultivation aimed at recommending appropriate measures necessary for sustainable and economically viable rubber cultivation

MATERIALS AND METHODS

The study was carried out in a land area covering an existing and proposed rubber plantation at Ikot Ukpora, Biase Local Government Area of Cross Rivers State. The area is a triangular shaped land area lying approximately within latitude 5°26.644' and 5°28.420' N

*Corresponding author. E-mail: Orimoloyej@yahoo.com.

and longitude 8°3.898' and 8°7.183' E covering about 1000 ha of land. Ikot Ukpora lies essentially in the humid rain forest zone with two distinct seasons (raining and dry seasons) which alternate annually with a mean annual rainfall of about 1950 mm distributed in about 9 months of the year with a peak between July and September. Relative humidity is approximately 89.9% while mean temperature is about 31.2°C. Cloud cover is about 80%, during the rainy season. The vegetation is mostly secondary forest re-growth. Young rubber plantations have undergrowths of broad leaves such as *Aspilla africana, Chromolaena odorata* etc with patches of grass species. Some perennial shrubs were scattered almost evenly while bamboo trees grow luxuriantly close to the streams and lowland areas. The cropping history is subsistence mixed plantain based arable cropping system with slash-and burn land clearing system along the contours and up to the peaks of the ridges.

The field study was carried out based on a field map of the entire landscape. A visual appraisal of the whole area was done and the field was mapped out for sampling based on the physiographic units, observed variabilities in landforms and geological materials. Spot elevations, slope gradients and aspects were obtained with a global positioning system (GPS) (Garmin etrex vista model). Soil samples were collected in various blocks shown in the map at Hillcrests, middle slopes and the valley bottoms. Samples were collected at two depths that is, 0 - 20 and 20 - 40 cm depths and a composite sample were collected for each depth at each of the sampling units. A total of 60 composite samples were collected, bagged, labelled and transported to the laboratory for analysis.

Soil pH was determined at 1:1 soil/water ratio using glass electrodes pH meter. Organic carbon was determined by chromic acid oxidation method. Available P was extracted using Bray 1 solution and the phosphate in the solution assayed by the Molybdenum blue colour method. Exchangeable cations were extracted with neutral normal ammonium acetate solution. Ca and mg were determined by EDTA titration, while Na and K were determined by flame photometry. Effective cation exchange capacity (ECEC) and base saturation were computed. Particle size analysis was by the modified Bouyoucous Hydrometer method as modified by Day (1965).

The interpretation of the results for fertility evaluation was based on the interpretation rules for tropical and subtropical soils as adopted by FDALR (1995). With respect to rubber cultivation, the information contained in Watson (1999), Orimoloye et al. (2004) and Orimoloye et al. (2005) was utilized at arriving at the conclusions and recommendations in conjunction with expert advice. Other criteria used for soil data interpretation and the sources of information are presented in Table 1.

RESULTS AND DISCUSSION

The area is underlain principally by quartz and quartzite schists characterized by ridges and steep slopes with no visible rock outcrop within the whole area though some remarkably high peaks were observed. Fragmented rocks of the mica and granite gneiss were encountered sporadically. In some exposed areas, it was observed that the quartzites exhibit lineaments that are North - South in orientation. The area is characterized by a remarkably undulating topography with slope gradients measuring between 5 - >25%. Global positioning system (GPS) elevation averaged about 89 m with the highest peak standing at 141 m. The area appears to be a watershed with numerous slopes ending invariably in a rivulet cutting deep into the hills. Principally, the area is drained by a tributary of the Cross River called Rubo river which flows westward in a very serpentine manner. The drainage pattern is dendritic in nature and most of the tributaries and rivulets joining the main stream almost at right angles. The courses of streams and rivers cut very sharp into the riverbeds and swampy narrow floodplains. The sharp angles and deep river beds are distinct features of resistance to weathering by underlying materials (Fagbami, 1981).

The soil analysis data were averaged across the land area as shown in Table 2 while the soil properties according to physiographic positions are presented in Table 3.

Soil physical properties

The soils of the study area ranged from loamy sand to sandy loam on the surface overlying a slightly heavier textured sub-soil ranging from sandy loam to sandy clay loam. This shows a degree of profile development and is desirable for rubber cultivation (Asawalam and Ugwa, 1993). Rubber has an extensive root system with taproots that could be as deep as 3 - 4 m and lateral roots that can extend from 15 - 20 m accordingly; a fairly deep, well drained soil with good physical structure on a gently sloping terrain with minimal soil erosion is required by rubber for optimum growth. Sandy clay loams and clay loams are among the most suitable textural classes. Hard pans and stoniness can greatly restrict root growth. Some of the soils of the study area have gradients that are higher than the recommended degrees for rubber but they are not under any serious threat of erosion due to the stability of the underlying parent material and a favourable orientation of the lineaments of the parent rock. However, most areas of the soils studied have gravelly layers less than 100 cm from the surface. In a matured plantation near the study area, an adaptation mechanism through modification in the tap root geometry of some rubber trees was observed (Plate 1). There was also a preponderance of extensive lateral root system that compensates for the tap root. It is however advised that clones that have high branching and dense canopy system should be avoided to minimize lodging and wind damage when rubber is fully established.

Soil chemical properties

The soil pH of the study area range from 4.8 - 5.9 which is within the suitable range for rubber without any need for liming (Watson, 1989; Ugwa et al., 2006). Rubber is essentially adapted to high rainfall areas which are often associated with leaching and acidity, the plant tolerates soils with an acid reaction (Akamigbo and Asadu, 1983). Most suitable soils for rubber are strongly to moderately acid with pH (in H_2O) ranging from 4.0 - 6.5. The

Table 1. Average soil physical and chemical properties at depth 0 - 20 and 20 - 40 cm in each block.

Block	Depth (cm)	Sand	Silt	Clay	Texture	pH (H2O)	Org C (%)	Aval P (mg kg^{-1})	Total N (%)	Exch Ac	Ca	Mg	Na	K	ECEC	BS (%)
			(%)									($Cmol\ kg^{-1}$)				
65	0 -20	71.9	7.99	20.11	SL	5.86	1.61	4.52	0.14	0.15	4.14	1.01	0.48	0.77	6.56	97.68
	20 - 40	67.23	8.25	24.52	SCL	5.31	2.12	2.42	0.05	0.34	2.06	0.33	0.30	0.56	3.61	90.58
66	0 - 20	58.18	14.92	26.90	CL	5.38	1.21	8.29	0.11	0.51	3.25	1.21	0.19	0.26	5.42	90.60
70	0 - 20	69.12	6.58	24.30	SCL	5.21	1.98	2.60	0.05	0.67	1.95	0.40	0.38	0.45	3.85	82.60
	20 - 40	55.78	14.92	29.30	SC	5.34	1.80	1.80	0.05	0.42	1.63	0.96	0.33	0.38	3.73	88.74
71	0 - 20	66.81	11.88	21.30	SL	5.85	1.99	11.33	0.04	0.32	4.14	1.01	0.48	0.77	6.76	95.26
	20 - 40	61.16	11.11	27.72	SCL	5.40	1.40	4.19	0.02	0.59	1.57	0.58	0.45	0.63	3.87	84.76
72A	0 - 20	60.48	13.22	26.30	SCL	5.28	1.64	6.34	0.10	0.26	1.95	0.29	0.23	0.08	2.85	90.88
	20 - 40	59.52	10.92	29.56	CL	5.13	1.04	3.94	0.08	0.67	1.62	0.18	0.18	0.05	2.73	75.47
75	0 - 20	67.48	13.22	19.30	SL	5.36	2.28	4.09	0.09	0.48	2.49	1.01	0.26	0.27	4.33	88.91
	20 - 40	61.48	16.22	21.30	SiCL	5.29	2.05	2.94	0.07	0.28	2.17	0.12	0.24	0.47	3.26	91.41
76	0 - 20	68.81	9.48	21.70	SL	5.42	1.77	5.77	0.12	0.28	1.46	0.52	0.38	0.61	3.27	91.43
	20 - 40	60.81	13.35	25.83	SCL	5.29	2.05	4.09	0.06	0.64	2.17	0.12	0.24	0.47	3.6	82.23
77	0 - 20	66.14	13.82	20.03	SL	5.28	1.65	6.32	0.10	0.26	1.46	0.20	0.16	0.19	2.74	90.51
	20 - 40	65.48	11.42	23.10	SCL	5.06	2.32	4.29	0.06	1.07	0.97	0.49	0.18	0.11	2.76	61.23
78A	0 - 20	66.81	15.22	17.63	SL	4.91	2.27	3.11	0.11	1.05	1.62	0.56	0.27	0.12	3.93	73.28
	20 - 40	57.83	14.79	27.38	SCL	4.84	1.74	1.51	0.08	0.95	2.06	0.32	0.43	0.12	2.73	65.50
Mean		63.83	12.19	23.90		5.30	1.82	4.56	0.07	0.52	2.15	0.54	0.31	0.37	3.88	84.77

exchangeable bases (Ca, Mg, Na and K) are generally low in the study area. The effective cation exchange capacity (ECEC) is also low this could be partly due to continuous cropping and leaching. Cox and Uribe (1992) observed soil cation decrease with cropping. An optimal value for CEC for rubber is 15 $cmolkg^{-1}$. Watson (1989) had suggested that soils that has less than 5 cmol kg-1 requires supplementary fertilizer programme for optimum rubber development. Base saturation mostly above 80% shows that the exchange complex of the soil particles are occupied by bases rather than reserved acidity (Al3+ and H+). The observed base saturation might be attributed to the fact that majority of the land area are well under vegetation cover and are just being opened up for cultivation. The status of the soil exchange complex is desirable for long term fertility management. The soil available P, total N and organic carbon contents of 4.71 mg/kg, 0.07 and 1.59, respectively are low to moderate in the soils. This is an indication of low fertility for rubber and characteristic of many of the rubber growing soils earlier studied (Eshett, 1991; Asawalam and Ugwa, 1993; Ugwa et al., 2006). These nutrients are mainly influenced by vegetation cover, bush burning and organic matter management. It is noteworthy here that some of the soils sampled are newly opened up from fallow and some in blocks 75, 76 and 77 have some fertilizers applied to some growing young rubber there. However, nitrogen is easily lost through several processes in the soil while P is easily immobilized by microbial activities as well as chemical reactions.

RECOMMENDATIONS

Based on field observations and soil analysis, the following recommendations are proposed.

(1) 50 g 0f NPK Mg with 100 g rock phosphate per plant should be thoroughly mixed with top soil and applied to the planting holes 1 - 2 weeks before planting. If the Mg component is not available, 15 g of dolomitic limestone may be applied.
(2) The rides of rubber should be kept clean of weeds before and after planting.

Table 2. Average soil chemical properties at three physiographic locations in all the field blocks sampled.

Physiography	Depth (cm)	Sand	Silt	Clay	Texture	PH (H_2O)	Org C	Av P (mg/kg)	Total N	Exch Ac	Ca	Mg	Na	K	ECEC	BS (%)
		($g\ kg^{-1}$)					($g\ kg^{-1}$)		($g\ kg^{-1}$)	($Cmol\ kg^{-1}$)						
	0 - 20	693.10	112.80	194.00	SL	5.51	15.70	8.45	01.20	0.45	3.79	0.49	0.29	0.71	5.73	92.15
Hill crest	20 - 40	622.00	132.40	242.40	SCL	5.29	10.30	6.12	00.60	0.61	2.02	0.37	0.30	0.35	3.65	83.28
	0 - 20	696.50	98.40	203.30	SL	5.40	16.60	5.04	00.40	0.22	3.92	0.56	0.40	0.55	5.65	96.11
Middle slope	20 - 40	625.40	100.40	274.20	SCL	5.28	10.00	2.13	00.20	0.34	2.80	0.55	0.35	0.53	4.57	92.56
Valley bottom	0 - 20	649.80	138.90	211.20	SL	5.39	24.90	3.31	01.00	0.44	2.12	0.27	0.34	0.49	3.66	87.97
	20 - 40	610.40	133.50	256.10	SCL	5.19	18.40	3.24	00.60	0.44	2.01	0.46	0.34	0.47	3.72	88.17
Mean	-	649.5	119.4	230.20	-	5.34	15.90	4.71	00.70	0.42	2.77	0.45	0.33	0.51	4.38	90.41

Table 3. Criteria for soil data interpretation.

Soil parameter	Very poor	Low	Moderate	Good	Authority
Ca ($Cmol\ kg^{-1}$)	1.0	1. - 2.3	2.3 - 5	7.0	FDALR (1995)
Mg ($Cmol\ kg^{-1}$)	0.4	0.5 - 1.0	1.1 - 1.5	3.0	FDALR (1995)
K ($Cmol\ kg^{-1}$)	0.01	0.01 - 0.2	0.2 - 0.4	0.4 - 0.8	FDALR (1995)
ECEC ($Cmol\ kg^{-1}$)	<4	4 - 9	9 - 12	>12	Ugwa et al. (2006)
Available P (mg/kg)	< 5	5 - 8	8 - 10	>10	FDALR (1995)
Total N (%)	<0.04	0.05 - 0.12	0.12 - 24	>0.24	Watson (1989)

(3)Application rate of 33 kg N/ha, 13.64 kg P2O5 /ha, 20.94 kg K2O per ha and 12.96 kg MgO /ha in two split applications is advised for the first year (Table 4) while a schedule for the first 5 years in Table 5 should be followed.

(4) To ensure maximum nutrient capture, fertilizers should be applied in a ring form a few centimetres from the base of rubber during the first year. The radius of the ring should be increased progressively to 80 - 90 cm at the end of the second year. From year 3 to 4, band placement along a clean weeded strip at 80 - 250 cm from the base of the tree till year 5.

(5) Frequent dressings and split application at 6 monthly interval is encouraged. Time of application should be adjusted to avoid periods of excessive rainfall or very dry periods of the year.

(6) Rooting assistance (hormones) such as indole acetic acid (IAA) or Indole butyric acid (IBA) to stimulate root development and nutrient capture should be done.

(7) Due to stoniness and the soil depth observed, regular pruning to avoid dense foliage should be carried out. In the alternative, clones with light crowns should be planted.

(8) Fertilizer application to mature rubber trees take 2 - 3 years before the effects are felt and are mainly for stability and yield maintenance. Application of 320 kg/ha (equivalent of 550 g/tree) per year of NPK Mg mixture of ratio 10: 7: 9: 4 respectively is recommended from year 7 - 15. Fertilizer application for matured tree could be modified based on responses obtained from the above and fertilizer regimes maintained at the immature phase.

Other soil management suggestions:

(1) Creeping legume cover crops may be beneficial to rubber in the study area. Mulching around the base of the rubber trees is well encouraged.

(2) Where applicable, intercropping with maize, cassava, cowpea, soybean and other common food crops including plantain that is a common

Plate 1. A taproot of rubber modified by stone line. Notice the root bent sideways instead of growing vertically into the soil.

Table 4. Soil nutrient observed and fertilizer recommendations for young rubber at Ikot-Okpora rubber estate at the first year of planting.

Nutrient element	Soil available (kg/ha)	Amount required (kg/ha)	Deficit/Surplus (kg/ha)	Fertilizer material	Application rate
				Urea/tree)	75 kg/ha or 125 g/tree
N	78.4	112	33.6	Ammonium sulphate	160 kg/ha or 266 g/ tree
				Ammonium phosphate	120 kg/ha or 200 g/tree
				SSP	75.78 kg/ha or 126 g/tree
P (P_2O_5)	12.06	25.70	13.64	Ammonium phosphate	68.20 kg/ha or 114 g/tree
K (K_2O)	197.06	218	20.94	Murriate of potash	34.90 kg/ha or 58.17g/tree
Mg (MgO)	108	120.96	12.96	Commercial MgO	2.4 g/tree
				Dolomite	120 g/tree

Table 5. Schedule of fertilizer application for the first 5 years of rubber planting the study area (g/tree).

Months after planting	N	P_2O_5	K_2O	MgO
At planting	80 g of NPK with 100 g rock phosphate per planting hole			
6	18	11.35	17.45	10.63
12	18	11.35	17.45	10.63
18	18	11.25	17.45	10.63
24	28.4	14.25	22.68	10.63
36	56	28.5	34.9	14.40
42	100	84	40	21
54	120	80	45	25
66	120	80	45	25

crop in the environment is encouraged during the first four (4) years of rubber. This will bring early return on investment and ensure proper maintenance of the plantation. Planting arrangement should be such that rubber is not choked up and fertilizer should be adjusted to accommodate the food crops.

(3) Contour terracing or construction of silt pits in areas where slope gradient is higher than 10% is suggested. (4) Termite control by destroying the termite hills and drenching with DEVAP or any of its variants at 2% concentration is highly recommended in the termite prone areas.

ACKNOWLEDGEMENTS

The management of Pamol Nigeria limited that funded this study, the Executive Director of who approved the study and the Laboratory staff of Soil and Plant nutrition Division RRIN who assisted in carrying out the soil analysis are gratefully acknowledged.

REFERENCES

Aigbekaen EO, Nwagbo EC (1999). An analysis of costs and returns of natural rubber production in Nigeria. Indian J. Nat. Rubber Res. 12 (1-2): 77 - 85.

Akamigbo FOR, Asadu CLA (1983). Influence of parent materials on the soils of South-eastern Nigeria. East Afr. Agric. For. J. 48(4): 81 - 91.

Anschel ER (1965). Economic aspects of present rubber production in Midwestern Nigeria. Ph.D Thesis. Michigan State University pp. 89 - 90.

Asawalam DOK, Ugwa IK (1993). Some soils of Northern Bendel State and their potentials for growing rubber. Indian J. Nat. Rubber Res. 12 (1-2): 137 - 142.

Cox FR, Uribe E (1992). Potassium in two humid tropical ultisol under corn ans oyabean cropping system: 11 Dynamics. Agron. J. 84: 485 - 496.

Day PR (1965). Particle fractionation and particle size analysis in C.A. Black (ed) Methods of Soil Analysis Part Agro. 9: 545 - 567.

Eshett ET (1991). General properties of iron and aluminium status of selected rubber growing soils of Southern Nigeria Indian J. Nat. Rubber. Res. 4(1): 8 - 15.

Fagbami A (1981). Soil formation processes in the Sub-humid tropics on basement complex. Niger. J. Soil Sci. 12: 131-146.

FDALR (1995). The reconnaissance soil map of Nigeria (1: 650,000) Soil report Vol. 3 Federal Department of Agricultural Land Resources, Lagos, Nigeria.

Juo ASR (1981). Mineralogy of 'acid sands' of Southern Soil Science Society of Nigeria (SSSN) Special Publication Monograph 11: 19 - 26.

Kang BT, Juo ASR (1986). Effect of forest clearing on soil chemical properties and performances. In: Lal R, Sanchez PA, Chumming RW Jr (Eds.) Land clearing and development in the Tropics. Balkema, Rotterdam, Botson.

Orimoloye JR, Omorusi VI, Ugwa IK, Idoko SO (2004). Effect of Dolomite, rock phosphate and organic manure on indigenous Mycorrhizal Fungi and rubber seedling in an acid sandy soil. J. Sustain. Trop. Agric. Res. 10: 38 - 89.

Orimoloye JR, Ugwa IK, Abolagba EO, Idoko SO (2005). Potentials of rubber cultivation in the Rainforest Savanna transition zone of mid - Western Nigeria. J. Sustain. Trop. Agric. Res. 14: 24 - 31.

RRIN (1998). Annual report. Rubber Research Institute of Nigeria, Benin City Nigeria.

Ugwa IK, Orimoloye JR, Esekhade TU (2006). Nutrient status of some soils supporting rubber in Mid-western Nigeria. Niger. Agric. J. 36:169 -176

Watson GA (1989). Nutrition in rubber C.C. Webster and W.J. Baulkwill (eds.) Longman Scientific and Technical, New York pp. 291 - 348.

18

Balanites aegyptiaca, a potential tree for parkland agroforestry systems with sorghum in Northern Ethiopia

Hailemariam Kassa[1], Kindeya Gebrehiwet[2] and Charles Yamoah[3]*

[1]Tigray Agricultural Research Institute, Humera Agricultural Research Centre, Humera, P. O. Box 230 Tigray, Ethiopia.
[2]Department of Land Resource Management and Environmental Protection, Faculty of Dryland Agriculture, Mekelle University, P. O. Box 231, Ethiopia.
[3]Faculty of Renewable Natural Resources, University of Development Studies, P. O. Box TL 1350 Tamale, Ghana.

***Balanites aegyptiaca* tree is preferred by farmers in Tigray region of Ethiopia for fuel wood, provision of shade, fodder and as a medicinal plant. This study was conducted to assess soil properties and sorghum yields on farm fields (Limat, Goblel and Korbebite and Endakeshe), where *B. aegyptiaca* are traditionally retained. At each site, six trees were randomly selected and soil properties and sorghum yields were compared 1) under the tree canopy (0 – 4 m), designated as zone D1; 2) near the tree, (4 – 6 m, D2) and 3) far away from the tree (8 – 12 m, D3). At the Limat site, results indicated that % clay and available P were significantly higher ($p < 0.05$) at D1 (under the canopy) than D3 (further away from the canopy). Also at the same site, % silt was significantly lower ($p < 0.05$) at D1 than D3. At Goblel and Korbebite sites, pH was significantly lower ($p < 0.05$) at D1 and D2 than D3. Apart from these, exchangeable Ca, exchangeable K, organic carbon (OC), cation exchange capacity (CEC), electrical conductivity (EC) and yields were non-significant among the three zones at the sites. Mean moisture levels of all sites at sowing period at D1 (14.36%) was significantly ($p < 0.05$) lower than D3 (16.32%), but yields were unaffected. Pot studies using soils from the above sites and zones demonstrated similar soil and yields' patterns. Thus, it may be appropriate to integrate sorghum crops with *B. aegyptiaca* trees on the same field in a parkland agroforestry system with minimal risk of crop failure.**

Key words: *Balanites aegyptiaca,* parkland agroforestry, sorghum.

INTRODUCTION

A major reason for practicing agroforestry land use systems is domestication of soil-improving trees for enhancing soil productivity through a combination of selected trees and food crops on the same farm field (ICRAF, 2000). Scattered trees are characteristics of a large part of the African agriculture landscape. The so-called parkland savanna may be defined as the regular presence of well-grown trees scattered on cultivated or recently fallowed fields. The trees are deliberately associated with the agricultural environment because of their specific use (ICRAF, 2000). *Sorghum bicolor* is the second dominant food crop cultivated in Tigray region (EARO, 2002).

Balanitus aegyptiaca is one of the dominant tree species deliberately retained on farm fields in the Tigray region of Ethiopia. It is useful as a medicinal plant, food, fodder, fuel wood, construction material and provision of shade (Kindeya, 2004).

Soil fertility depletion and low crops' yields are widespread in the Tigray region (EARO, 2002). The use of supplementary inorganic fertilizers has become less affordable for many Ethiopian farmers, following the removal of fertilizer subsidies. Consequently, improved fallows such as *Sesbania sesban,* is used to address low fertility problems of small holders fields in the region (EARO, 2002). Additionally, it is important to combine plant biomass with inorganic fertilizers so as to increase fertilizer-use efficiency or both (Juo and Franzluebbers,

*Corresponding author. E-mail: yamoahcf@yahoo.com.

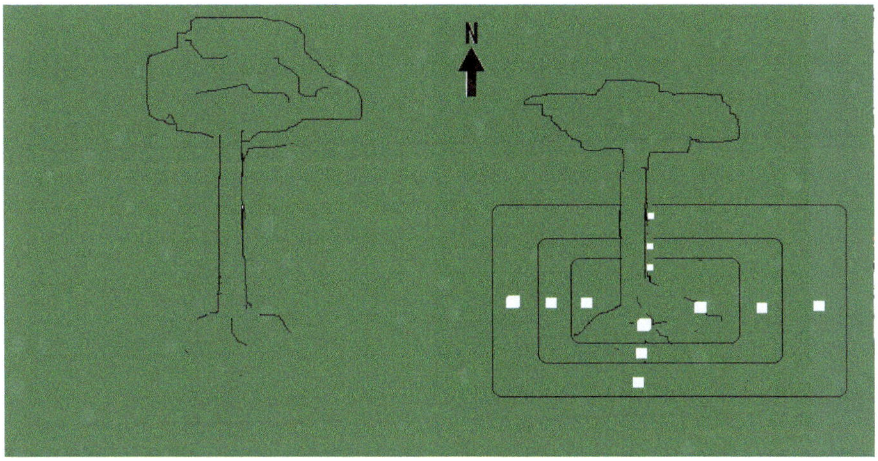

Figure 1. Design of soil sampling and measurement of sorghum plant height and yield from *B. aegyptiaca* tree.

2003). Usually, the rapid-decomposing biomass with low carbon to nitrogen ratios (C:N < 10) may be used as fertilizers to supply nutrients in a shorter time frame. On the contrary, the slow-decomposing biomass could be used as mulch to improve soil physical conditions (Tian et al., 1995).

There are a number of agroforestry options in drylands of Eastern Africa that are available for the purpose of both resources conservation and economic development for sustainable livelihoods (Jama and Zeila, 2005). There are many indigenous tree species including *B. agyptiaca* that have a potential of maintaining the ecology and economic development (Kindeya, 2004). Dispersed trees on farms, that is, parkland agroforestry systems, are typical in Tigray region of Ethiopia (Tesfay, 2004) covering large tract of agricultural land. However, despite their significance in contributing to the livelihood of rural people, they have been largely ignored in the research domain. Rather, the impact of trees like *Faidherbia albida*; *Cordia africana,* and *Crotton macrostachyus* (Poschen, 1986) on soil properties and yields have been extensively researched. Our objective is to determine the effects of *B. agyptiaca* on soil properties and sorghum yield in order to assess the tree's potential and suitability for use in parkland agroforestry systems.

MATERIALS AND METHODS

Description of the study sites

This study was conducted at Humera district located at (13°42' - 14°28'N, 36°20' - 37°31'E), Tigray region of Ethiopia, bordering Eritrea. Specific villages selected were Lemat, Goblel, Korbebit, and Endakashe, because of the sorghum-based cropping systems and *B. aegyptiaca* trees present on farm fields at these villages. The dominant soil type is Vertisol (USDA, 2006) characterized by deep soils (> 150 cm), clay content of 40 - 60%, electrical conductivity of 0.047 - 0.179 mmohs/cm, organic matter content of <2% and CEC ranging from 37 – 77 meq/100 g soil (Mitiku et al., 2006). The mean annual rainfall is 581.2 mm mostly occurring from June to September. The maximum temperature varies from 42°C in April to 33°C in August, while minimum temperature is from 22.2°C in May to17.5°C in July (EARO, 2002).

Experimental design

At each village, six *B. aegyptiaca* trees were randomly selected based on DBH (diameter at breast height), crown diameter and heights. Soil samples (0 - 20 cm) were taken at three distances from *B. aegyptiaca* designated as zones (D) were: 1) 0 to 4 m (D1), 2) 4 m to 6 m (D2) and 3) 8 to 12 m (D3) all from the base or zero cm of the tree. The six trees acted as replications. The test crop was an improved *S. bicolor* variety "Gobye" ,which is released to farmers by government to resist striga weeds (EARO, 2002). Data collection included height and biomass and grain yields from 8 quadrants of 0.5 m × 0.5 m taken from four directions from the tree as shown in Figure 1. The data were analyzed using a one-way analysis of variance.

Soil analyses

Soil samples were analyzed for total nitrogen (N) by Keldahl method (Jackson, 1958); available phosphorus (P) (Olsen and Sommers, 1982); exchangeable K by ammonium acetate (Jackson, 1958); organic carbon (OC) (Walkley and Black, 1934), electrical conductivity (EC) and cation exchange capacity (CEC) by (Houba et al., 1989); and pH, (1:2.5 soils to water ratio) by Jackson (1958). Texture was determined by hydrometer method (Gee and Bauder, 1982) and bulk density by the core method. Gravimetric moisture contents (Blake and Hartge, 1986) were determined at the beginning of rains to the flowering stage of sorghum that is, two months later.

Pot experiment

Soil samples collected from six trees on four sites (24 trees) and three distances or zones were used for the plastic pot experiments. The plastic pots were 25 cm high and a diameter of 10 cm and contained 2.5 kg soil. The experimental design was complete randomized (CRD) with 5 replications. Sorghum was the test crop. Three seeds were sown and thinned to one plant after germination. Plant height was measured at two weekly intervals and yield was

Table 1. Effect of *Balanites aegyptiaca* tree on soil properties at Limat site.

Soil property	Under the canopy (D1)	Near the canopy (D2)	Far from the canopy (D3)
OC %	$0.60^{a} \pm 0.027$	$0.55^{a} \pm 0.068$	$0.47^{a} \pm 0.6$
Total N %	$0.07^{a} \pm 0.004$	$0.07^{a} \pm 0.004$	$0.06^{a} \pm 0.004$
Available P (ppm)	$1.28^{ab} \pm 0.08$	$1.37^{a} \pm 0.23$	$0.75^{b} \pm 0.10$
Exchangeable K (ppm)	$320.21^{a} \pm 20.10$	$305.47^{a} \pm 23.85$	$289.00^{a} \pm 14.15$
EC ds/m	$0.29^{a} \pm 0.04$	$0.17^{a} \pm 0.03$	$0.31^{a} \pm 0.07$
CEC meq/100 g soil	$51.45^{a} \pm 5.23$	$49.62^{a} \pm 1.29$	$53.59^{a} \pm 1.83$
pH	$8.17^{a} \pm 0.08$	$8.15^{a} \pm 0.09$	$8.40^{a} \pm 0.03$
Clay %	$45.24^{a} \pm 3.96$	$37.09^{ab} \pm 2.36$	$30.90^{b} \pm 3.12$
	LSD = 11.81		
Silt %	$16.76^{b} \pm 4.45$	$22.91^{ab} \pm 0.63$	$29.09^{a} \pm 3.57$

Rows with the same superscript letters are not significantly different at $P < 0.05$. OC % = Organic carbon percent. EC ds/m = electrical conductivity; Total N% = total nitrogen; CEC meq/100 g = Cation exchange capacity; Available P (ppm) = available Phosphorus; Exchangeable K (ppm) = exchangeable Potassium.

Table 2. Effect of *Balanites aegyptiaca tree* on soil properties at Goblel site.

Soil property	Under the canopy (D1)	Near the canopy (D2)	Far from the canopy (D3)
OC %	$0.81^{a} \pm 0.04$	$0.77^{a} \pm 0.03$	$0.77^{a} \pm 0.01$
Total N %	$0.047^{a} \pm 0.01$	$0.05^{a} \pm 0.00$	$0.04^{a} \pm 0.01$
Available P (ppm)	$1.77^{a} \pm 1.04$	$1.06^{a} \pm 0.39$	$0.33^{a} \pm 0.13$
Exchangeable K (ppm)	$295.88^{a} \pm 29.66$	$276.91^{a} \pm 13.9$	$262.59^{a} \pm 11.19$
EC ds/m	$0.38^{a} \pm 0.03$	$0.34^{a} \pm 0.07$	$0.28^{a} \pm 0.023$
CEC meq/100 g soil	$50.56^{a} \pm 2.91$	$50.63^{a} \pm 1.54$	$50.89^{a} \pm 2.56$
pH	$7.96^{b} \pm 0.04$	$8.33^{a} \pm 0.06$	$8.22^{a} \pm 0.06$
	LSD = 0.188		
Clay %	$36.51^{a} \pm 2.41$	$31.75^{a} \pm 1.86$	$32.81^{a} \pm 2.23$
Silt %	$23.64^{a} \pm 1.66$	$25.20^{a} \pm 1.59$	$25.44^{a} \pm 2.41$

Rows with the same superscript letters are not significantly different at $p < 0.05$. OC %= Organic carbon percent EC ds/m = electrical conductivity; Total N% = total nitrogen; CEC meq/100 g = Cation exchange capacity; Available P (ppm) = available Phosphorus; Exchangeable K (ppm) = exchangeable Potassium.

determined at harvest. Plant height, yield and soil analytical data were subjected to statistical analysis using JMP version 0.5. Duncan multiple range test (DMRT) was used for treatment mean separation.

RESULTS AND DISCUSSION

Effects of *B. aegyptiaca* tree on soil properties at the Limat site

Available phosphorus, clay and silt percentages were significant ($P < 0.05$) at Limat. Available P was higher at D1 (1.28 ppm) and D2 (1.37 ppm) than D3 (0.75 ppm), representing an increase by 82.7%. With available P of less than 10 ppm (Sanchez et al., 1982; Marx et al., 1999), the soils at Limat were deficient in available P. Percent clay was higher at D1 (45%) than D3 (31%), showing an increase of 45%. However, silt percent was the reverse that is,17% at D1 and 29% at D3, showing a decrease of 41 % (Table 1). Contrary to our expectation, there was no significance ($P > 0.05$) difference in OC among D1 (0.60%), D2 and D3 (0.47%). This range of OC is often what is encountered in soils of the semiarid areas, which is >1% (Bationo et al., 1993). Similarly, there was no significance difference in total nitrogen, exchangeable potassium, EC, CEC and pH among the three zones, that is, D1, D2 and D3 (Table 1).

Effect of *B. aegyptiaca* on soil properties at Goblel, Korbebit and Endakashe sites

Soil pH is the only parameter that was significantly different at Goblel and Korbebit sites (Tables 2 and 3). At

Table 3. Effect of *Balanites aegyptiaca tree* on soil properties at Korbebit.

Site parameter	Under the canopy	Near the canopy	Far from the canopy
OC%	$0.75^a \pm 0.03$	$0.69^a \pm 0.02$	$0.72^a \pm 0.03$
Total N %	$0.02^a \pm 0.00$	$0.03^a \pm 0.01$	$0.03^a \pm 0.02$
Available P (ppm)	$1.47^a \pm 0.42$	$1.10^a \pm 0.14$	$0.92^a \pm 0 .15$
Exchangeable K (ppm)	$199.34^a \pm 8.83$	$217.12^a \pm 9.58$	$212.83^a \pm 10.91$
EC ds/m	$0.24^a \pm 0.044$	$0.180^a \pm 0.03$	$0.18^a \pm 0.03$
CEC meq/100 g soil	$34.32^a \pm 1.97$	$37.60^a \pm 4.01$	$43.15^a \pm 4.61$
pH	$8.46^a \pm 0.07$	$8.08^b \pm 0.11$	$8.49^a \pm 0.02$
Clay %	$41.03^a \pm 1.12$	$41.03^a \pm 0.84$	$40.36^a \pm 0.77$
Silt %	$29.28^a \pm 0.73$	$28.28^a \pm 2.24$	$28.61^a \pm 1.12$

Rows with the same superscript letters are not significantly different at $p < 0.05$. OC %= Organic carbon percent EC ds/m = electrical conductivity; Total N% = total nitrogen; CEC meq/100 g = Cation exchange capacity; Available P (ppm) = available phosphorus; Exchangeable K (ppm) = exchangeable potassium.

Table 4. Effect of *Balanites aegyptiaca tree* on soil properties at Endakeshe.

Site parameter	Under canopy (D1)	Near canopy (D2)	Far from canopy (D3)
OC %	$0.77^a \pm 0.06$	$0.77^a \pm 0.04$	$0.77^a \pm 0.05$
Total N %	$0.03^a \pm 0.00$	$0.03^a \pm 0.00$	$0.02^a \pm 0.00$
Available P(ppm)	$0.66^a \pm 0.11$	$0.46^a \pm 0.06$	$0.79^a \pm 0.15$
Exchangeable K(ppm)	$214.79^a \pm 15.30$	$186.11^a \pm 16.17$	$194.80^a \pm 7.91$
EC (ds/m)	$0.22^a \pm 0.02$	$0.21^a \pm 0.03$	$0.22^a \pm 0.03$
CEC (meq/100 g soil)	$53.50^a \pm 1.59$	$53.56^a \pm 1.44$	$51.65^a \pm 3.29$
pH	$8.14^a \pm 0.16$	$8.44^a \pm 0.04$	$7.95^a \pm 0.19$
Clay %	$32.693^a \pm 1.74$	$32.08^a \pm 1.41$	$32.36^a \pm 2.06$
Silt %	$22.28^a \pm 0.86$	$22.20^a \pm 0.66$	$24.61^a \pm 1.69$

Rows with the same superscript letters are not significantly different at $p < 0.05$. OC %= Organic carbon percent EC ds/m = electrical conductivity; Total N % = total nitrogen; CEC meq/100 g = Cation exchange capacity; Available P (ppm) = available phosphorus; Exchangeable K (ppm) = exchangeable potassium.

Goblel, pH was lower at D1 (7.96) than D3 (8.22) showing a decrease of 3.1% (Table 2). The pH range reflects the calcareous nature of the $CaCO_3$ parent material (Sharma and Gupta, 1989). Like the Limat site, there was no significant difference in OC among the three zones. The non-significant difference of organic carbon supports the assumption regarding the rapid mineralization of OC in the semiarid environment in Ethiopia (Tesfy, 2004). At Korbebit site, soil pH was significantly lower at D2 (8.08) than D1 (8.46) and D3 (Table 3).

According to Marx et al. (1999) and Sanchez et al. (1982), the soils were slightly alkaline, and that would not normally create an unfavorable condition for growth of common cereals and legumes. The same holds true for exchangeable potassium and EC. Cation exchange capacity of > 40 meq/100 g is above average. Total nitrogen value was >0.05 in the study area which is of the lower side of common soils. Even though total N is not a good measure of plant available N, typical agricultural soils has percent total N of 0.1 - 0.15 (Marx et al., 1999). Since OC, available P and total N are limiting (Tables 1, 2, 3 and 4), there is a need for supplementary inorganic fertilizers (e.g. DAP (diammonium phosphate) in addition to litter fall of *B. aegyptiaca* and possibly biomass of *S. sesban* to sustain increase sorghum yields. Diammonium phosphate is the recommended fertilizer by the Ministry of Agriculture for the Tigray region and *B. aegyptiaca and S. sesban* are indigenous shrubs in the region. At Endakashe, there was no significant difference in pH, organic carbon, total nitrogen, available phosphorus, exchangeable potassium, EC, CEC, clay and silt among the three zones (Table 4).

Combination soil chemical properties of the four sites

The combined OC of the four sites was not significantly different ($p > 0.05$) among the three zones from the base

Table 5. Combined all sites soil properties.

Site parameter	Under canopy (D1)	Near canopy (D2)	Far from canopy (D3)
OC %	0.72[a] ±0.02	0.69[a] ±0.03	0.68[a] ±0.03
Total N %	0.04[a] ±0.00	0.04[a] ±0.00	0.04[a] ±0.01
Available P(ppm)	1.29[a] ± 0.28	0.99[a] ±0.13	0.70[a] ±0.08
Exchangeable K(ppm)	257.55[a] ±14.21	245.76[a] ±12.57	239.81[a] ±9.46
EC (ds/m)	0.28[a] ±0.02	0.22 [a] ±0.02	0.25[a] ±0.02
CEC (meq/100 g soil)	47.46[a]±2.20	47.96[a] ±1.67	49.82[a] ±1.72
pH	8.17[a]±0.01	8.30[a] ±0.050	8.24[a] ±0.06
Clay %	38.87[a] ±1.54	35.40[ab] ±1.12	34.11 [b] ±1.29
	LSD 4.4987		
Silt %	22.99[a] ± 1.47	24.40[a] ±0.83	26.94 [a] ±1.18

Rows with the same superscript letters are not significantly different at $p < 0.05$. OC% = Organic carbon percent EC ds/m = electrical conductivity; Total N% = total nitrogen; CEC meq/100 g = Cation exchange capacity; Available P (ppm) = available Phosphorus; Exchangeable K (ppm) = exchangeable Potassium.

of *B. aegyptiaca* tree (Table 5). This may be due to low litter fall during the period of the experiment and the existing organic matter might have readily decomposed due to elevated temperatures often observed in the arid ecozone. The increase in soil organic matter under the canopy of different tree species has also been reported. For example, Abebe et al. (2001) reported an average increase of about 13% organic matter under canopy of Cordial Africana trees. Similarly, Felker (1978), Bernard-Reversat (1982), Tadesse (1997) and Yeshanew et al. (1999), all reported significantly higher organic matter under different trees' canopy than the corresponding open area. In addition, Jonsson (1995) found nearly three times higher percent organic carbon near the bole of the *Vitvaria paradoa* tree than at 150 cm away from the bole of the tree. Correspondingly, Tadesse (1997) and Zebene (2003) all reported significantly higher organic carbon under different trees canopy than the corresponding open area, which disagrees with the result of the present study. This study is similar to that of Kindeya's (1995) who found no difference in OC under and outside the canopy of *B. aegyptiaca* sp. as reported in the present study.

Also, total nitrogen content (%) of the topsoil did not show significant differences among the three zones (Table 5). The highest was observed near the canopy zone showing an increase of 8.2% over the control, that is, further away from the tree. Low N under the canopy may be due to low organic matter and rapid decomposition of already low OC that release N which were tapped by the trees roots or might have leached down the soil profile. In on-farm field experiment in Niger under the *F. albida* canopy, nitrogen availability under trees was estimated to be more than 200% higher than in the open field causing a 26% production increase (Elseed et al., 2002).

The concentration of available phosphorus was 2.29 ppm at D1, 0.99 ppm at D2 and 0.7 ppm at D3, and was statistically insignificant (Table 5). The highest concentration of available phosphorus under the canopy of *B. aegyptiaca* may be due to the release of available organic P during the decomposition of animal manure in addition to low native organic matter. In the semi-arid environment, livestock like cattle, sheep, goats and chicken etc hide under trees to avoid the scorching sun. The livestock litter and urinate whilst resting under the trees. The concentration of exchangeable potassium in the soil was higher under the canopy (257.55 ppm,) than far from the canopy 239.81 ppm (Table 5). However, exchangeable K was not significant ($p > 0.05$). The same pattern was noted for EC and CEC.

As regards to soil physical properties, percent clay of the four sites was higher under canopy (38.37%) was significantly different ($p < 0.05$) than further away from the canopy (34.11%). This agrees with the study of Pandey et al. (2000). In the Indian subcontinent they noted that under *Acacia nilotica,* canopy sand particles declined by 10 and 9% whereas clay particles increased by 14 and 10% respectively. Silt percent did not show significant difference between the treatments, but decreased 14.6% under canopy than far from canopy (Table 5).

Bulk density and soil moisture content

The moisture content at the beginning of rainy season when crops are sown was significantly lower than that at flowering stage two months later. The moisture content at both periods showed a gradual decrease towards the trees. However, sorghum yields were unaffected by the differential decrease in moisture levels (Tables 6). Findings of Sursh and Rao (1998) substantiated our observation by reporting that soil moisture content at different growth stages of sorghum was lower under *A. lebbeck* at all depths, indicating rapid depletion of soil moisture under this species. Sorghum growth under

Table 6. Effect of *Balanites aegyptiaca tree* on soil moisture and bulk density.

Parameter	Under canopy (D1)	Near canopy (D2)	Far from canopy (D3)
Soil moisture (%)			
At sowing	14.36^{b} ± 0.79	15.87^{ab} ± 0.35	16.32^{a} ± 0.42
After 2 months	21.08^{a} ± 2.58	24.40 ^{a} ± 0.83	26.61^{a} ± 3.20
Bulk density (gm/cm^{3})	1.36^{a} ± 0.05	1.24^{a} ±0.1	1.36^{a} ±0.06

Rows with the same superscript letters are not significantly different at $P < 0.05$.

Table 7. Effects of *Balanites aegyptiaca* on sorghum plant height, grain and biomass yields at Limat's field and potted soils from the same site.

Type of experiment	Parameter	Under canopy (D1)	Near canopy (D2)	Far from canopy (D3)
Field	Plant height	1.20^{a} ±0.05	1.36^{a}±0.06	1.27^{a} ±0.04
	Grain yield	172.80^{a} ±32.65	210.3^{a} ±36.77	221.1^{a} ±38.40
	Biomass yield	464.34^{a}±49.52	508.94^{a} ±64.86	492.86^{a}±72.94
Pot	Plant height	0.65^{a} ±0.03	0.58^{a}±0.04	0.63^{a} ±0.03
	Grain yield	135.4 ^{a}±26.55	117.3 ^{a}±30.8	159.8 ^{a}±27.4
	Biomass yield	481.5 ^{a}±84.9	403.55 ^{a}±90.25	434.6 ^{a}±53.25

Rows with the same superscript letters are not significantly different at $P < 0.05$

Table 8. Effect of *Balanites aegyptiaca tree* on sorghum plant height (m), grain and biomass weight (gm^{-2}) on farm field at Goblel .

Type of experiment	Parameter	Under the canopy (D1)	Near the canopy (D2)	Far from the canopy (D3)
Field	Plant height	1.10^{a} ± 0.05	1.09^{a} ± 0.02	1.09^{a} ± 0.03
	Grain yield	202.55^{a} ± 32.4	135.64^{a} ± 21.04	195.01^{a} ± 22.55
	Biomass weight	414.70^{a} ± 44.03	385.85^{a}±75.45	377.53^{a}±47.49
Pot	Plant height	0.60^{a} ±0.04	0.64^{a} ±0.04	0.57^{a} ±0.02
	Grain yield	134.6 ^{a} ±19.25	186.5 ^{a} ±41.10	149.15 ^{a} ±45.35
	Biomass weight	436.05 ^{a} ±53.7	485.8 ^{a} ±67.15	352.7 ^{a} ±15.85

Rows with the same superscript letters are not significantly different at $P < 0.05$.

A.ferruginea faced similar effects but of a lower magnitude compared to *A. lebbeck* (Sursh and Rao, 1998).

There was no significance difference in bulk density among the three zones, probably because of lack of differences in organic carbon levels (Tables 5 and 6). It is well known that incorporation of organic matter in soil improves physical (aggregate stability, bulk density, water retention, etc.) and biological properties (nutrients availability, cation exchange capacity, reduction of toxic elements etc.) of soils (Lal, 1989).

Effects of *B. aegyptiaca* on sorghum plant height (m), grain and biomass yields (gm^{-2}) at Limat, Goblel, Korbebit and Endakashe sites

Sorghum height, grain and biomass yields were not significantly different among the three zones (Tables 7, 8, 9, 10). Sorghum yields at Limat were 172.80 gm^{-2} at D1 (under canopy) and 221.1 gm^{-2} at D3 (further from canopy), that is, a decrease of 22% relative to D3. However, at Goblel, there was an increase of 3.87 and 9.85% in grain and biomass yields, respectively under the canopy. At Korbebit, there was an increase of 16.8% in plant height and decrease of 9% in biomass yields under canopy. Endakashe experienced an increase of 4.6% in plant height, a decrease of 2.0% in grain and 6.82% in biomass yields. Sorghum plant height, grain and biomass yields for the pot experiment at Limat are presented in Table 7. Evidently, there is no difference in sorghum performance, that is, height, biomass and grain yields when all the four sites were combined (data not shown). The patterns of sorghum plant height, grain and biomass yields in the field and the pots experiments were comparable in the four sites (Tables 7, 8, 9, 10). It is suspected that shading and nutrients competition may be

Table 9. Effects of *Balanites aegyptiaca tree* sorghum plant height (m), grain and biomass weight (gm^{-2}) on farm field at Korbebit .

Type of experiment	Parameter	Under the canopy (D1)	Near to the canopy (D2)	Far from the canopy (D3)
Field	Plant height	$1.46^a \pm 0.06$	$1.33^a \pm 0.05$	$1.25^a \pm 0.1$
	Grain yield	$210.25^a \pm 24.79$	$195.22^a \pm 25.23$	$210.81^a \pm 22.37$
	Biomass weight	$408.22^a \pm 28.66$	$404.88^a \pm 57.96$	$445.32^a \pm 31.98$
Pot	Plant height	$0.65^a \pm 0.04$	$0.66^a \pm 0.05$	$0.64^a \pm 0.06$
	Grain yield	$184.8^a \pm 26.7$	$163.55^a \pm 29.65$	$160.4^a \pm 23.35$
	Biomass weight	$508.75^a \pm 34.95$	$572.5^a \pm 42.4$	$487.7^a \pm 21.65$

Rows with the same superscript letters are not significantly different at $P < 0.05$.

Table 10. Effects of *Balanites aegyptiaca tree* on sorghum plant height (m), grain and biomass yields (gm^{-2}) on farm field at Endakeshe.

Type of experiment	Parameter	Under the canopy (D1)	Near to the canopy (D2)	Far from the canopy (D3)
Field	Plant height	$1.14^a \pm 0.17$	$1.13^a \pm 0.05$	$1.09^a \pm 0.06$
	Grain yield	$117.48^a \pm 10.46$	$107.39^a \pm 16.04$	$119.88^a \pm 11.17$
	Biomass weight	$268.24^a \pm 21.23$	$242.03^a \pm 29.63$	$287.86^a \pm 14.32$
Pot	Plant height	$0.55^a \pm 0.03$	$0.57^a \pm 0.06$	$0.56^a \pm 0.04$
	Grain yield	$36.95^a \pm 5.3$	$32.7^a \pm 5.90$	$32.1^a \pm 4.65$
	Biomass weight	$562.3^a \pm 103.05$	$533.75^a \pm 64.4$	$597.3^a \pm 110.$

Rows with the same superscript letters are not significantly different at $P < 0.05$.

the primary factors that are responsible for the observed yield reduction in the fields (Yamoah et al., 1986; Yamoah, 1991).

CONCLUSIONS AND RECOMMENDATIONS

The study concluded that there were no significant differences in total N, EC, CEC, exchangeable K and organic carbon among the three zones. However, available P on Limat site, pH at Goblel and Korbebite were significantly different. Apart from the Limat site, clay, silt and bulk density were not significant. *B. aegyptiaca* depletes soil moisture under canopy during the first two months, however, sorghum performance was not significantly affected. Indeed, *B. aegyptiaca* did not affect plant height, grain and biomass yields of sorghum. It is recommended to retain and protect *B. aegyptiaca* trees on sorghum farm fields in parkland agroforestry systems so as to attract multiple benefits namely additional incomes from the sale of wood and improved sorghum yields. Furthermore, farmers are advised to maintain *S. sesban* which is native to the highland alongside *B. aegyptiaca* on their farm fields of Ethiopia in order to capture the multiple benefits of agroforestry.

REFERENCES

Abebe Y, Fisseha I, Olson M (2001). Contribution of Indigenous tree to soil Properties: The case of scattered trees of *Cordia Africana Lam.* In crop land of western Oromia. Ethiopian J. Natural Res. 3(2): 245-270.

Bationo A, Christianson CB, Klaij MC (1993). The effect of crop residue and fertilizer on pearl millet yields in Niger. Fertilizer Res. 34: 251-258.

Blake GR, Hartge KH (1986). Bulk Density, Methods of Soil Analysis, Part 1, Physical and Mineralogical Methods. 2nd edn., Soil Science Society of America, Madison, WI, pp: 363–367.

EARO (2002). An Assessment of the agricultural Production base, technological Packages and innovation strategies for commercial farmers in Kafta Humera wereda of Tigray Regional State Ethiopia.

Elseed AM (2002). Nutritive evaluation of some fodder tree species during the dry season in Central Sudan Asian-Australasian. J. Anim. Sci., 15(6): 844-850.

Gee GW, Bauder JW (1982). Particle Size Analysis in Methods of Soil Analysis. Chemical and Microbiological Properties.Agronomy No.9.2nd Edition. Soil Sci. Soc. Am. Medison Wiskonsin, USA. pp. 383-411

Houba VGJ, Vander L, Novazamsky I, Walinga I (1989). Plant and Soil Analysis procedures. Department of Soil Science and Plant Nutrition Agric. Univ. Wageningen, the Netherlands.

Jackson ML (1958). Soil Chemical Analysis. Prentice; Halls, Inc., Englewood cliffs, New Jorsey. Sixth printing. p. 498.

Jama B, Zeila A (2005). Agroforestry in the dry lands of eastern Africa: A call to action. ICRAF Working Paper – no. 1. Nairobi: World Agroforestry.

Jonsson K (1995). Agroforestry in dry savannah areas in Africa: interaction between trees ,soils and corps. dissertation bound with a collection of 4 prints summaries. p. 31.

Juo ASR, Franzluebbers K (2003). Tropical Soils, Properties and Management for Sustainable Agriculture. Oxford Press. pp. 281.

Kindeya G (1995). Scope for improving land productivity in northern Ethiopia using native tree species Msc. Thesis University Collage of Northern Walas, Bangor, United Kingdom.

Kindeya G (2004). The development of Agro forestry in the dry lands of

Ethiopia. Paper presented in the dry lands agroforestry Workshop,1-3 September 2004, Nairobi, Kenya. pp. 1-45.
Lal R (1989). Agroforestry systems and soil surface management of a tropical Alfisol. Agrofor. Syst. 8: 97-111
Marx ES, Hart J and Stevens RG (1999). Soil test interpretation guide. Oregon State Extension Service. p. 7.
Mitiku H, Herweg K, Stillhardt B (2006). Sustainable Land Management: A new approach to Soil and water conservation in Ethiopia, Mekelle, Ethiopia. Land Resource Management and Environmental Protection Department, Mekelle university: Bern, Switzerland center for Development and Environment, University of Bern. and Swiss National center of competence in Research, NCCR) north-south. p. 20.
Olsen SR, Summer LE (1982). Methods of soil analysis; part 2. Chemical and microbiological properties of Phosphorus. ASA Monograph number. 9: 403-430.
Pandey CB, Singh AK, Sharma DK (2000). Soil properties under *Acacia nilotica* trees in a traditional agroforestry system in central India Agroforestry Syst. 49(1): 53-61.
Poschen P (1986). An of the *Acacia albida*- based agroforestry practices in the Hararghe highlands of eastern Ethiopia. Agroforestry Syst., 4: 129-143.
Sanchez PA, Couto W, Boul SW (1982). The fertility capability soil classification system: interpretation, applicability and modification. Geoderma 27: 283-309.
Tadesse H (1997). Effect of *Melletia ferruginea* on soil fertility and growth performance of maize. Msc. Thesis, SLU .Sweden.
Tesfay T (2004). Organic Inputs from Agroforestry Trees on Farms for Improving Soil Quality and Crop Productivity in Ethiopia. Doctoral thesis No.2005: 122.
Tian G, Bruusard L, Kang BT (1995). An index for assessing the quality of plant residues and evaluating their effects on soil and crop in the (sub-) humid tropics. Appl. Soil Ecol. 2: 25-32.
USDA Soil Survey Staff (2006). Keys to Soil Taxonomy, 10th ed. USDA-Natural Resources Conservation Service, Washington, DC.
Walkley A, Black C (1934). Examination of the Degtjareff method for determining soil organic matter and a proposed modification of the chromic acid titration method. Soil Sci., 37: 29-38.
Yamoah CF, Agboola AA, Wilson GF (1986). Nutrient contribution and maize performance in alley cropping systems Agroforestry Syst., 4(3): 255.
Yamoah CF (1991). Choosing suitable intercrops prior to pruning Sesbania in an alley cropping configuration. Agroforestry Syst., 13: 87-94.
Yeshanew A, Tekalign M, Olsson M (1999). Changes in soil chemical Properties under scattered *Croton macrostachyus* trees in the traditional agroforestry system in north-western Ethiopia. Ethiopian J. Natural Res., 1(2): 215-233.
Zebene A (2003). Tree species Diversity, Topsoil Conditions and Arbuscular Mycorhizaal Association in the Sidama traditional agroforestry land Use, Southern Ethiopia. Ph.D. Thesis SLU, Department of forest management and products, Uppsala, Sweden.

The physical and chemical characteristics of soils of Northern Kenya Aridlands: Opportunity for sustainable agricultural production

E. M. Muya[1]*, S. Obanyi[1], M. Ngutu[2], I. V. Sijali[1], M. Okoti[2], P. M. Maingi[1] and H. Bulle[2]

[1]National Agricultural Research Laboratories, P.O. Box 14733, Nairobi, Kenya.
[2]Kenya Agricultural Research Institute (KARI) P.O. Box 147, Marsabit, Kenya.

Biophysical characterization was carried out in the mountain and oasis areas within the Northern Kenya Arid Lands with a view of identifying, soil constraints and opportunities for sustainable agricultural production in the area. The soil aspects were studied through desk-top analysis of the existing databases and collection of secondary data at regional scale, site evaluation surveys at site level and detailed soil survey at farm level. Based on biophysical data, the mountain and oasis area of the region was divided into three major eco-zones, namely (1) upper regions: mountains, hills and uplands, (2) middle level: footslopes and (3) low-lying areas: riverine, plains and bottomlands, which were found to occupy 20, 5 and 54% of the total land area of Kenya Arid and Semi-Arid Lands (KASALs) respectively. In these areas, soil structural degradation has taken place at varied rates through pulverization in the upper regions, compaction in the middle level and dispersion in the low-lying areas. The mean productivity index for the upper zone, middle slopes and the lowest zone was found to be 18.5, 19.6 and 1.3%, the most limiting factors being high acidity, increased compaction and high sodicity/salinity respectively. The opportunities for sustainable agriculture was found to be elimination of acidity and increased water saving for supplementary irrigation in the upper zone; harnessing run-off water and improving water holding capacity through subsoiling on the footslope; and precision and market oriented irrigated farming for improved water use efficiency in the lowest zone.

Key words: Biophysical characterization, soil quality and productivity index.

INTRODUCTION

The challenge to any project with an objective to improve the productivity of the area is to have baseline on land productivity and identify soil-related constraints in different zones or ecosystems. This can be the basis of formulating appropriate intervention packages and identifying areas with reasonable agricultural potentials for the selected areas. In most of these ecosystems, efficient use of limited water and nutrient resources is required for both livestock and crop production. In this context, efforts for improving the productivity through sustainable approaches will bear fruits if they are accompanied by participatory methods and initiatives to improve the capacity of the pastoral communities to sustain agricultural production in these fragile and resource limited environments. This is because these communities have a pastoral background with limited farming skills and knowledge to face the challenges of sustainable agriculture – for example crop diversification, use of locally available resources (for example organic inputs), water harvesting techniques and conservation.

In addition, such research efforts should also come up with tangible outputs and data on environment-agriculture interactions, which has been one of the major impediments in formulating appropriate land use and environmental policy for dry lands. Therefore, land resources inventory through systematic characterization of the area using appropriate soil quality indicators is one of the major steps required to trigger of the agricultural development processes. In this regards, stepwise and hierarchical approaches are required at different scales of

*Corresponding author. Email: edwardmuya@yahoo.com.

Table 1. Climatic zones of drylands.

Zones	r/Eo (%)	r mm	Eo mm	Climate designation
III	50 to 55	900 to 960	1750 to 1800	Semi-humid
IV	40 to 50	750 to 900	1800 to 1880	Semi-humid to semi-arid
VI	25 to 40	525 to 750	1880 to 2095	Semi-arid
VII	15 to 25	320 to 525	2095 to 2150	Arid
VIII	<15	170 to 320	2150 to 2280	Very arid

r=rainfall, Eo=potential evapotranspiration. Source: Van Kekem (1986).

biophysical data collection in order to capture adequate information which are commensurate with different levels of determinations. This can provide decision support tools in formulating the appropriate intervention packages to be implemented at the farm level. Against this background, the objective of carrying out site characterization was to identify, prioritize and document key constraints with a view of formulating the most promising management options to be tested and up scaled in broad geographical scales.

MATERIALS AND METHODS

Characterization of the soils of Northern Kenya Arid Lands focused on the mountain and oasis areas, which are unique micro-climatic ecosystems that can support arable farming at subsistence or small-scale, commercial level. The areas covered were those in which high degree of land degradation has taken place, thereby impacting negatively on the quality of land resources. The study sites are located in very arid and semi-arid lands, with annual rainfall varying from 170 to 750 mm. In this area low soil quality is one of the main causes of poor crop performance at farm level. The decrease in vegetation cover and diversity towards settled areas indicates that settlement had impacted negatively on environmental quality and biodiversity, hence loss in ecosystem functions, low biophysical sustainability and poor crop performance (Muya et al., 2008). The study area is located in a typical dryland with arid to very arid climate (Table 1). These areas occurred mainly in Marsabit, Wajir North, Garissa, Turkana and Ijara districts.

Biophysical site characterization was carried out at three levels, namely regional, site and farm level. At the regional scale, description and maping of the major areas of Kenya Arid and Semi-Aridlands (KASAL), were done by clipping on the exploratory soil maps (Sombroek et al., 1981) to indicate major soils of the region. This was to provide an overview of the possible areas to be up scaled and the potential impacts of the developed technologies. The type and quantity of data collected depended on the scale. At the regional scale, the secondary data were studied to gather biophysical information from the existing databases. The biophysical information from the secondary data included climatic characteristics of the study area, geomorphic processes, the degree of land degradation and general soil conditions. At the site level, the time and space dynamic soil degradation processes such as aggregate collapse, compaction, pulverization, salinization, sodification and nutrient depletion were assessed. At the farm level, detailed soil and topographical soil survey were carried on the basis of physical, hydraulic and chemical characteristics of soil.

The chemical characteristics of soils were used to determine their productivity, based on the thresholds values indicated in Table 2, after Aune and Lal (1997) and Kamoni and Wanjogu, (2006). The response functions applied were regressed relationships between each of the soil quality attributes selected and yield of maize. The soil quality attributes selected were those with known influence on crop growth such as nitrogen, phosphorus, potassium and organic carbon. The productivity index of different soils was calculated by the method provided by Aune and Lal (1997). Base on this method, the critical limits were derived and applied in developing the productivity index. According to Aune and Lal (1997), the critical limit is defined as the numerical value of soil quality where crop yield is 80% of the maximum. The productivity index of a given soil mapping unit (PI) was calculated by the following equation:

$$PI = SQI_1 + SQI_2 + SQ_3 + SQ_4 + SQ_5 + \ldots\ldots$$

Where:

PI = Productivity index

SQI_1, SQI_2, SQI_3, SQI_4, SQI_5 is the relative yield read off from the yield response curves under specified soil quality indicator.

RESULTS AND DISCUSSIONS

Climatic characteristics and status of land degradation

The results showed that the area has three major ecological zones, namely mountains, hills and uplands; foot slopes; and plains, riverine and bottomlands. The three zones have different rainfall regimes and degree of land degradation (Table 3).In the mountain, hills and uplands, most land forms are as a result of geological and geomorphological processes such as faulting, folding, volcanic activities and erosion. The mountains have a relief intensity of 300 to 1000 m and slopes ranging from 8 to 30%. The mountains are important water catchment areas. The volcanic mountains consist of various volcanic rocks, and the dominating volcanic mountains are mount Kulal in the North-East and mount Marsabit in the East. They rise to 1740 and 2230 m above the sea level respectively.

Hills have relief intensity varying from 50 to 300 m, with slopes ranging from 8 to 30%, while uplands have relief intensity varying from 10 to 100 m. On the mountain tops, the soils are deep to very deep. They are reddish brown to brown clay. In some areas the soils are stony and gravelly. In places the soils are compact between 0 and 20 cm depth. The footridges on the southern slopes of mount Marsabit have moderately deep to deep clay loam to clay soils with favorable physical and chemical properties.

Table 2. Soil quality attributes and their threshold levels.

Indicators	Threshold values	Source
K me/100g	0.2 to1.5	Kamoni and Wanjogu (2006)
K me/100g	0.83	Aune and Lal (1997)
N%	0.2	Kamoni and Wanjogu (2006)
C%	2 to 4	(Kamoni and Wanjogu (2006)
C%	1.08	Aune and Lala (1997)
P ppm	20 to 80	Kamoni and Wanjogu (2006)
P ppm	7.6	Aune and Lal (1997)
Cation exchange capacity	15 to 25	Kamoni and Wanjogu (2006)
pH	5.5 to 7.0	"
Electrical conductivity (EC)	4.0 mS/cm	"
Ca me/100g	2.0 to 10.0	"
Mg me%	1.0 to 3.0	"
Mn me%	0.1 to 2.0	'

Table 3. Climatic characteristics and status of land degradation.

Eco-zones	% Coverage	Altitude (m a.s.l.)	Annual rainfall (mm)	Annual temperature in degree C	Relative degree of land degradation
Mountains, hills and uplands	20	1,500 to 2,000	500 to 800	17 to 20	High
Footslopes	5	1,000 to1,500	400 to 750	27 to 29	Moderate
Riverine, plains and bottomlands	54	<1000	150 to 600	28 to 31	Very high

The footridges of mount Kulal have association of three soil types, depending on topographical positions. On the lower slopes are shallow to moderately deep calcareous soils with texture of stony, gravelly, sandy clay loam to sandy clay. At the middle upper slopes, the soils are clay loam. On the lowest slopes, the soils are brown, friable clay and in some areas they are stony and shallow.

Physical and chemical characteristics of soils

The physical and chemical characteristics of soils in these landforms are indicated in Tables 4 and 5 respectively. Over 50% of these soils were found to be clay and soil moisture holding capacity tends to increase with the clay content. The bulk density generally increased with the soil depth. The hydraulic conductivity followed a similar trend. A good number of soils in the mountains, hills and uplands are brown to red, pulverized clay loam to clay at the depth of 0 to 20 cm. Increased bulk density with the soil depth explained increased compaction, hence reduction in macrospores. However, the general increase in hydraulic conductivity was explained by the general increase in microspores (due to increase in clay content) which provide flow pathways in saturated conditions under which hydraulic conductivity is determined.

The soils in most of the sites surveyed are acidic (pH<6.5), hence high possibility of aluminium and other heavy metal toxicity. The nitrogen level is lower than the critical limit (0.2%) in all the sites except site No 10. Similarly, organic carbon level in all the sites is less than the critical limit of 2% except for the site No 5. From these observations, it can be established the soil fertility status of the mountains, hills and uplands is generally low. However, phosphorous level is above the critical limit (20 ppm) for all the sites except for the site No 9. But its availability to plants, however, could be impeded through fixation by aluminium ions under acid conditions, rendering it unavailable to plants. The highly pulverized topsoils are also an impediment to soil moisture availability.

In this eco-region, the opportunity for sustainable agriculture lies on improved tillage tillage techniques, saving the spring water originating from the steep slopes for supplementary irrigation and eliminating aluminium and other heavy metal toxicity and releasing fixed phosphorous for improved nutrient availability. The footslopes are developed from colluvium from crystalline limestone. They are well drained, very deep, dark reddish brown, firm, moderately to strongly calcareous, clay. The soils have very compact subsoils with relatively high bulk density (>1.2 g/cc) and low available water holding capacity (Table 6). This implies high volume of run-off that causes

Table 4. Physical and hydraulic properties of different soils.

Observation site No	Soil depth (cm)	Texture	Bulk density (g/cc)	Hydraulic conductivity (cm/h)	Water holding capacity (mm)
	0 to 20	Sandy clay to clay	1.05 to 1.10	25 to 50	25
1	20 to 100	Clay	1.20 to 1.22	5.0 to10.0	80
	80 to 100	Clay	1.31 to 1.41	8.0 to15.0	100
	0 to 20	Clay loam to clay	1.00 to1.10	20.0 to 55.0	18
2	20 to 80	Clay	1.25 to 1.30	2.0 to 5.0	75
	80 to 100	Clay	1.22 to 1.30	12.0 to21.0	17
	0 to 20	Clay	1.00 to 1.10	12.0 to18.0	22
3	20 to 80	Clay	1.20 to 1.28	23.0 to 33.0	68
	80 to 100	Clay	1.21 to 1.27	20.0 to 34.0	25
	0 to 20	Silty clay loam/clay	1.10 to 1.15	16.0 to 27.0	18
4	20 to 80	Clay	1.28 to 1,30	1.0 to 5.0	10
	80 to 100	Silty clay to clay	1.33 to 1.44	2.0 to 8.0	70
	0 to 20	Clay	0.9 to 1.05	22.0 to 25.0	16
5	20 to 80	Clay	1.5 to 1.55	0.5 to 5.5	11
	100	Sandy clay/clay	1.23 to 1.24	10.0 to 15.7	87
	0-20	Silty clay loam	0.80 to 095	14.8 to 21.9	22
6	20-80	Clay loam to clay	1.1 to1.15	0.56 to 1.9	70
	80-100	Clay	1.33 to1.44	5.8 to12.5	17
7	0-20	Clay loam to clay	1.11 to 1.18	29.2 to 32'8	19
8	20-80	Clay	1.12 to 1.28	6.7 to14.8	77
	100	Clay	1.05 to 1.10	12.5 to 18.6	56
	0-20	Clay	0.88 to 1.00	35.0-50.5	14
9	20-80	Clay	1.36 to 1.44	0.28-5.6	68
	80-100	Clay	1.24 to1.28	10.5-17.6	12
	0-20	Silty clay	1.1 to 1.18	23.6-24.5	20
10	20-80	Clay	1.10 to1.21	15.8-19.6	28
	80-100	Clay	1.10 to1.20	33.6-43.8	40

Table 5. Chemical properties of soils.

Observation site No	pH	N%	C%	Pppm	K m.e%	Productivity index (PI) %
1	6.1	0.02	1.95	62	1.29	9.75
2	5.7	0.07	1.71	63	0.68	13.46
3	5.9	0.01	1.43	30	0.81	1.93
4	5.8	0.02	1.88	33	0.93	5.82
5	5.9	0.02	2.15	50	1.47	10.00
6	5.0	0.02	1.08	30	1.64	47.5
7	5.0	0.21	1.05	36	1.54	47.5
8	6.0	0.18	0.92	66	2.00	41.4
9	5.0	0.18	1.59	2	2.04	3.47
10	6.0	0.36	1.65	84	3.29	14.85
11	5.5	0.1	0.88	28	2.1	22.00
12	5.9	0.022	0.67	20	1.87	3.69
						Mean=18.45

Table 6. Physical and hydraulic properties of soils of the footslopes.

Observation No	Soil depth (cm)	% 2 to 5 mm stable aggregates	Texture	Bulk density (g/cc)	Hydraulic conductivity (cm/day)	Water holding capacity (mm)
	0 to 20	5.0	Sandy clay/clay	1.34	3.5	32
	20 to 80	3.0	Clay	1.39	2.1	23
	80 to 100	15.0	Clay	1.27	5.6	36
	0 to20	65.0	Clay loam/clay	1.10	26	33
1	20 to 80	2.0	Clay	1.42	0.5	12
	80 to 100	10.0	Clay	1.22	12.5	45
2	0 to 20	2.0	Clay	1.45	1.2	11
	20 to80	2.5	Clay	1.34	1.5	24
	80 to100	5.0	Clay	1.28	3.6	56
	0 to 20	15.0	Sandy clay loam	1.00	22.5	38
3	20 to 80	20.0	Clay loam	1.05	13.7	55.9
	80 to 100	28.0	Clay	1.13	15.8	25.8
	0 to 20	48.6	Sandy loam/sandy clay	1.09	29.7	16.1
4	20 to 80	43.4	Sandy clay loam	1.08	18.9	45.7
	80 to100	36.8	Clay loam/clay	1.12	30.8	27.6

Table 7. Fertility gradients across the farmers' field.

Topographical position	Nutrient levels			
	% C	% N	P ppm	K me%
Topmost slope	0.54	0.11	14.5	1.34
Middle slope	1.24	0.23	5.8	1.08
Lowest slope	2.15	0.31	19.3	0.76

land degradation. Organic carbon and nitrogen is generally above the critical levels, except for phosphorous which is less than 20 ppm (Table 6). The dominantly compact soils occurring in the footslopes have physical and hydraulic properties that result into poor soil moisture regimes.

The sharp contrast in soil aggregate stability between the topsoils and subsoils explains the degree to whichsoil aggregate formation and stabilization has been adversely affected in these areas (Table 6). As shown in Table 7, the dynamics of soil nutrients are indicated by the differences in nutrient levels across the slopes, which exhibit an increasing trend from the topmost slope, through the middle slope to the lowest. Soil organic carbon and nitrogen increase, while P reduces at the middle slope and then increases. On the contrary to these general trends, K decreases with the slope. Productivity index (PI) for the topmost slope, middle slope and lowest slope is 10.77, 16.9 and 31.00 respectively. The mean PI is 19.55 and the major limiting factor in the eco-zone is high compaction of the subsoil resulting into inappropriate partitioning of the rainwater into soil water storage and run-off, the former being much less than the latter. Therefore, the opportunity for sustainable agriculture lies on the improvement of soil water uptake and storage as well as harnessing the run-off for improved water use efficiency.

Riverines, plains and bottomlands occur generally in lower lying area with an altitude ranging from 500 to 1000 m above the sea level. The soils vary, as influenced by the soil forming processes. On the riverine ecosystem, the soils are developed on sediments derived from various sources (recent flood plians). In some sites (e.g. Garissa), they are developed from fluvatile sediments that receive fresh materials at regular time intervals. The soils are well drained to imperfectly drained, very deep, dark brown to yellowish brown, stratified, non to strongly calcareous, loamy sand to silty clay; in places firm and

Table 8. Physical and hydraulic characteristics of soils.

Soil depth (cm)	% 2 to 5 mm stable aggregates	Texture	Bulk density (g/cc)	Hydraulic conductivity (cm/h)	Water holding capacity (mm)
0 to 20	10.5	Silty clay	1.05	48.9	34
20 to 50	4.8	Sandy clay loam	1.01	55.7	22
50 to 80	2.6	Loamy sand	1.09	70.8	25
100 to120	3.2	Sandy clay loam	1.10	66.7	24
120to 150	11.8	Clay	1.07	45.7	30
120 to150	7.7	Silty clay	1.05	46.7	35

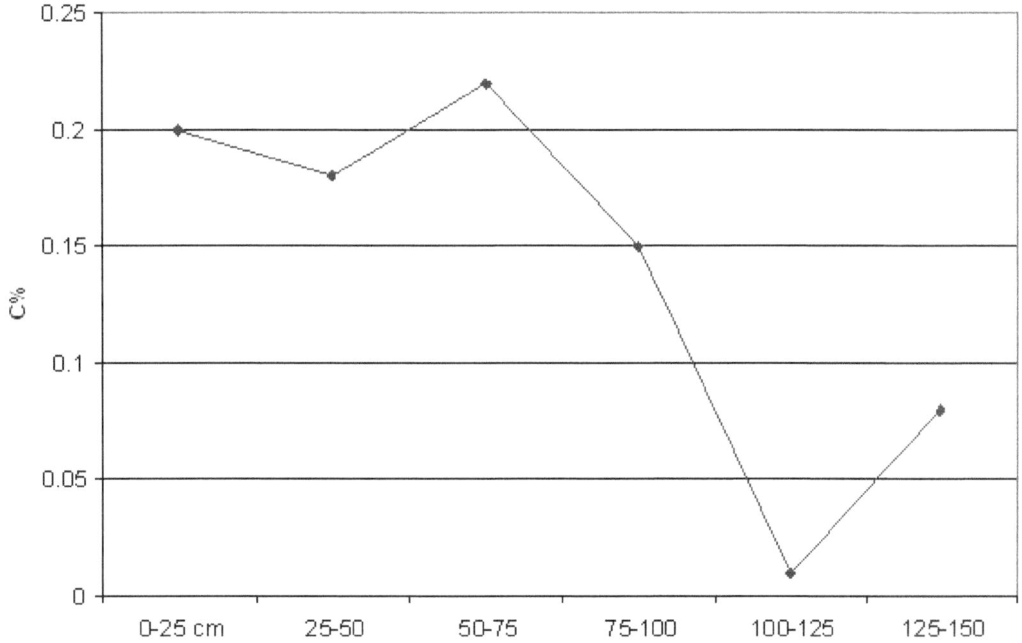

Figure 1. Distribution of organic carbon with depth.

abruptly overlying dense clay. Most of these soils have very good physical conditions being highly friable, permeable and reasonably high available water holding capacity. However, the degree of aggregation is very low and this explains relatively low percentage of aggregates with size ranging from 2 to 5 mm in diameter (Table 8). These soils are stratified with irregular distribution of organic carbon across the soil depth (Figure 1).

The typical physiographical processes of riverine ecosystems were observed in Shantole scheme as evidenced by the formation of the physiographic features such as river levees on the upper part, grading down gently to the lowest slopes, which form the backswamps. These soils have been formed through differential transportation and deposition, resulting into vertical and horizontal stratification, hence the existence of the profile and spatial textural gradients. The soils in this area were found to be in a sequence that reflected these processes (Tables 9, 10 and 11). These soils were found to be sand to loamy sand on the upper top slopes of the river levees; sandy loam to sandy clay loam on the middle slopes; and clay loam to clay on the lower middle slopes and silty clay to clay on the lowest slopes. The quality and productivity of most of these soils are lower than the optimum values because of low organic matter, adverse nutrient rations and high salinity. Based on organic carbon, salinity and ESP as the most limiting factors, productivity index of the levees, middle slope and backsawmp was found to be only about 0.5, 2.5 and 3.0% respectively. In general, the mean value of the productivity index for the area is 1.25%.

The soil pH is generally high and tending to be too alkaline (>7.0) in all the physiographical locations. According to Landon (1984), adverse chemical conditions associated with high soil pH include toxic environment due to increased availability of heavy metals, reduced

Table 9. Soils on the top slopes of levees under fruit trees.

Soil characteristics	Values under in different soil depth (cm)		
Soil depth	0 to 20	20 to 60	60 to150
Soil pH-H_2O (1:2.5)	8.9	9.1	9.4
Electrical conductivity	8.0	8.0	5.5
Carbon%	0.1	0.1	0.1
Texture	Sandy loam	Loamy sand	Loamy sand
Cation exchange capacity me/100g	6.7	4.5	4.2
Calcium me/100g	12.7	3.4	7.3
Magnesium me/100g	0.6	0.3	0.4
Potassium me/100g	1.5	0.5	0.3
Sodium me/100g	1.8	1.1	0.9
Sum me/100g	16.6	5.3	8.9
Exchangeable sodium percentage	10.8	20.8	10.1

Table 10. Soil on the middle slopes under maize and beans.

Soil characteristics	Values under in different soil depth (cm)		
Soil depth	0 to 20	20 to 60	60 to 150
Soil pH-H_2O (1:2.5)	7.6	9.0	8.9
Electrical conductivity	26.0	26.0	0.3
Carbon%	0.5	0.2	0.2
Texture	Sandy loam	Sandy clay loam	Sandy clay loam
Cation exchange capacity me/100g	12.9	11.3	11.4
Calcium me/100g	8.1	8.2	11.8
Magnesium me/100g	0.7	0.7	0.6
Potassium me/100g	1.4	0.4	0.6
Sodium me/100g	1.7	1.7	1.2
Sum me/100g	11.9	11.0	14.2
Exchangeable sodium percentage	14.3	15.5	8.5

Table 11. Soil on the lowest slopes.

Soil characteristics	Values under in different soil depth (cm)		
Soil depth	0 to 20	20 to 40	40 to 60
Soil pH-H_2O (1:2.5)	7.7	7.8	7.7
Electrical conductivity	1.5	2.8	3.5
Carbon%	0.6	0.3	0.3
Texture	Clay loam	Clay	Clay loam
CEC me/100g	17.3	17.4	9.4
Calcium me/100g	4.7	31.0	7.7
Magnesium me/100g	0.4	1.9	0.7
Potassium me/100g	0.9	2.3	1.9
Sodium me/100g	0.8	3.9	2.1
Sum me/100g	6.8	39.1	12.4
ESP	11.8	9.97	16.94

availability of some of the important micronutrients and nutrient imbalances. The electrical conductivity is high on the levees and middle slope, the threshold value being 4.0 mmhos/cm. The values in the middle slope are extreme at the depth of 0 to 60 cm being 26.0 mmhos/cm. Exchangeable sodium percentage is higher than the threshold value (6) for all the physiographic positions. According to Withers and Vipond (1974) ESP value of more than 6 causes soil structural degradation through clay dispersion.

The cation exchange capacity is lower than the threshold value (24 me/100 g) for all the physiographic positions.

Conclusions and Recommendations

The three major eco-zones identified in the mountains and oasis areas of Kenya arid and semi-arid lands were mountains, hills and uplands in the upper zones, footslopes in the middle-level zone; and riverine, plains and bottomlands in the lowest zone. The soils in the upper zones are reddish brown to red moderately deep to deep clay loam to clay, in places stony and gravely, sandy clay loam to sandy clay. The clay content generally increases with the depth. The bulk density varies from 0.9 to 1.44 g/cc, while the available moisture holding capacity is from 5 to 100 mm. The hydraulic conductivity is in the order of 0.28 to 43 cm/h. The mean productivity index is 19%. The major soil-related constraints are high acidity and pulverized topsoils. The opportunity for sustainable agriculture lies on the improvement of nutrient availability and water availability.

The soils of the middle slopes are well drained, very deep, dark reddish brown, firm, moderately to strongly calcareous, sandy clay loam to clay. The bulk density is generally very high (>1.2 g/cc), while the available water holding capacity varies from 12 to 56 mm. The hydraulic conductivity is generally low (<35 cm/h) and the mean productivity index is 19.55%. The major limiting factor is high compaction in the subsoil. The opportunity for sustainable agriculture lies on the improved water uptake and storage through subsoiling and harnessing run-off water for enhanced water use efficiency. The soils in the lowest zone are well drained to imperfectly drained, very deep, dark brown to yellowish brown, stratified, micaceous, non to strongly calcareous, sand to silty clay; in places firm and abruptly overlying dense clay. The bulk density is generally low (<1.2 g/cc) and the available water holding capacity ranges from 20 to 35 mm, while the hydraulic conductivity is in the order of 40 to 75 cm/h. The mean productivity index is 1.25%.

The major limiting factors are low organic matter content, high pH, high salinity and extremely high exchangeable sodium percentage. The opportunity for sustainable agriculture lies on the development of precision and market oriented irrigated farming for improved economic water use efficiency.

REFERENCES

Aune JB, Lal R (1997). Agricultural productivity and critical limits of properties of Oxisols and Alfisols. Trop. Agric. (Trinidad), 74(2): 96-102.

Van K (1986). The soils of Mount Kulal area. Reconnaissance soil survey report. Landon, J.R. 1984. Bookers Tropical Soil Manual, pp. 1010-1018.

Muya EM, Esilaba AO, Lelon JK, Gachini GN (2008). Combating land degradation in irrigated areas through systematic characterization of soils for improved irrigation efficiency. Proc. Drylands degradation and desertification conference, Ben-Gurio Unversity of the Negev, Isreal, 14th – 17th December.

Kamoni PT, Wanjogu SN (2006). Land evaluation processes. Miscellaneous paper. Withers, B. and Vipond, S. 1974. Irrigation design and practice.

Manure application setback effect on phosphorus and sediment in runoff

Ahmed Al-wadaey[1,6], S. Charles Wortmann[2*], A. Charles Shapiro[3], G. Thomas Franti[4] and E. Dean Eisenhauer[5]

[1]Department of Agronomy and Horticulture, University of Nebraska, Lincoln NE 68583-0915, USA.
[2]Department of Agronomy and Horticulture, University of Nebraska, 58C Filley Hall, Lincoln NE 68583-0951, 402 472 2902, USA.
[3]Department of Agronomy and Horticulture, University of Nebraska, 279 Plant Science, Lincoln NE 68583-0951, 402 584 3803, USA.
[4]Department of Biosystems Engineering, University of Nebraska, 242 Chase Hall, Lincoln NE 68583-0726, 402 472 9872, USA.
[5]Department of Biosystems Engineering, University of Nebraska, 238 Chase Hall, Lincoln NE 68583-0726, 402 472 1637, USA.
[6]ICARDA, P. O. Box 5466, Aleppo, Syrian Arab Republic.

Surface application of animal manure increases nutrients at the soil surface with increased potential for nutrient loss in runoff. Non-application setbacks are often required with the intent to reduce nutrient loss to surface water. The objective of this research was to determine the effect of setback distance on phosphorus and sediment in runoff. The research was conducted in eastern Nebraska on 24 ha of a terraced field with a 4 to 7% slope and predominantly Yutan silty clay loam Mollic Hapludalf soil. Experimental units were drainage areas served by risers for the tile-outlet terrace system. The seven treatments included no manure applied and setbacks with radii of 0, 5, 10, 20, 30, and 40 m; the setback distances were limited on the terrace side by the top of the terrace bund. Composted feedlot manure was applied at approximately 74 Mg ha^{-1} with 222 kg Pha^{-1} applied. Over a two-year period, the mean precipitation was 770 mm per year and six major runoff events occurred. Setback distance did not affect dissolved P, particulate P, total P, and sediment concentrations or losses except that sediment loss was greater by 51% with increased setback distance, confirming the value of manure in reducing soil erodibility. Sediment and P concentrations were related to time since the on-set of a runoff event and peak intensity of rainfall events. The results show that manure application setbacks around risers on fields protected with tile-outlet terraces are ineffective in reducing P and sediment runoff.

Key words: CAFO, compost, concentrated animal feeding operations, ISCO, terraces, water quality.

INTRODUCTION

Application of organic or inorganic phosphorus (P) to cropland can increase the risk of P loss to surface waters. Agricultural runoff is considered a major nonpoint P source of pollution for many lakes, rivers, estuaries, and coastal waters, often causing accelerated eutrophication (Carpenter et al., 1998). Transport of P from agricultural areas to surface waters is primarily by runoff and erosion, and P concentration may be affected by surface soil P content and the method, rate, and timing of fertilizer and manure P application (Sharpley et al., 1993). The loss of P to surface water may be reduced by management

*Corresponding author. E-mail: cwortman@unlnotes.edu

Abbreviations: CFM, composted feedlot manure; DP, dissolved P; PP, particulate P; TP, total P.

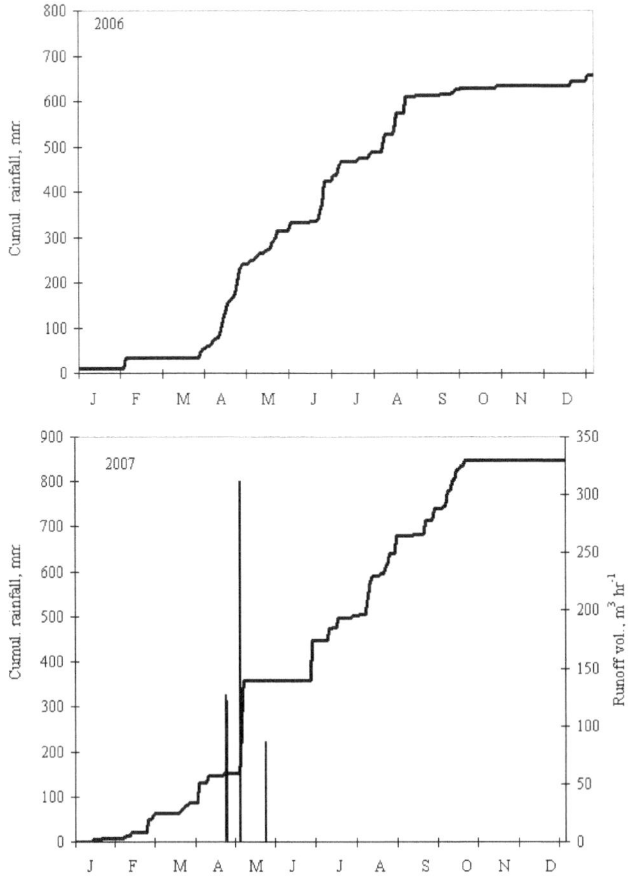

Figure 1. Cumulative rainfall for 2006 and 2007 and mean runoff volume (m^3 ha^{-1}), as indicated by the bars, of four runoff events collected using ISCO samplers in eastern Nebraska in 2007. Runoff volume was not recorded for runoff events in 2006.

practices such as non-application setbacks.

Setbacks, with or without perennial vegetation, are distances between pollutant sources and an aquatic ecosystem (National Research Council, 2000). The distance from the point of P application to concentrated flow can affect the risk of P reaching surface water (McDowell and Sharpley, 2002). A 30-m setback of manure application from surface waters and conduits (channels, open tile line-intake structures, sinkholes, and agricultural wellheads) to surface waters is required for regulated confined animal feeding operations (CAFOs) (USEPA, 2003) to minimize "the potential runoff of pathogens, hormones such as estrogen, and metals and to reduce nutrient and sediment runoff" (Federal Register page 3054). An 11-m setback in perennial vegetation is an alternative. In some situations, a CAFO may be able to demonstrate to the permitting authority that a setback or vegetative buffer can be reduced or is unnecessary.

Setback effectiveness for reducing runoff P has not been well evaluated in field situations. The setback concept may not adequately consider the value of manure application to enhanced formation of water-stable soil macro-aggregates and reduced runoff and erosion (Wortmann and Shapiro, 2008; Gilley and Risse, 2000). This effect can persist for several years after manure application (Wortmann and Walters, 2007). Withholding application of manure from setback areas means loss of this beneficial effect in the setback area. Liu et al. (1997) found that P applications did and did not significantly affect dissolved P (DP) concentrations in runoff for 10 and 5% slopes, respectively. Setbacks may be of little value on fields with well-designed and maintained terrace systems that effectively reduce soil erosion and enhance sedimentation within the field. Mickelson et al. (1998) reported that 20.1 m setbacks around a riser of a tile-outlet terrace did not reduce off-site herbicide transport, at least in part because the setback area is the ponding area where water may be over 1 m deep. Franti et al. (1998) reported that setbacks reduced atrazine and cyanazine losses only in proportion to the area not sprayed.

Setbacks can complicate manure application and soil fertility management. A 30-m setback from a point, such as a well or tile intake, can be 0.28 ha that is not available for manure application but still available for crop production (Koelsch, 2005). It is likely a producer will apply needed nutrients on the setback areas using fertilizer to maintain productivity and this may negate any possible positive effects of the setbacks for reducing P loss. The inconvenience of the setbacks may often causes producers to apply manure to other land that is not protected by tile-outlet terraces and with higher runoff P risk.

Research was conducted to evaluate the USEPA setback requirement for tile-outlet terraces by determining the effect of setback distance on runoff P and sediment loss following surface application of composted feedlot manure (CFM). The effects of rainfall intensity and time during runoff events on runoff P and sediment loss were evaluated.

MATERIALS AND METHODS

Site characteristics

The study was conducted on 24 ha (59 acre) of a field protected with underground-outlet terraces located northeast of Lincoln, NE, at 41°3'43"N and 96°29'51"E. The mean elevation was 374 m. The mean precipitation was 767 mm per year with 75 to 80% falling from April to September, often at intensities of > 25 mm per day, and occasionally, 125 to 150 mm per day (Figure 1). The slope was 4 to 7%. The soil was deep, moderately well-drained Yutan silty clay loam formed in loess (fine-silty, mixed, superactive, mesic Mollic Hapludalf). The soil was sampled for the 0 to 20 cm depth before 48 h after runoff ceased; the remaining sample was discarded.

Samples were collected from runoff events that had runoff for most manure and fertilizer application using a 1.9 cm (0.75 in) diameter soil probe for basic chemical analyses. The soil chemical analysis was as follows: pH, 6.8; organic matter, 29.2 g kg^{-1}; Bray-P, 52 mg kg^{-1} (ppm); NO_3-N, 7.25 mg kg^{-1}; and K, 509 mg kg^{-1}.

Experimental design, treatments, and crop management

The experimental units were the drainage areas of 21 tile-outlet terrace risers grouped into three blocks of seven treatments. Blocking was according to slope aspect with approximately east, west, and north to northeast aspects for blocks 1, 2, and 3, respectively.

The treatments were CFM applied with non-application setbacks of 0, 5, 10, 20, 30, and 40 m, and no CFM applied. The CFM was applied to bare, frozen soil in February 2006 to minimize compaction. The CFM contained 3.71, 0.04, 0.89, 3.00 kg mg^{-1} of organic N, nitrate N, ammonium N, and P, respectively. This resulted in the application of 222 kg P ha^{-1}. The intended CFM application rate was 56 Mg ha^{-1} but application was by the animal feeding operation and calculation from the total amount applied to the 24 ha research site determined the mean application rate to be 74 Mg ha^{-1}. This is a high but not excessive rate as feedlot manure is commonly applied at heavy rates once in 3 to 6 *year* to minimize compaction, with much of the P removed in harvests between the applications. Winter application of feedlot manure is common in this part of Nebraska where repeated snow melts during the winter are common with little or no runoff. Winter application on bare soil is considered to pose less P runoff risk than spring application but more than late summer and fall application (Wortmann and Walters, 2007). Considering availability factors of 15, 0, and 100% for organic N in CFM, ammonium N with no incorporation, and nitrate N, respectively, 107 kg ha^{-1} of manure N was estimated to be available to the 2006 corn (*Zea mays* L.) crop (Eghball and Power, 1999).

The CFM was first carefully applied around flagged setback areas reducing risk of CFM application in the setback while applying to the rest of the field. No fertilizer P was applied to the setback area or the rest of the field. The mean P application rate for the drainage area of the experimental units ranged from 220 kg ha^{-1} for the 0 m setback compared with 152 kg ha^{-1} for the 40 m setback.

Crop management was at the farmer's discretion but uniform across the trial area with the understanding that no fertilizer P was to be applied and no tillage was conducted. Corn was planted on 3rd May 2006 and soybean (*Glycine max* L.) was planted on 21st May 2007. No tillage was conducted during these years. Urea-ammonium nitrate (32-0-0) was applied at 146 kg N ha^{-1} in 2006. This was more N than was needed where CFM was applied as CFM was expected to supply 107 kg ha^{-1}, but the farmer opted to apply the fertilizer N uniformly to setback and non-setback areas. No fertilizer was applied in 2007. A tipping-bucket rain gauge (Onset Data Logging Rain, Global Water Instrumentation, INC, Gold River, CA) was positioned on the east side of the field to record rainfall time and intensity in increments of 0.25 mm.

Sampling

At each of 21 risers, a 1-L bottle was installed to collect runoff samples with the inlet tube 75 mm above the soil surface. The inner and outer diameters of the inlet tubes were 6 and 9 mm, respectively. A 0.5 L sub-sample was collected for analysis within risers on 30th March, 6th April, and 8th August in 2006, and on 25th April, 7th May, and 25th May in 2007. The 25th May 2007 event actually began on the 24th May and ISCO samples were collected on both 24th and 25th May; therefore, the ISCO data was analyzed as four events in 2007. Two other runoff events with much missing data were excluded from statistical analysis.

Automated samplers (ISCO Inc., Lincoln, NE) were installed in 2007 in experimental units with the 0 and 30 m setback distances to measure the depth of ponded runoff and collect runoff samples. The ISCO samplers held 12 jars and collected samples every 10 min once the ponded runoff reached the sampler's intake.

The runoff samples were kept at 4°C until analyzed. Sediment concentration was gravimetrically determined by drying 100 mL of unfiltered runoff at 104°C. Total P (TP) was determined on unfiltered runoff samples using perchloric/nitric acid extraction ($HClO_4$–HNO_3). Dissolved P in runoff was determined after filtration of a 100 mL sample through a <0.45 μm filter (Pote and Daniel, 2000). Particulate P (PP) was determined as the difference between TP and DP. Phosphorus was measured colorimetrically at 880 nm with a Spectronic 601 (Milton-Roy, Rochester, NY) according to Murphy and Riley (1962).

Calculations

The drainage area of each sampler was calculated using survey equipment to calculate the stage storage curve that defines the relationship between the depth of water and the associated storage volume in the drainage area. This step involves initially establishing contours for the ponding area and determining the elevation-storage relationship for the pond.

Discharge rate through the riser, Q ($m^3 s^{-1}$), was calculated using ISCO data as a function of the riser orifice diameter, orifice depth, and ponding depth (Martin et al., 1997):

$$Q = C_d A_o (2 g H)^{0.5}$$

Where the discharge coefficient (C_d) = 0.65, A_o = orifice area (m^2), g = gravitational constant = 9.8 ms^{-2}, and H = depth from the pond surface to the orifice (m). Runoff volume (m^3ha^{-1}) was calculated from the total volume discharged.

Mean runoff or inflow (I) rate (m^3 ha^{-1} hr^{-1}) was calculated from change in storage volume and estimated discharge to the underground tile for each 10 min interval:

$$I = (S_{j+1} - S_j) / \Delta t + (O_{j+1} - O_j) / 2$$

Where I is inflow rate (m^3 ha^{-1} hr^{-1}), S is storage (m^3), O is the discharge from the riser ($m^3ha^{-1}h^{-1}$) calculated from Q, and t is time step (h).

Statistical analysis

The analysis of variance (α = 0.05) was conducted using Proc Mixed (SAS Institute, 2003) to determine the effects of setback treatments on sediment, TP, DP, and PP concentrations for samples collected in the 1-L bottles with setback distance as a fixed effect and block as a random effect. ISCO runoff volume, concentration, and load data from the 30 m and no setback treatments were analyzed using the Proc Mixed model for the effects of setback treatments with rainfall event as a repeated measure.

Table 1. The mean effect of 74 Mg ha^{-1} composted feedlot manure applied with 30 m compared with no setback on runoff volume and concentrations and loads of total P (TP), particulate P (PP), dissolved P (DP), and sediment for four runoff events in 2007 based on ISCO sampling.

Setback distance	Runoff volume	Sediment	TP	DP	PP	Sediment	TP	DP	PP
(m)	(m^3 ha^{-1-})	(g L^{-1})	(mg L^{-1})			(kg ha^{-1})			
0	171	0.84	2.74	1.03	1.72	182	0.47	0.28	0.21
30	175	1.24	2.78	1.16	1.69	279	0.46	0.26	0.23
Setback	NS	NS	NS	NS	NS	*	NS	NS	NS
Setback × event	NS	NS	***	NS	**	NS	NS	NS	NS

*, **, *** Significant at P ≤ 0.1, 0.05, and 0.001 respectively. NS, not significant.

Drainage area, amount of P applied, and mean P application rate per experimental unit were significant for some variables as co-variants in the ANOVA but did not increase the results that were statistically significant. All reported results are from ANOVAs without co-variants.

Various relationships were evaluated. Pearson correlation coefficients and means with confidence intervals were determined for the effect of sampling method on sediment and P concentrations. Polynominal regression analysis was used to relate the 10-min peak rainfall intensity with sediment and nutrient concentrations in runoff using the 1-L bottle data. Regression analysis (exponential decay) was used to relate sediment and P concentrations to sampling time (min) during runoff events.

RESULTS AND DISCUSSION

Effect of setback distances on P and sediment concentration

Application of CFM and setbacks did not significantly affect P concentrations when measured from the 1-L sample bottles for the mean of five rainfall events. The means with confidence limits (P = 0.05) were 1.3 ± 0.20, 1.82 ± 0.31, and 3.20 ± 0.42 mg L^{-1}, respectively, for DP, PP, and TP. There was not a significant relationship of P concentrations with setback distance and no difference in P concentrations for CFM applied compared with no CFM applied. Sediment concentration, averaged across the five runoff events, was not affected by setback distance, but was significantly greater (P = 0.003) with no CFM applied (3.69 g kg^{-1}) compared with the mean for all setback distances with CFM applied (1.98 g kg^{-1}). Another very intense and erosive runoff event was excluded from this analysis because of heterogeneity of variance, but also because of much variation independent of the treatments.

Runoff volume and concentrations of sediment and DP were not different for the 0-m compared with the 30-m setback as determined from the ISCO sampling data collected from four runoff events in 2007 (Table 1). The setback distance by runoff event interaction was significant for total and PP concentrations because of relatively high concentrations with the 30-m setback for the 24th April 2007 compared with other runoff events. Sediment load was significantly less with 0-m compared with 30-m setback.

Setbacks did not reduce dissolved and TP concentrations and runoff. Application of CFM did reduce sediment concentration compared with no CFM applied when measured with the 1-L sample bottles. Sediment load was greater with setbacks compared with no setback. The results confirm the value of manure application in reducing erosion, possibly because of an increase in water-stable aggregates (McDowell and Sharpley, 2001; Wortmann and Shapiro, 2008). Wortmann and Shapiro (2008) also reported higher P concentration in water-stable aggregates compared with the whole soil, with the aggregates protecting much soil P from runoff. When repeated manure application results in excessive levels of soil test P, however, the soil amendment effect is inadequate to compensate for the high levels of soil P, and runoff P is increased (Wortmann and Walters, 2006, 2007). In this study, even with surface application of 222 kg ha^{-1} P in the winter, CFM application did not add significantly to P in runoff and setbacks were not beneficial in reducing P and sediment runoff.

Another factor that affected P but not sediment in runoff was that manure P was applied at a lower mean rate to experimental units with greater setbacks. The drainage areas ranged from 0.39 to 2.13 ha and setback area comprised, on average, 0, 0.7, 2.9, 21.4, 23.3, and 35.8% of the drainage area for the 0, 5, 10, 20, 30, and 40 m setbacks, respectively. Therefore, 35.8% less manure P was applied with the 40 m compared with the 0 m setback, which is expected to reduce runoff P concentration and load, and yet setback distance did not affect P concentrations in runoff.

Sediment, PP, and DP react differently in erosion, runoff, and ponding events. The results of this study are in

Table 2. Sample means and confidence limits (P = 0.95) comparing use of 1-L bottles for sampling with ISCO samples. Means of six experimental units and three runoff events.

Pollutant	1-L bottle	ISCO
Sediment, g L^{-1}	1.71 ± 0.60	1.14 ± 0.37
Total P, mg L^{-1}	3.02 ± 0.50	2.72 ± 0.43
Dissolved P, mg L^{-1}	1.37 ± 0.29	1.47 ± 0.18
Particulate P, mg L^{-1}	1.66 ± 0.46	1.30 ± 0.31

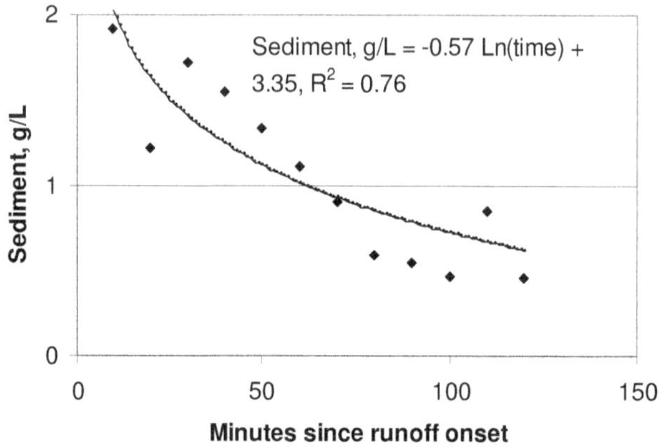

Figure 2. Effect of time since on-set of runoff ponding on sediment concentration based on samples collected at 10-min intervals with ISCO samplers. Each point represents the mean of eight observations from runoff of both 0 and 30 m manure application setbacks and for four runoff events.

in agreement with findings that setbacks were not effective in reducing herbicide runoff (Mickelson et al., 1998), except in proportion to the area not sprayed (Franti et al., 1998).

Sampling method comparison

The 1-L bottle sampling method is an inexpensive means of obtaining samples from drainage areas but gives only one sample per drainage area per event and is useful only for pollutant concentrations. The ISCO sampling has the advantage of collecting multiple samples per event while measuring ponding depth to provide a means to calculate runoff rate, but at much higher cost. Mean concentrations with ISCO were not outside the confidence intervals of the mean concentrations with the 1-L bottle method, but the coefficient of correlation between the two methods was significant for DP concentration only (r = 0.6; P = 0.003) (Table 2). Some of the difference in results from the two methods was probably due to time during the event when the sample was collected, at least for sediment concentration which dropped steadily during the runoff event (Figure 2). Using the 1-L bottles, samples were collected within 30 min after the ponding reached the inlet tube while ISCO samples were collected throughout the event. Concentrations of P were not consistently related to time during runoff events.

Rainfall properties and sediment and phosphorus concentration

Particulate P was 70% of the TP load in agricultural runoff. Sediment and P concentrations were increased with higher peak rainfall for spring runoff events (Figure 3). The peak rainfall intensities, determined as the maximum intensity for an event over a 10 min period, were 72 mmh^{-1} on 30th March, 52 mmh $^{-1}$ on 6th April, and 58 $mmha^{-1}$ on 8th August in 2006 and 29 mmh $^{-1}$ 25th April, 17 mmh $^{-1}$ on 6th May, and 12 mmh^{-1} on 24th May in 2007.

The relationship of sediment and P concentrations with

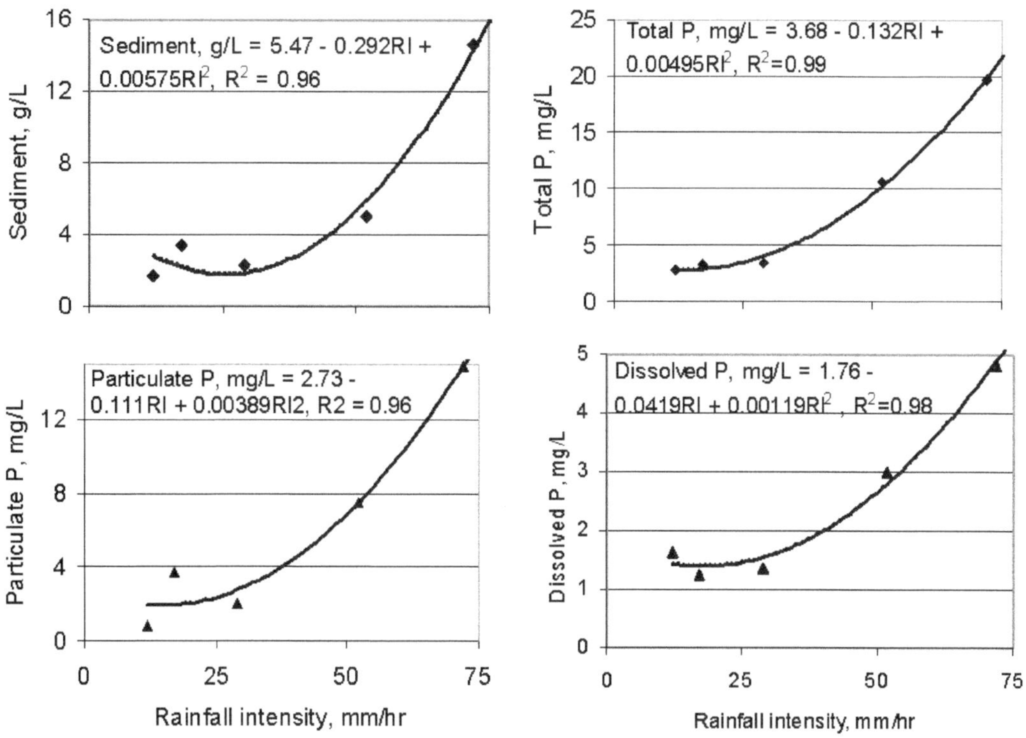

Figure 3. The effect of peak 10-min rainfall intensity on concentration of (A) sediment, (B) total P, (C) dissolved and (D) particulate P based on 1-L samples collected by risers of tile outlet terrace systems during six spring runoff events in eastern Nebraska in 2006 and 2007.

rainfall intensity has been reported by others (Owens and Shipitalo, 2006; Cassell et al., 1998; Quinton et al., 2001). Detachment of soil particles during intense rainfall results in increased sediment and P concentrations in runoff (Fraser et al., 1999). Most runoff events occurred in March to May in the current study as well as in other studies conducted in eastern Nebraska (Wortmann and Walters, 2006, 2007). Soil water content is often high during these months, thus reducing the rate of water infiltration and increasing the potential for runoff. Maintaining ground cover during this period before corn or soybean crop canopies are formed may be important to minimizing sediment and P loss. As the season progresses, the crop canopy and the extraction of water from the soil tend to reduce runoff and erosion potential erosion potential.

Conclusion

Setbacks around risers of tile-outlet terrace systems were not effective in reducing P loss in runoff. Application compared with no application of CFM resulted in reduced sediment concentration by 51%. Sediment loss in runoff was less with 0-m (182 kg ha) compared with 30-m setback (279 kg ha^{-1}). Rainfall intensity is a major factor influencing sediment and P concentration because it affects soil detachment by raindrop impact and transport of detached particles by runoff. These research findings do not support the use of setbacks around risers of tile-outlet terraced land as a means for reducing P loading of surface waters. This study did not address manure-borne antibiotics or pathogens, but the failure of setbacks to reduce DP in runoff, as well as the works of Mickelson et al. (1998) and Franti et al. (1998) which found that setbacks did not reduce herbicide in runoff, indicates that setbacks around risers are not effective for reducing other manure contaminants. Sediment loss can be reduced with manure application. Setbacks around risers of tile-outlet terrace systems are ineffective in reducing P in runoff.

ACKNOWLEDGMENTS

We thank Jerry Newsham, the field owner, and Mark Strnad, Maribeth Milner, Steven Katchman, Allen Boldt, Ken Kotera, and Roger Renken for technical support. This research was conducted under the USDA CSREES Managed Ecosystems project 2005-55101-16369.

REFERENCES

Braun, DC (1998). Modeling phosphorus dynamics in ecosystems: mass balance and dynamic simulation approaches. J. Environ. Qual. 27:293-298.

Carpenter SR, Caraco NF, Correll DL, Howarth RW, Sharpley AN, Smith VH (1998). Nonpoint pollution of surface waters with phosphorus and nitrogen. Ecol. Applic. 8:559-568.

Cassell EA, Dorioz JM, Kort RL, Hoffman JP, Meals DW, Kirschtel, D,

Eghball B, Power JF (1999). Composted and noncomposted manure application to conventional and no-till systems: corn yield and nitrogen uptake. Agon. J. 91: 819-825.

Franti TG, Peter CJ, Tierney DP, Fawcett RS, Myers SA (1998). Reducing herbicide losses from tile-outlet terraces. J. Soil Water Conserv. 53: 25-31.

Fraser AI, Harrod TR, Haygarth PM (1999). The effect of rainfall intensity on soil erosion and particulate phosphorus transfer from arable soils. Water Sci. Technol. 39:41-45.

Gilley JE, Risse LM (2000). Runoff and soil loss as affected by the application of manure. Trans. ASAE 43:1583-1588.

Koelsch R (2005). Evaluating livestock system environmental performance with whole-farm nutrient balance. J. Environ. Qual. 34: 149-155.

Liu F, Mitchell CC, Hill DT, Odom JW, Rochester EW (1997). Phosphorus recovery in surface runoff from swine lagoon effluent by overland flow. J. Environ. Qual. 26:995-1001.

Martin W, Kersten R, Eaglin R (1997). Hydrology water quantity and quality control. John Wiley & Sons 2nd ed. p. 283

McDowell RW, Sharpley AN (2001). Approximating phosphorus release from soils to surface runoff and subsurface drainage. J. Environ. Qual. 30: 508-520.

McDowell RW, Sharpley AN (2002). Phosphorus transport in overland flow in response to position of manure application. J. Environ. Qual. 31: 217-227.

Mickelson SK, Baker JL, Melvin SW, Fawcett RS, Tierney DP, Peter CJ (1998). Effects of soil incorporation and setbacks on herbicide runoff from a tile-outlet terraced field. J. Soil Water Conserv. 53: 18-25.

Murphy J, Riley JR (1962). A modified single solution method for the determination of phosphate in natural waters. Anal. Chim. Acta 27: 31-36.

National Research Council (2000). Watershed management for potable water supply: assessing the New York City strategy. Commission on Geosciences, Environment and Resources (CGER) ISBN 10: 0-309-06777- 4:427-428. Available at http://books.nap.edu/openbook.php?record_id=9677&page=427 (verified 1 Oct 2009).

Owens LB, Shipitalo MJ (2006). Surface and subsurface phosphorus losses from fertilized pasture systems in Ohio. J. Environ. Qual. 35:1101-1109.

Pote DH, Daniel TC (2000). Analyzing for dissolved Phosphorus in water samples. p. 91-93. *In* G.M. Pierzynski (ed.) Methods of phosphorus analysis for soils, sediments, residuals, and waters. Southern Coop. Ser. Bull. 396. North Carolina State University, Raleigh.

Quinton JN, Catt JA, Hess TM (2001). The selective removal of phosphorus from soil: is event size important? J. Environ. Qual. 30:538-545.

SAS Institute. 2003. The SAS system for Windows. Release 8.2. SAS Inst., Cary, NC.

Sharpley AN, Daniel TC, Edwards DR (1993). Phosphorus movement in the landscape. J. Prod. Agric. 6: 492-500.

USEPA (2003). National pollutant discharge elimination system permit regulation and standards for concentrated animal feeding operations (CAFOs). Final rule. Fed. Regist. 29: 7176-7274.

Wortmann CS, Shapiro CA (2008). The effects of manure application on soil aggregation. J. Nutr. Cycl. Agroecosys. 80: 173-180.

Wortmann CS, Walters D (2006). Phosphorus runoff during four years following composted manure application. J. Environ. Qual. 35:651-657.

Wortmann CS, Walters D (2007). Residual effects of compost and plowing on phosphorus and sediment in runoff. J. Environ. Qual. 36: 1521-1527.

Soil classification and micromorphology: A case study of Cholistan Desert

Farooq Ahmad

Department of Geography, University of the Punjab, Lahore, Pakistan. E-mail: F.Ahmad@sheffield.ac.uk.

Cholistan Desert is an extension of the Great Indian Desert and covering an area of 26,330 km^2. The desert can be divided into two main geomorphic regions: Northern region, known as Lesser Cholistan, constitutes the desert margin and consists of a series of saline alluvial flats alternating with low sand ridges/dunes; and Southern region, known as Greater Cholistan, a wind resorted sandy desert comprising of a number of old Hakra River terraces with various forms of sand ridges and inter-ridge valleys. Cholistan Desert presents quite a complex pattern of alluvial and aeolian depositions. The westerly drifting of the river combined with the lowering of the base level of erosion caused the formation of a number of terraces, which represent different depositional stages. Soil morphology is the field observable attributes of the soil within the various soil horizons and the description of the kind and arrangement of the horizons.

Key words: Cholistan, Hakra River, sand ridges, sediments, soil horizon, soil morphology

INTRODUCTION

Cholistan Desert lies within the Southeast quadrant of Punjab Province, placed between 27° 42' and 29° 45' North latitude and 69° 52' and 73° 05' East longitude (Ahmad, 2002; 2006; 2008; 2010). The word 'Cholistan' is possible derived from a Turkish word, 'Chol', which means a desert (Figure 1); while some historians believe the name has been distorted from Iraqi (Kurdish) word, 'Chilistan' meaning waterless waste land (Ahmad et al., 1992; FAO/ADB, 1993; Auj, 1995; Ahmad, 1999; 1999a, 2005; Ahmad et al., 2005). Cholistan is popularly known as *'Rohi'.* In a dialect still spoken in some parts of Tibet, '*roh'* means a hill, from which the name *Rohilla* has been attributed. In fact, *Rohi* has been derived from the Pushto word *'roh'*, meaning a sandy desert. The man from *Rohi* is called *Rohilla* (Auj, 1987; 1991). However, a reference has been made in Geography of Sindh, that the word Cholistan has been derived from the 'Seraiki' word *'Wacholo'* meaning 'in between'. Since this desert is situated between the Thar and Rajasthan deserts, it was called *'Wacholo'*, which later became *'Cholo'* and then Cholistan. This appears to be a more probable explanation for the origin of the word (Khan, 1992; Ahmad, 2002).

Around 4000 BC, Cholistan was a cradle of civilization commonly known as Hakra valley civilization, when Hakra River flowed through the region. The river supplied water until 1200 BC and about 600 BC it became irregular in flow and consequently vanished (Auj, 1987). Cholistan forms a part of the vast Indo-Gangetic Plain lying in a great tectonic trough, which lies between the foothills of the Himalayan Mountains and the central core of South-Asia. This trough was subsequently filled up with the thick mass of alluvium derived from the Himalayan and deposited by the Indo-Gangetic system. This was later transformed into a vast plain (FAO/ADB, 1993).

RESEARCH DESIGN AND METHODS

The work involves investigation on both constructional and depositional aspects of the archaeological record and the micro-environmental conditions. It is often possible to relate a soil type to a particular ecological niche in the landscape (Retallack, 1994; Buol et al., 2003). Soils and their properties is the product of different soil-forming factors (Jenny, 1941) and the parent material. As soil-forming factors also govern geomorphic processes, landscape evolution is intimately related to soil development (McFadden and Kneupfer, 1990; Kapur and Stoops, 2008; Kapur et al., 2011). Soil samples from surface to 150 cm depth were collected at different locations from Cholistan desert and were subsequently analyzed. The methods included both fieldwork and laboratory analysis. Field survey was conducted with the collaboration of soil survey of Pakistan for micro classification of soil. Soils occurring on

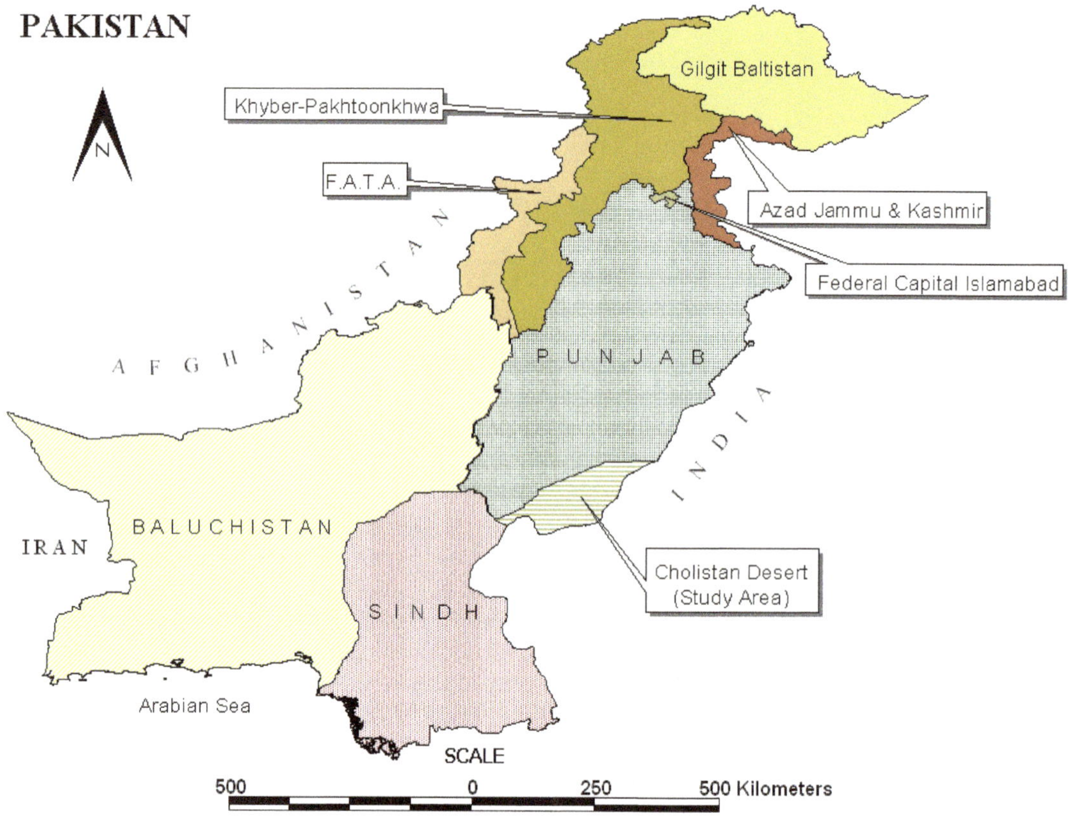

Figure 1. Location map of Cholistan Desert (Pakistan).

different topographic positions within each unit were studied and described according to the Soil Survey Manual (USDA, 1951) and the Guidelines for Soil Profile Descriptions (FAO, 1965).

GEOMORPHOLOGY

Geomorphologically the area presents quite a complex pattern of alluvial and aeolian depositions (Figures 2 and 3), which was followed by: (a) wind resorting and further deposition of the sediments into various forms and sand ridges; (b) resorting and further deposition in spill channels; (c) deposition of clayey sediments in flats; and (d) present day wind resorting and dune formation (Ahmad, 2002; 2008). The alluvium consists of mixed calcareous material, which was derived from the igneous and metamorphic rocks of the Himalayas and was deposited by the Sutlej and abandoned Hakra Rivers most probably during different stages in the sub-recent periods (FAO/ADB, 1993; Ahmad, 2008). The aeolian sands have been derived mainly from the Rann of Kutch and the sea coast and partly from the lower Indus Basin. Weathered debris of the Aravalli has also contributed. The material was carried from these sources by the strong South-western coastal winds (FAO/ADB, 1993; Ahmad, 2002; 2008). The westerly drifting of the river combined with the lowering of the base level of erosion caused the formation of a number of terraces, which represent different depositional stages.

Based on differences in topographic form, parent material and soils, Cholistan Desert can be divided into two main geomorphic regions: Northern region, known as Lesser Cholistan (12,370 km^2), constituting the desert margin and consisting of a series of saline alluvial flats alternating with low sand ridges/dunes; and Southern region, known as Greater Cholistan (13,960 km^2), a wind resorted sandy desert comprising of a number of old Hakra River terraces with various forms of sand ridges and inter-ridge valleys (FAO/ADB, 1993; Tahir et al., 1995; Ahmad, 2008).

The area consists of four main soil types, like dune land with topography ranging between undulating and steep slopes. The sand dunes lie parallel to each other and are connected by small streamers. They are very excessively drained, coarse textured, and their structure less derived from aeolian material deposited by strong winds. Sandy soils are nearly level to gently sloping, deep to very deep, excessively drained, calcareous, coarse textured. Loamy soils are level to nearly level with hummocks of fine sand on the surface, moderately deep, somewhat excessively drained to well drained, calcareous, moderately coarse textured to medium textured. Clayey soils are mostly level, moderately deep, poorly drained, calcareous, saline-sodic, fine moderate textured to fine textured (Baig et al., 1975).

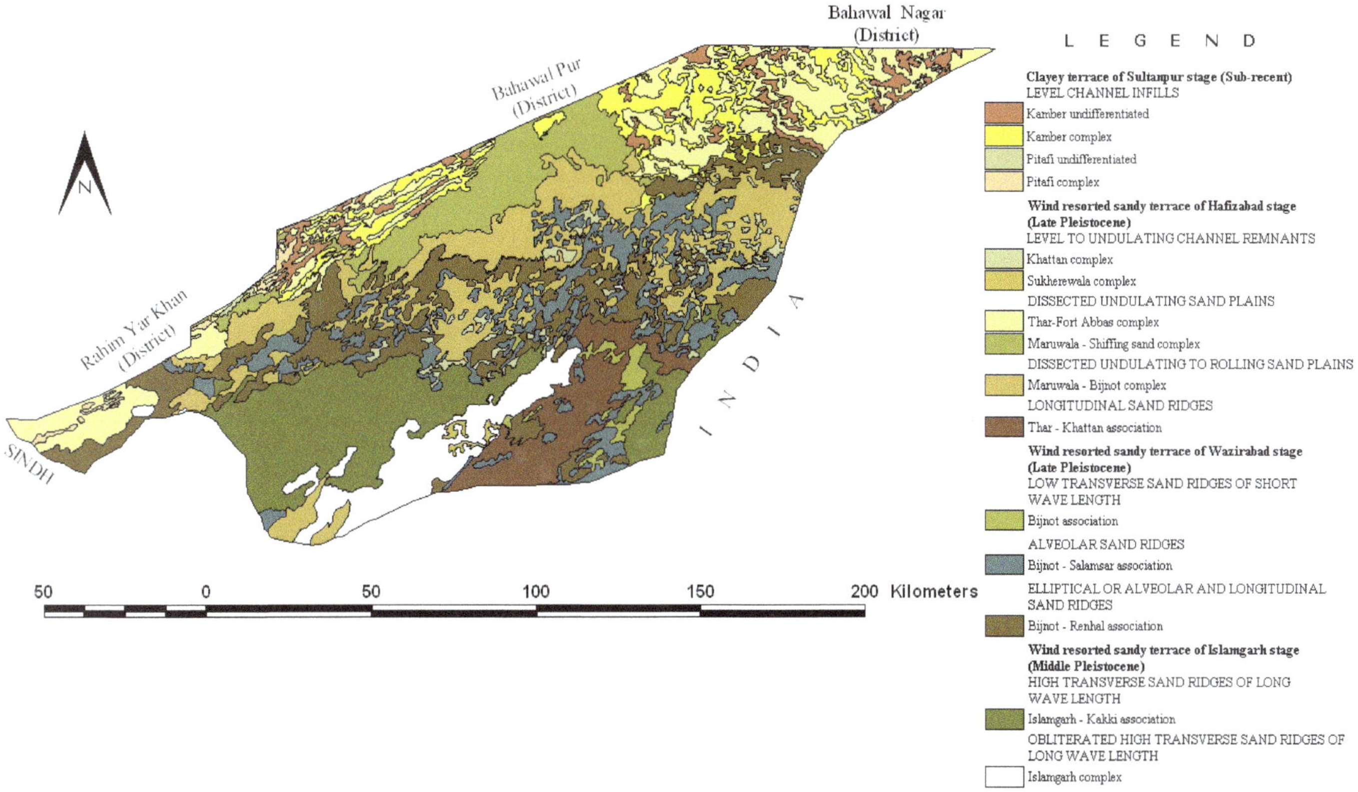

Figure 2. Cholistan: Soils and landforms. Source: After soil survey of Pakistan 1975, modified by the author.

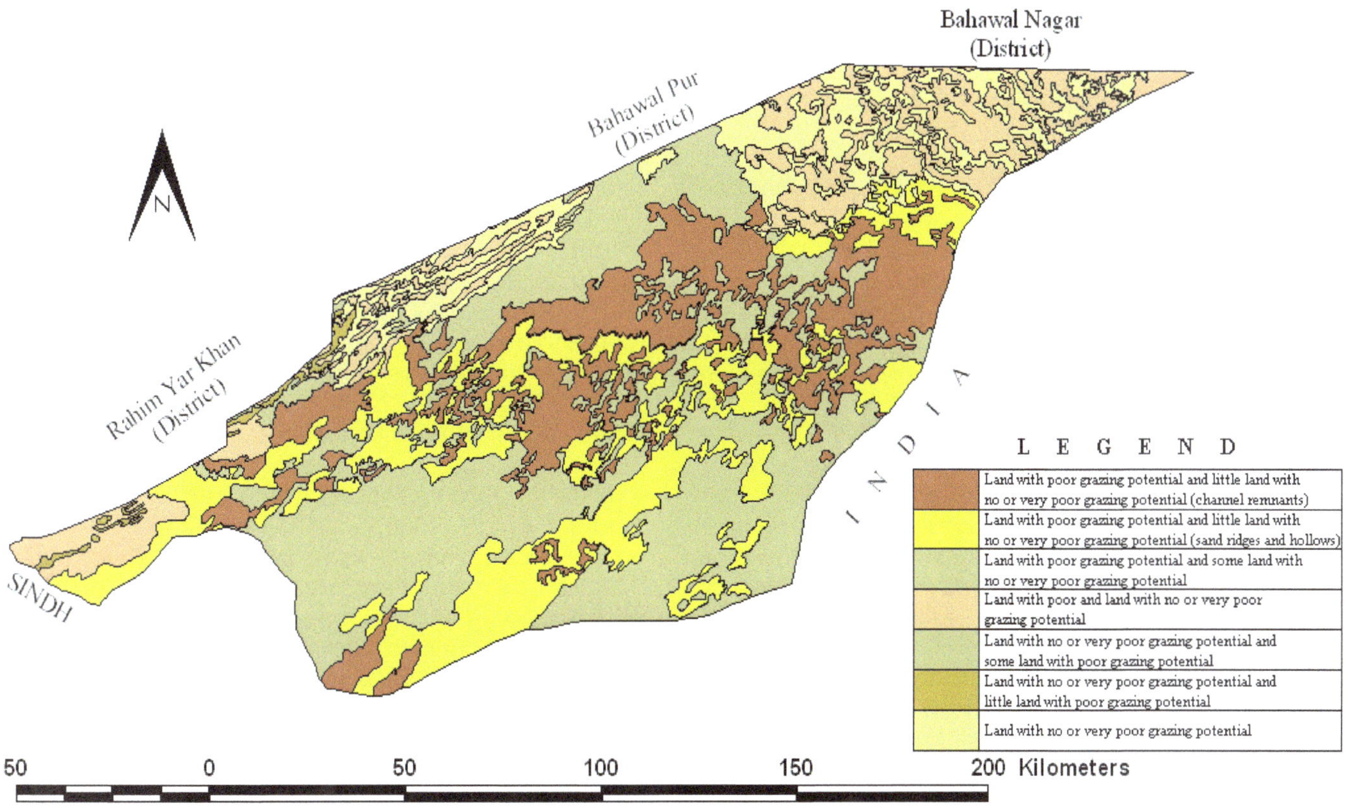

Figure 3. Cholistan: Land capability. Source: After soil survey of Pakistan 1975, modified by the author.

Desert margin (Lesser Cholistan)

This region lies 2 to 4 m below the wind resorted sandy desert (Greater Cholistan). It consists of a complex of two distinct terraces separated by a bluff ranging in height from about half a meter to three meters (FAO/ADB, 1993). The terraces are:

(a) Sub-recent clayey terrace (10,000-1,000 years ago).
(b) Dissected wind resorted sandy terrace of Hafizabad stage (17,000-10,000 years ago).

The sub-recent terrace is lower and comprises the saline alluvial flats causing the sub-recent channels of the drying Hakra River (Tahir et al., 1995). The sediments of this terrace are predominantly clayey and the soils are invariably saline or saline-sodic (Baig et al., 1980). The vegetation cover, which consists of xeromorphic halophytic plant species, is a true indicator of the saline soils of the region. The wind resorted sandy terrace is higher and older than the former and is represented by patches of undulating to rolling sand ridges/dunes occurring as remnants of Pleistocene terraces. The sediments of this terrace are sandy and bear a vegetative cover consisting true desert shrubs similar to those of the Greater Cholistan (FAO/ADB, 1993).

Wind resorted sandy desert (Greater Cholistan)

The three terraces have been distinguished in the wind resorted sandy desert formed during different stages of Pleistocene period. Each terrace is separated from the other by a bluff ranging in height from about one to six meters (FAO/ADB, 1993). The terraces are:

(a) Wind resorted sandy terrace of Hafizabad stage (Late Pleistocene, 17,000-10,000 years ago).
(b) Wind resorted sandy terrace of Wazirabad stage (Late Pleistocene, 50,000-20,000 years ago).
(c) Wind resorted sandy terrace of Islamgarh stage (Middle Pleistocene, 250,000-50,000 years ago).

Each terrace is associated with one or more type(s) of sand ridges and depressions. The form and orientation of these ridges and depressions control the soil moisture and eventually the vegetation and provide the basis for their development as rangelands (Baig et al., 1980). Four types of sand ridges and depressions have been recognized on these terraces. They are: (i) Longitudinal sand ridges, (ii) Alveolar sandy ridges, (iii) Low transverse sand ridges, and (iv) High transverse sand ridges (FAO/ADB, 1993).

The longitudinal sand ridges are found within the first and the lowest terrace of Hafizabad stage, the alveolar sand ridges and low transverse sand ridges are confined to the second intermediate terrace of Wazirabad stage whereas the high transverse sand ridges are connected with the third, the highest and the oldest terrace of Islamgarh stage (Baig et al., 1980).

In addition to the above mentioned physiographic units, courses of old abandoned river occur in patches as covered channel remnants throughout the wind resorted sandy desert. These channel remnants are 3 to 10 m below the general level of surrounding landscape and are 1.5 to 5 km wide (Baig et al., 1975; FAO/ADB, 1993). The soil of the wind resorted sandy desert are predominantly sandy ranging in texture from sand to sandy loam (Shirazi et al., 2004), although small proportion of loam soils are also found. In addition, some clayey soils occur as small scattered patches. The ridges are invariably made up of sands. The windward faces of the ridges contain stabilized sands, whereas the leeward faces and the crest carry shifting sands and semi-stabilized sands respectively. The inter ridge depressions contain soils ranging in texture from loamy sand through sandy loam parts of the depressions are occupied by loam and the intermediate parts by sandy loam. The soils are brown in colour, deeply homogenized with gradual gradation of colour and texture, which become lighter with depth. The soils are moderately to strongly calcareous containing very few fine to common, fine to medium scattered lime nodules in the profile but without a definite zone of lime accumulation (Mohammad and Mirza, 1985; Mohammad, 1989). The soils are non-saline and non-sodic with pH values ranging 8.2 to 8.4 (Baig et al., 1975).

MICROMORPHOLOGICAL CHARACTERISTICS AND SOIL PROFILE

The soil profile of Table 1, located in the Northwest of Ramewala *toba* in the vicinity of Dingarh Fort, shows profile texture to be clay loam to silty clay up to 34 cm depth, while the remaining depth of the profile up to 150 cm consists of loamy very fine and sand textures (PADMU, 1986; Malik, 1995). The soil horizon (cb) from depth 34 to 67 cm is buried and soils horizon (cg) from depth 90 to 110 cm is clayed (PADMU, 1986; Malik, 1995; Ahmad, 2002). The structure of the profile is between massive and sub-angular blocky characteristics. The pores in the top horizon (A) of the profile from soil surface to 6 cm depth are common fine vesicular types. These pores are not continuous; therefore, infiltration of water will not go deep. The pores in other horizons of the profile are absent. The pH of the profile is between 8.8 and 9 (PADMU, 1986; Malik, 1995). Due to high sodicity in the profile, the soil horizons are devoid of porosity. Mottles and roots are also absent in the profile. The soil profile up to 110 cm consists of $CaCO_3$ in the form of common very fine to fine lime nodules and a few lime *kankers* that also cause hardness (PADMU, 1986; Malik, 1995; Ahmad, 2002). The profile from soil surface to 34 cm depth is hard to very hard. The horizon from 90 to 110 cm is clayed indicating once it was under water; as a result there is no pore space. The characteristics of the profile indicate that the soils are non-porous and

Table 1. Profile characteristics as a catchment for rainwater harvesting located about 125 meters north-west of Ramewala *toba* in Dingarh vicinity.

Soil horizon	A	B	Cb	C1	Cg	C2
Soil depth (cm)	0 – 6	6 – 34	34 – 67	67 – 90	90 – 110	110 – 150
Texture	Clay loam	Silty clay	Loamy very fine sand	Loamy fine sand	Loamy fine sand	Loamy very fine sand
Structure	Massive	Sub angular blocky	Massive	Massive	Massive	Massive
Pores	Common fine vesicular	No	No	No	No	No
pH	8.8	9.0	9.0	8.8	8.8	8.8
Mottles	No	No	No	No	No	No
Roots	No	No	No	No	No	No
$CaCO_3$	Common very fine nodules	Common fine nodules	Common fine and a few lime *cankers*	Common very fine lime nodules	Common very fine lime nodules	Nil
Hardness	Hard	Very hard	Soft	Friable	Friable	Very friable

Source: PADMU - Pakistan Desertification Monitoring Unit 1986, modified by the author.

Table 2. Profile characteristics as a catchment for rainwater harvesting located about 2.37 km from Dingarh Fort on Barianwala *toba* track.

Soil horizon	A	B	Cb	C
Soil depth (cm)	0 – 8	8 – 35	35 – 90	90 – 150
Texture	Clay loam	Silty clay	Loamy very fine sand	Loamy fine to fine sand
Structure	Massive	Sub angular blocky	Massive	Massive
Type of pores	Vesicular	No	No	No
No. of pores	Common fine	Nil	Nil	Nil
pH	8.8	9.0	9.0	9.2
Mottles	No	Yes	Yes	Yes
No. of mottles	Nil	Common fine to Very fine	Many fine	Many fine
Roots	No	No	No	No
$CaCO_3$	Common fine lime nodules	Common fine lime nodules	Common fine to a few medium *kankers*	Common very fine lime *kankers*
Hardness	Hard	Very hard	Friable	Friable

Source: PADMU - Pakistan Desertification Monitoring Unit 1986, modified by the author.

non- drained to very poorly drain and indicate that this area is very suitable for rain runoff collection.

The profile of Table 2 is located at 2.37 km away from Dingarh Fort on Barianwala *toba* track towards west. The characteristics of this profile mostly resemble to profile under Table 1 except some features. The depth of buried horizon is more and profile does not contain clayed horizon. The profile is consisting of common very fine to fine mottles indicating that these soils once had been under water (PADMU, 1986; Malik, 1995; Ahmad, 2002). This profile also indicates that the area which is non to very poorly drained is therefore very appropriate for runoff harvesting and collection.

The profile of Table 3, located at 0.5 km North-west of Dingarh Fort, shows that the texture of the profile is between clay loam to silty clay up to 125 cm depth and the remaining profile which is up to 150 cm is buried, thereby having coarse textured fine sand (PADMU, 1986; Malik, 1995; Ahmad, 2002). The major part of the profile consists of B-horizon indicating developed soils. The.

Table 3. Profile characteristics as a catchment for rainwater harvesting located about 500 m north-west of Dingarh Fort.

Soil horizon	**A**	**B21**	**B22**	**B23**	**Cb**
Soil depth (cm)	0 – 9	9 – 75	75 – 108	108 – 125	125 – 150
Texture	Clay loam	Silty clay	Silty clay	Silty clay	Fine sand
Structure	Massive	Sub angular blocky	Sub angular blocky	Sub angular blocky	Sub angular blocky
Type of pores	Vesicular	Tubular	Tubular	No	No
No. of pores	Common fine	A few very fine	A few very fine	Nil	Nil
pH	9.0	9.0	9.0	9.0	9.2
Mottles	No	Yes	Yes	Yes	Yes
No. of mottles	Nil	Common fine	Common fine	Many fine	Common medium
Roots	No	Yes	No	No	No
No. of roots	Nil	A few fine	Nil	Nil	Nil
$CaCO_3$	Nil	Nil	Nil	Nil	Nil
Hardness	Hard	Very hard	Very hard	Very hard	Friable

Source: PADMU - Pakistan desertification monitoring Unit 1986, modified by the author.

Table 4. Profile characteristics as a catchment for rainwater harvesting located about 200 m south-west from Bachewala *toba*.

Soil horizon	**A**	**B1**	**B2**	**Cb**
Soil depth (cm)	0 – 8	8 – 50	50 – 70	70 – 150
Texture	Silty clay	Silty clay	Silty clay	Loamy sand
Structure	Massive	Sub angular blocky	Sub angular blocky	Sub angular blocky
Type of pores	Few fine vesicular	A few very fine tubular	A few very fine tubular	No
pH	9.0	9.2	9.0	9.2
Mottles	No	Yes	Yes	Yes
No. of mottles	Nil	Common fine	Common fine	Common fine
Roots	No	No	No	No
No of roots	Nil	Nil	Nil	Nil
$CaCO_3$	Nil	Nil	Nil	Common fine to common medium *Kankers*
Hardness	Hard	Very hard	Very hard	Friable

Source: PADMU - Pakistan desertification monitoring Unit 1986, modified by the author.

structure of the profile is massive to sub-angular blocky The profile up to 9 cm depth is consisting of common fine vesicular pores, while up to depth 108 cm are a few fine to very fine tubular pores. The pH of the profile is between 9.0 and 9.2 indicating very high sodicity (PADMU, 1986; Malik, 1995). The profile is consisting of common fine to many fine to medium mottles indicating that once this area was under water and it caused reduction of iron and manganese. Mottles also indicate that these soils are non-porous. The profile is devoid of roots except a few fine roots in the horizon B21 from depth 9 to 75 cm. The profile is hard to very hard up to depth 125 cm. The features of the profile indicate that the area is very poorly drained and suitable for rainwater harvesting and collection.

The profile of Table 4, located at 200 m South-west of Bachewala *toba*, is mostly similar to profile located about 500 m North-west of Dingarh Fort except for some minor features (PADMU, 1986; Malik, 1995). This profile indicates that the area is very poorly drained and very suitable for rainwater harvesting and collection. The profile of Table 5, located at about 4.15 km away from Dingarh Fort on Gappenwala *toba* track and about 500 m towards East, is almost similar to the profiles located at about 200 m South-west from Bachewala *toba* and the profile located at about 500 m North-west of Dingarh Fort, except minor difference. Therefore, this area is also a good catchment for rainwater harvesting and collection. The characteristics of mentioned profiles in Table 5 show that fine textured soil lies in Cholistan having level to nearly levelled topography is dense, impervious, non porous to very poorly porous, non/very poorly drained, saline-sodic, capable of generating maximum runoff after absorbing minimum water (PADMU, 1986; Malik, 1995). The features and characteristics of the soil indicate that Cholistan is the best catchment area for rainwater

Table 5. Profile characteristics as a catchment for rainwater harvesting located about 4.15 km from Dingarh Fort on Gappenwala *toba* track.

Soil horizon	A1	B1	B2	B3	Cb
Soil depth (cm)	0 – 12	12 – 45	45 – 75	75 – 125	125 – 150
Texture	Clay loam	Clay loam	Silty clay	Silty clay	Very fine sand
Structure	Massive	Sub angular blocky	Sub angular blocky	Sub angular blocky	Massive
Type of pores	Vesicular	Interstitial	Interstitial	No	No
No. of pores	Common very fine	A few fine	A few very fine	Nil	Nil
pH	9.0	9.0	9.2	9.2	9.2
Mottles	No	Yes	Yes	Yes	Yes
No. of mottles	Nil	A few fine	Common fine	Many fine	A few medium
Roots	No	Yes	Yes	No	No
No. of roots	Nil	A few fine & medium	A few very fine	Nil	Nil
$CaCO_3$	Common very fine nodules	Common fine nodules	Common fine nodules	Nil	Nil
Hardness	Hard	Hard	Very hard	Very hard	Very friable

Source: PADMU - Pakistan desertification monitoring unit 1986, modified by the author.

Table 6. Land system.

Mega	Macro	Percentage
Lesser Cholistan (12,370 km^2): Wind resorted dissected terrace remnants	1. Flats with sandy terrace remnants	31.7
	2. Sandy terrace remnants with some flats	15.8
Greater Cholistan (13,960 km^2): Wind resorted sandy terraces	3. Low sand dunes and hollows	15.3
	4. Longitudinal ridges and valleys	6.4
	5. Low transverse ridges and valleys	0.7
	6. Alveolar ridges and valleys	15.3
	7. High transverse ridges and valleys	13.0
	8. Partly covered channels	1.8

Source: Based on field survey, January 1998, January 1999 and June 2000.

harvesting and collection.

Conclusions

Mega land system (Lesser and Greater Cholistan) are split into macro land system (Table 6) which controls soils, moisture and eventually vegetation. Landforms map of Cholistan is compiled from the integration of information on geomorphology, soil, wind erosion, and vegetation. The units are biophysically homogeneous and versatile and such can be used confidently for land use planning of crops, range/livestock, irrigation and road alignment.

ACKNOWLEDGEMENTS

The author wish to thank Dr. Mohammad Arshad, Cholistan Institute of Desert Studies, Islamia University Bahawalpur, Pakistan for providing technical assistance during the field study of Cholistan Desert and valuable comments on a draft-version of this paper.

REFERENCES

Ahmad F (1999). Eco-regeneration and runoff collection in Cholistan, UNEP. Desertif. Control Bull., 35: 50-54.

Ahmad F (1999a). Ecological restoration in Cholistan. J. Geograph., 2(1): 34-38.

Ahmad F (2002). Socio-economic dimensions and ecological destruction in Cholistan, Ph.D. dissertation (unpublished), Department of Geography, University of Karachi, Pakistan.

Ahmad F (2005). Historical and archaeological perspectives of soil degradation in Cholistan". J. Geogr., 10: 31-35.

Ahmad F (2006). Soil classification and micromorphology: A case study of Cholistan desert, Abstract in 11th Congress of Soil Science, 28-31 March 2006, Islamabad, Pakistan, p.133.

Ahmad F (2008). Runoff farming in reducing rural poverty in Cholistan desert, Revista Sociedade & Natureza, Instituto de Geografia, Universidade Federal de Uberlândia, Brazil, 20(1): 177-188.

Ahmad F (2010). Leptochloa Fusca cultivation for utilization of salt-

affected soil and water resources in Cholistan desert. Revista Sociedade & Natureza, Instituto de Geografia, Universidade Federal de Uberlândia, Brazil, 22(1): 141-149.
Ahmad F, Gulzar F, Shirazi SA, Farooq S, Ali Z (2005). Agropastoral systems in Cholistan, Proceedings of Silvopastoralism and Sustainable Land Management International Congress, 18-24 April 2004, Lugo, Spain, CAB International, United Kingdom, pp.346-347.
Ahmad TF, Akbar G, Tahir MB, Ahmad I (1992). Developing Cholistan desert – A perspective. Prog. Farm., 12(6): 35-40.
Auj N, Zaman A (1987). Ancient Bahawalpur, Caravan Book Centre, Multan, Pakistan.
Auj N, Zaman A (1991). Cholistan: land and people, Caravan Book Centre, Multan, Pakistan. Kindly provide the page number.
Auj N, Zaman A (1995). Legacy of Cholistan, Caravan Book Centre, Multan, Pakistan.
Baig MS, Akram M, Hassan MA (1980). Possibilities for range development in Cholistan desert as reflected by its physiography and soils, The Pakistan. J. For., pp.61-71.
Baig MS, Khan, EH, Zaheer MR, Ahmad M (1975). Reconnaissance soil survey of Cholistan, Soil Survey of Pakistan, Lahore, Pakistan.
Buol SW, Southard RJ, Graham RC, McDaniel PA (2003). Soil genesis and classification, Blackwell Publishing Company, p. 494.
FAO (1965). Guidelines for soil profile descriptions, Soil Survey and Fertility Branch, Land Water Development Division, FAO, Rome, Italy.
FAO/ADB (1993). Cholistan area development project, Food and Agriculture Organization (FAO), Rome, Italy. Report No. 59/53 ADB-PAK 58,
Jenny H (1941). Factors of soil formation, McGraw-Hill, New York, p.146.
Kapur S, Eswaran H, Blum WEH (2011). Sustainable land management, Springer, Verlag Berlin.
Kapur S, Stoops G (2008). New trends in soil micromorphology, Springer, Verlag Berlin.
Khan SRA (1992). Agricultural development potential of Cholistan desert. Linkers Publishers, Lahore, Pakistan.
Malik MK (1995). Surface water development in Cholistan desert through rainfall harvesting and conservation technologies. J. Pure. Appl. Sci., 14(1): 49-61.
McFadden LD, Kneupfer PLK (1990). Soil geomorphology: The linkage of pedology and superficial processes, Geomorphology, Proceedings of the 21st Annual Binghamton Symposium in Geomorphology, 3: 197-205.
Mohammad N (1989). Rangeland management in Pakistan. International Centre for Integrated Mountain Development, Katmandu, Nepal.
Mohammad N, Mirza Sarwat N (1985). Range management and forage research in Pakistan". Prog. Farm., 5(5): 44-51.
PADMU (1986). Desertification process in Cholistan desert", PCRWR Publication No. PADMU 7/86.
Retallack GJ (1994). The environmental factor approach to the interpretation of Paleosols in Factors of Soil Formation, Fiftieth Anniversary Retrospective, Special Publication: Soil Science Society of America, Madison, p. 33.
Shirazi SA, Ahmad F, Gulzar F, Farooq S (2004). Soil classification and micromorphology: A case study of Cholistan desert, Proceedings of 12th International Meeting on Soil Micromorphology, 20-26 September, Adana, Turkey, pp. 61-63.
Tahir MA, Naim MA, Ahmad M (1995). Landforms and soils of Bahawalpur division, Proceedings of the 6th All Pakistan Geographical Conference, 26-29 December 1993, The Islamia University of Bahawalpur, Pakistan, pp. 311-325.
USDA (1951). Soil survey manual hand book, Government Printing Office, USDA, Washington, USA, p. 18.

The potential of four non traditional legumes in suppressing the population of nematodes in two Ghanaian soils

K. Osei[1]*, J. O. Fening[2], S. R. Gowen[3] and A. Jama[3]

[1]Crops Research Institute, Box 3785, Kumasi, Ghana.
[2]Soil Research Institute, Academy Post Office, PMB, Kwadaso Kunmsi, Ghana.
[3]University of Reading, School of Agriculture, Policy and Development, Box 237, Earley Gate, RG6 6AR, Reading, UK.

Most African farmers could not afford the purchase of pesticides to control crop pest, alternatives such as the use of phytochemicals are most useful. For best results, sound recommendations are needed on the use of different antagonistic plants in controlling pest. In Ghana few reliable data are available on the potential of antagonistic plants to control nematode population. A field trial was conducted in the forest and forest savanna transitional zones of Ghana to evaluate the potential of four leguminous crops in suppressing nematodes population. *Mucuna pruriens, Crotalaria spectabilis* and *C. retusa* were antagonistic but *Phaseolus vulgaris* was a favourable host to the pest. *M. pruriens* did not gall and no egg masses found on roots. *C. spectabilis* and *C. retusa* recorded insignificant galling indices of 2.6 and 1.8 respectively. However, *P. vulgaris* was significantly affected by *Meloidogyne incognita* as expressed by the significantly higher gall and egg mass indices. In addition, nodulation was greatly reduced in *P. vulgaris*. Consequently, significant higher population of nematodes (an average of 600 J2/g^{-1}root) was recovered from *P. vulgaris* whilst lower population (an average of 50 J2/g^{-1}root) was recovered from *M. pruriens*. The growth and yield of okra following *M. pruriens* resulted in a significant 48% yield increase over the control and *P. vulgaris* treatments. *M. pruriens* recorded 31 and 38% yield increase over *C. spectabilis* and *C. retusa* respectively.

Key words: Agricultural production, antagonistic plants, legumes, *Meloidogyne incognita.*

INTRODUCTION

Agricultural production, world wide, is beset with a myriad of constraints. In sub Saharan Africa, the two major constraints are low and declining soil fertility and pest infestation. The low and declining soil fertility arises from continuous cultivation where the levels of soil replenishment are too low to mitigate the process of nutrient mining (IFDC, 1998). Plant parasitic nematodes are important pests that cause economic yield losses of crops, (Orion, 2000; Fourie et al., 2001 and Koenning et al., 2001) especially in the tropics where environmental conditions and cropping systems favour their development (Luc et al., 2005). They have extensive host ranges (Kratochvil, 2004; Dickson and De Waele, 2005; Quénéhervé, 1995), thus making their control extremely difficult. Root-knot nematodes cause root galling that is often accompanied by stunting, chlorosis, and wilting of the host plants. Total crop failure can occur when nematodes population exceed economic threshold level (ETL) (Sikora and Fernandez, 2005). Nematicide application is the single most effective management strategy against nematodes (Thomas, 1996). However, chemical usage presents environmental problems (Bell, 2000). The use of antagonistic plants in agricultural production offers tremendous potential as nematicidal principles have been identified in many higher plants (Chitwood, 2002). Legume family (Fabacea) contains members with considerable agricultural utility. They are desirable rotational and cover crops. By fixing atmospheric nitrogen, these legumes play a key role in

*Corresponding author. E-mail: kofifening@yahoo.com.

restoration of degraded soils (Peoples and Craswell, 1992; Thomas et al., 1997). Evaluating legumes for nematode management stimulate greater exploitation of the legumes in cropping systems.

MATERIALS AND METHODS

The studies were conducted in 2004 and 2005 at two sites, Ejura (07°24'N, 01°21'W) in the forest savanna transition zone and Fumesua (01°28'N, 06°41'W) in the forest zone. Composite soil samples from rhizosphere region at a depth of 15 cm were taken from each site at the beginning and end of the study. At each site, soil was collected and bulked from three locations, air dried and sieved to pass through a 2 mm mesh. Soil reaction (pH) was determined by using the glass electrode pH meter (Radiometer Analytical, Lyon, France) in water at 1:1 (soil: water) ratio. Organic carbon (%C) was determined by the dichromate-oxidation method of Walkley-Black (Walkley and Black, 1934). From these values, the organic matter (%OM) content of soil was computed using a correction factor of 1.72 (Nelson and Sommers, 1996). Total nitrogen (%N) was determined by the Kjeldahl method (Pereira et al., 2006).

Tomato bioassay

Tomato bioassay was performed prior to the establishment of the legumes. Soil samples at a depth of 20 cm were taken randomly from twenty plots of 5 x 4 m by walking in a zigzag fashion. From each plot, four samples were taken making a total of eighty. The soil samples were put into eighty plastic pots each measuring 5 kg. One seedling each of a two–week old tomato, *Solanum lycopersicum* L. (cv. Tiny Tim) which had been germinated on tissue paper in a 9 cm petri dish was transplanted. The pots were arranged in a randomized complete block design in a green house. The plants were grown for 8 weeks and root gall index was determined (Netscher and Sikora, 1990).

Treatments and experimental design

Experiments were conducted with five treatments namely; *C. retusa, C. spectabilis, M. pruriens, P. vulgaris* and weeded fallow. The experiments were replicated four times and arranged in a randomized complete block design (RCBD). Seeds of these legumes were directly sown in plots of four, 5 m rows 1 m apart. Hand weeding was done three times during the growth period. Fallow plots were weeded whenever the other plots were weeded. Three months after sowing, shoot biomass of the leguminous crops was determined using the quadrat sampling method. Shoots in four 50 x 30 cm quadrats were taken per plot. Root galling, nodule score, root length, number of egg mass index per 5 g and J2 g 200 cc of soil were also recorded.

Cultivation of okra

Three healthy seeds of *Abelmoscus esculentus* (L.) Moench were sown 50 cm apart in four 5 m rows on the plots previously where legume was grown. Seedlings were thinned to one three weeks after planting. Hand weeding was done three times during the crop season. A second crop was sown after the harvest of the current crop. Observation was recorded on: fresh weight of root, fresh and dry weight of stem, yield, gall index, number of egg mass per 5 g root and J2 per 200 cc soil root. Okra roots, one per plant were randomly sampled and extracted for eggs according to the procedure of Hussey and Barker (1973).

Statistical analyses

Where experiments were conducted and repeated at separate locations and in different years, the mixed model (REML) approach was used in the analysis of data. Nematode count data was log transformed, $\ln(x + 1)$ and indices based data were square root transformed, $\sqrt{(x + 0.5)}$ to comply with the assumption of normal distribution. Continuous data such as yield of treatments was however not transformed. Two-way and one-way analyses of variance (ANOVA) were used for two or more factor and single factor experiments for randomized complete block design. Means were separated using the Standard Error Difference (SED) test ($P \leq 0.05$). All statistical analyses were performed using GenStat Release 8.1 (Lawes Agricultural Trust, VSN International).

RESULTS

Chemical analysis of soil

There were no differences in the initial levels of soil chemical properties at the two locations (Table 1). Significant interaction was however observed between the treatments and the soil chemical properties after the cultivation of the legumes. At both locations, *M. pruriens* significantly increased the levels of organic carbon, nitrogen, and organic matter. The weeded fallow on the other hand recorded decreased levels of organic carbon and nitrogen but increased the organic matter content of the soil at Ejura. At Fumesua, weeded fallow treatment decreased only the nitrogen content of the soil. *P. vulgaris, C. spectabilis* and *C. retusa* did not change the chemical status of the soil after cultivation (Table 2).

Tomato bioassay

Root-knot nematodes were distributed in soil from all the plots at both locations. There were significant differences between the plots assigned to the different treatments. At Ejura, the highest gall index recorded was 4.5 while the lowest was 1.9. Similarly, at Fumesua, the highest gall index was 3.5 and the lowest was 1.3. In a related experiment, a mixed population of root-knot nematodes (*Meloidogyne arenaria-incognita* complex) was identified from the plots using the Sequence Characterized Amplified Region - Polymerase Chain Reaction (SCAR – PCR) method.

The reaction of leguminous cover crops to root-knot nematode infection

The response of the cover crops to infection by *Meloidogyne* species was measured by root galling and egg mass intensity. Highly significant differences ($P < 0.01$) were observed in these variables at both sites

Table 1. Initial chemical properties of soil.

	Ejura				Fumesua			
Plot no.	pH	C %	N %	OM%	pH	C %	N%	OM%
1	0.85	6.0	0.69	0.04	5.2	0.97	0.06	0.91
2	0.80	5.9	0.56	0.05	5.5	0.92	0.06	0.78
3	0.73	6.1	0.58	0.05	5.2	0.89	0.05	0.65
4	0.69	5.9	0.66	0.08	5.5	0.93	0.07	0.76
5	0.75	6.1	0.65	0.07	5.0	0.68	0.07	0.80
P- value		0.748				0.276		
SED		0.09				0.14		

Table 2. Effect of leguminous crops on some chemical properties of soil[†].

Treatment	Ejura				Fumesua			
	pH	C %	N %	OM%	pH	C %	N %	OM %
M. pruriens	6.1	1.03	0.09	1.77	5.1	1.64	0.14	2.82
P. vulgaris	6.0	0.74	0.07	1.27	5.4	1.37	0.12	2.37
C .retusa	6.2	0.92	0.08	1.58	5.2	1.03	0.09	1.87
C. spectabilis	6.0	0.81	0.07	1.39	5.5	1.09	0.10	2.12
Fallow	5.5	0.61	0.05	1.06	5.1	1.00	0.06	1.47
P- value		< 0.01				< 0.01		
SED		0.01				0.02		

[†] Figures are changes in initial measurements after the cultivation of the cover crops, except pH values which are absolute values.

Table 3. Gall index GI (0-10), egg mass index EI (0-10) and susceptibility index GI x EI of leguminous crops at Ejura and Fumesua.

Treatment	GI‡	EI‡	GI × EI	GI‡	EI‡	GI × EI
M. pruriens	0*	0*	0*	0*	0*	0*
P. vulgaris	7.0 (2.7)	7.9 (2.8)	55.2 (7.4)	6.9 (2.6)	8.0 (2.8)	55.6 (7.4)
C. retusa	2.6 (1.6)	3.7 (1.9)	9.7 (3.1)	2.5 (1.6)	3.0 (1.7)	7.6 (2.7)
C. spectabilis	1.8 (1.3)	2.7 (1.6)	4.9 (2.2)	2.0 (1.4)	3.3 (1.8)	7.0 (2.6)
P- value		<0.001			<0.001	
SED		(0.1)			(0.2)	

‡ √ (x + 0.5) transformed data used in analysis in parenthesis.Figures not used in analysis.

respectively (Table 3). Similarly, a strong interaction was observed between the treatments (leguminous crops) and gall, egg mass and susceptibility indices. *M. pruriens* did not gall and no egg mass was found. The susceptibility indices of the *Crotalaria* species were significantly low at the two locations. However, *P. vulgaris* recorded significantly higher susceptibility indices; 11 times and 8 times greater than *C. spectabilis* at Ejura and Fumeusa respectively. Populations of second stage juveniles (J2) per gram root of the cover crops were significantly different ($P < 0.01$). Juveniles per gram of root in *P. vulgaris* were approximately 13 times more than in *M. pruriens* (Table 4).

The fresh, dry shoot and fresh root weights of the cover crops were not significantly different ($P > 0.05$) at the different ecological zones. The fresh, dry shoot and fresh root weights of *M. pruriens* (76.4, 54.1 and 50.8 g) at Ejura were not different from (82.0, 58.6 and 51.9 g) at Fumesua. Similarly, *P. vulgaris* recorded (76.2, 53.1 and 84.2 g) at Ejura which were not different from (81.7, 59.7 and 79.0 g) at Fumesua. The weights of the *Crotalaria* species were also not different.

Root length of the cover crops was measured to determine the extent to which the active ingredients in the root could affect nematodes in the soil system. The cover crops ability to nodulate freely at both locations was also

Table 4. Numbers of second stage juveniles (J2) g^{-1} root of leguminous crops at Ejura and Fumesua.

Treatment	J2	J2§	J2	J2§
M. pruriens	64	(1.8)	47	(1.7)
P. vulgaris	657	(2.8)	628	(2.8)
C. retusa	158	(2.2)	124	(2.1)
C. spectabilis	114	(2.0)	100	(2.0)
P- value	<0.001			
SED	(0.05)			

Data are means of four replications.
§ln(x + 1) transformed data used in ANOVA in parenthesis.

Table 6. Nodule score (0-5) of leguminous crops at Ejura and Fumesua.

Treatment	Nodule score	Nodule score‡
	Ejura	Fumesua
M. pruriens	2.9 (1.7)	2.7 (1.6)
P. vulgaris	0.4 (0.5)	0.4 (0.4)
C. retusa	1.2 (1.0)	1.7 (1.3)
C. spectabilis	1.8 (1.2)	1.3 (1.1)
P- value	<0.01	
SED	0.2	

Data are means of four replications
‡ $\sqrt{(x + 0.5)}$ transformed data used in ANOVA in parenthesis.

Table 5. Root length of leguminous crops at Ejura and Fumesua.

Treatment	Root length (cm)	Root length (cm)
	Ejura	Fumesua
M. pruriens	96.3	87.5
P. vulgaris	82.0	63.8
C. retusa	64.3	24.0
C. spectabilis	51.5	19.8
P- value	<0.01	
SED	5.4	

Data are means of four replications.

studied as an index to nitrogen fixation. Significant differences in treatments (P < 0.01) were observed at both locations (Tables 5 and 6). However, there was no interaction between treatments and location. There was no difference in nodulation between *C. spectabilis* and *C. retusa* but the *Crotalaria* varieties recorded significantly higher nodulation than *P. vugaris* and lower than *M. pruriens*.

Effect of leguminous cover crops on yield of okra

There were highly significant differences (P < 0.01) regarding gall, egg mass and susceptibility indices of okra which followed the cultivation of leguminous cover crops at both locations in both years of experiment (Table 7). Okra cultivated on the plots previously sown to *P. vulgaris* recorded significantly higher (P < 0.01) gall, egg mass, susceptibility indices and J2 g^{-1} root of 59, 52, 79 and 77% than *M. pruriens* which recorded the lowest 33, 25, 51 and 46% than the weeded fallow treatment. Gall, egg mass, susceptibility indices and J2 g^{-1} root of okra from *M. pruriens* plots were not significantly different from *C. spectabilis* plots. However, okra grown on plots previously sown to *C. spectabilis* recorded significantly lower indices and population per gram of root than the weeded fallow treatment.

Highly significant differences were observed in fresh and dry shoot and fresh root weights of okra cultivated on plots previously sown to leguminous cover crops (P < 0.01) at both locations in both years. Fresh shoot weight (77.1g) of okra planted on plots previously cultivated with *M. pruriens* was significantly higher than (54 and 59 g) recorded from plots previously sown to *C. retusa* and *C. spectabilis* respectively. Similarly, dry shoot weight of 53.2 g from *M. pruriens* plots was significantly higher than 35.3 and 42 g from *C. retusa* and *C. spectabilis* plots respectively. A similar trend was observed in fresh root weight of okra as a weight of 19.4 g from *M. pruriens* plots was significantly higher than 15.5 and 15 g from *C. retusa* and *C. spectabilis* plots respectively.

Highly significant differences were observed amongst treatments regarding the yield of okra (Table 8). The yield of okra from plots previously cultivated with *M. pruriens* was the highest and exceeded the yield of okra from weeded fallow plots by 48%. Similarly, yield from plots previously grown with *C. spectabilis* out yielded that of weeded fallow by 25%. There was no difference between the yield of okra from weeded fallow and that of *P. vulgaris* plots. The performance of *C. retusa* and *C. spectabilis* were the same.

DISCUSSION

M. pruriens consistently increased the levels of nitrogen and organic matter at both locations probably due to its high biomass and higher mineralization rate. The decline in organic carbon and nitrogen levels in the weeded fallow is not unexpected. A number of tropical legumes have been used to suppress root-knot nematode population densities. For instance *Mucuna deeringiana* was used to control a variety of root-knot nematode species and races in green house and micro plots in the United States (Rodriguez et al., 1992; McSorley and Gallaher, 1992) and in Puerto Rico (Vicente and Acosta, 1987). In Martinique, *M. pruriens* was effectively used to

Table 7. Effect of earlier cover crop treatment on *Meloidogyne* spp. gall index (GI), egg mass index (EI), susceptibility index (GI x EI) and J2 g^{-1} root of okra at Ejura and Fumesua in 2004 and 2005.

Treatment	GI‡	EI‡	GI × EI	J2 $g^{-1§}$
M. pruriens	2.5 (1.5)	3.4 (1.8)	9.0	98 (1.9)
P. vulgaris	6.1 (2.5)	7.1 (2.7)	43.5	432 (2.6)
C. retusa	3.9 (1.8)	4.2 (2.0)	13.6	217 (2.2)
C. spectabilis	3.0 (1.7)	4.2 (2.0)	13.2	141 (2.1)
Weeded fallow	4.1 (2.0)	5.3 (2.3)	21.3	234 (2.3)
P- value	(<0.001)	(<0.001)	<0.001	(<0.001)
SED	(0.06)	(0.06)	2.1	(0.03)

Data are means of four replications.
‡ $\sqrt{(x + 0.5)}$ transformed data used in ANOVA in parenthesis.
§ ln (x + 1) transformed data used in ANOVA in parenthesis.

Table 8. Effect of cover crops on the yield (kg/ha) of okra in 2004 and 2005 at Ejura and Fumesua

Treatments	Predicted mean
Mucuna pruriens	3648
Phaseolus vulgaris	1872
Crotalaria spectabilis	2527
Weeded fallow	1885
Crotalaria retusa	2257
P- value	<0.001
SED	334.1

Data are means of four replications.

control *M. incognita* and *Rotylenchulus reniformis* in vegetables in polytunnels (Quénéhervé et al., 1998). In the current study, *M. pruriens* significantly reduced *Meloidogyne arenaria-incognita* complex. In addition, roots of *M. pruriens* did not gall neither were egg masses found in the root system. This observation is consistent with the work of McSorley et al., (1994). This antagonistic activity of *Mucuna* might be due to the production of phytoalexins by the roots (Vargas et al., 1996). The *Crotalaria* species were also effective in reducing nematode population. According to Fassuliotis and Skucas (1969) the nematicidal compound in the root of *C. spectabilis*, monocrotaline, is responsible for suppression of nematode density. *P. vulgaris* on the contrary stimulated population increase, roots were severely galled and significantly higher egg mass index was recorded. *P. vulgaris* was considered highly susceptible to the pest under the conditions of this study which was consistent with the observation of (Sikora et al., 2005). Okra cultivated on *M. pruriens* plots recorded significantly low gall, egg mass and susceptibility indices. The most remarkable potential became evident in J2 g^{-1} root where populations were 77 and 58% lower than those on *P. vulgaris* and weeded fallow plots respectively. The residual effect of *M. pruriens* did not only control mixed population of root-knot nematodes and increased plant growth but also resulted in significant yield increase. *M. pruriens* out yielded the control and *P. vulgaris* treatments by 48 and 49% respectively. There was no difference in yield between *P. vulgaris* which was considered highly susceptible to the *Meloidogyne arenaria-incognita* complex and the weeded fallow (control) treatment.

Conclusion

The study has shown that *M. pruriens, C. spectabilis* and *C. retusa* were all effective in suppressing nematode population densities. *M. pruriens* however could be the most preferred management candidate for the following reasons: In addition to all the benefits accrued from inclusion of legumes in cropping systems, *M. pruriens* provides excellent hay for livestock, and its seeds are used as feed supplement after processing (Maasdorp et al., 2001). More importantly, *M. pruriens* is used for food in some African countries (Rachie and Roberts, 1974). Fallowing also causes some challenges in nematode management, hence a pulse crop without antagonistic effect can be included to maintain the population in the field.

ACKNOWLEDGEMENTS

The authors are grateful to the Agricultural Services Sub-sector Investment Project for the fellowship and also indebted to the Council for Scientific and Industrial Research (CSIR) Ghana for the study leave granted to Kingsley Osei for his PhD studies at the University of Reading, England.

REFERENCES

Bell CA (2000). Fumigation in the 21st century. Crop Protection 19: 563-569.

Chitwood DJ (2002). Phytochemical based strategies for nematode control. Annual Review of Phytopathol., 40: 221-249

Dickson DWD, Waele D (2005). Nematode Parasites of Peanut. In: Luc M, Sikora RA and J Bridge (eds). Plant parasitic nematodes in subtropical and tropical agriculture. Wallingford, UK: CABI Wallingford, UK. Pp 393-436.

Fassuliotis G, Skucas GP (1969). The effect of pyrrolizidine alkaloid ester and plants containing pyrrolizidine on *Meloidogyne incognita* acrita. J. Nematol., 1: 287-288.

Fourie H, McDonald AL, Loots GC (2001). Plant parasitic nematodes in field crops in South Africa. 6, Soybean. Nematolo., 3: 447-454.

Hussey RS, Barker KR (1973). A comparison of methods of collecting inocula of *Meloidogyne* spp. including a new technique. Plant Disease Reporter 57: 1025–1028.

IFDC (1998). Soil Nutrient Depletion. In: Sarfo EY and EA Dennis (eds). Report of the Sub-Committee on Fertilizer use for the National Agricultural Research Project (NARP), Ghana. AM Peanut Res. Edu. Soc.pp. 139-163.

Koenning SR, Barker KR, Bowman DT (2001). Resistance as a tactic for management of *Meloidogyne incognita* on cotton in North Carolina. J. Nematol. 33: 126-131.
Kratochvil RJ, Sardanelli SI, Everts K, Gallagher E (2004). Evaluation of crop rotation and other cultural practices for management of root-knot and lesion nematodes. Agron. J. 96: 1419-1428.
Luc M, Sikora RA, Bridge J (2005). Plant parasitic nematodes in subtropical and tropical agriculture. Wallingford, UK: CABI Wallingford, UK. p. 871
Maasdorp BV, Jiri O, Titterton M (2001). Production and utilization of *mucuna* in Zimbabwe with emphasis on its use as a feed resource. In: Flores M, Eillitta M,
McSorley R, Gallaher RN (1992). Comparasion of nematode population densities on six summer crops at seven sites in north Florida. J. Nematol., 24: 699 –706.
McSorley R, Dickson DW, De Brito JA, Hewlett TE, Frederick JJ (1994). Effect of tropical rotation crops on *Meloidogyne arenaria* population densitiesand vegetable yields in micro plots. J. Nematol. 26: 175-181.
Nelson DW, Sommers LE (1996). Total carbon, organic carbon and organic matter. Agron., 9: 961-1010.
Netscher C, Sikora RA (1990). Nematode parasites of vegetables. **In:** Luc M, Sikora RA and J Bridge (eds). Plant parasitic nematodes in subtropical and tropicl agriculture. Cambrian Printers, Ltd, Aberystwyth. pp.237-283.
Orion D (2000). Nematodes of agricultural importance in Israel. Nematolo., 2: 735-736.
Peoples NB, Craswell ET (1992). Biological nitrogen fixation: investments, expectations and actual contributions to agriculture. Plant and Soil 141:13- 40.
Pereira MG, Espindula Jr A, Valladares G S, Cunha dos Anjos LH, Benites V and Schultz N (2006). Comparison of total nitrogen methods applied for Histosols and soil horizons with high organic matter content. Soil Science and Plant Analysis 37: 939-943.
Quénéhervé P, Drob F, Topart P (1995). Host status of some weeds to *Meloidogyne* spp., *Pratylenchus* spp., *Helicotylenchus* spp and *Rotylenchulus reniformis* associated with vegetables cultivated in polytunnels in Martinique. Nematropica 25: 149-159.
Quénéhervé P, Topart P, Martiny B (1998). *Mucuna pruriens* and other rotational crops for control of *Meloidogyne incognita* and *Rotylenchulus reniformis* in vegetables in polytunnels in Martinique. Nematropica 28:19- 30.
Rachie KO, Roberts LM (1974). Grain legumes of the lowland tropics. Advances in Agronomy 26: 1-132.
Rodríguez-Kábana R, Pinochet J, Robertson DG and Wells L (1992). Crop rotation studies with velvetbean (*Mucuna deeringiana*) for the management of *Meloidogyne* spp. J. Nematol., 24: 662-668.
Sikora RA, Fernandez E (2005). Nematode parasites of vegetables. In: M. Luc M, Sikora RA and J Bridge (eds). Plant parasitic nematodes in subtropical and tropical agriculture. CAB International, Wallingford, UK. 319-392.
Sikora RA, Greco N, Silva JFV (2005). Nematode parasites of food legumes. **In:** M. Luc M, Sikora RA and J Bridge (eds). Plant parasitic nematodes in subtropical and tropical agriculture. CAB International, Wallingford, UK. 2005; 259-318.
Thomas WB (1996). Methylbromide: effective pest management tool and environmental threat. Supplement J. Nematol. 28:586-590.
Thomas RJ, Asakawa NM, Rondon MA, Alarcon HF (1997). Nitrogen fixation by three tropical forage legumes in an acid-soil Savanna of Columbia. Soil Biol. Biochem., 29: 801-808.
Vargas R, Rodríguez A, Acosta N (1996). Components of nematode suppressive activity of velvet bean, *Mucura deeringiana.* Nematropica pp. 26: 323[Abstr.]
Vicente NE, Acosta N (1987). Effects of *Mucuna deeringiana* on *Meloidogyne incognita*. Nematropica 17: 99-102.
Walkley A, Black IA (1934). An examination of the Degtjareff method for determining organic carbon in soils: Effect of variations in digestion conditions and of inorganic soil constituents. Soil Sci., 63: 251-263.

The effects of soil density on the vegetation of the Umfolozi catchment in South Africa

Clifford Tafangenyasha*, Amos T. Mthembu, Hector Chikoore, Nothile Ndimande, Sifiso Xulu and Nonkululeko Gcwensa

Department of Geography and Environmental Science, University of Zululand, P, Bag X10003, KwaDlangezwa 3886, South Africa.

The effects of soil compaction on plant productivity were studied in the dry season of September 2008 in Umfolozi catchment, South Africa. This study used 10 paired plots each measuring 20 x 20 m and 1 x 1 m quadrats randomly located within the study plots in both degraded and undegraded sites. Soil samples were collected on the outer edge of the quadrats within the study plots. Soil moisture (%) in the soil samples collected from the degraded and undegraded sites was 8.94 and 13.48%, respectively. Organic matter (%) was 3.17 and 5.46% on degraded and undegraded sites, respectively. Average soil bulk density (g/cm^{-3}) was 0.99 g/cm^{-3} (CV=1.81 %) and 0.83 g/cm^{-3} (CV=5.02 %) on degraded and undegraded sites, respectively. Porosity was greater on undegraded sites (36.2%) than on degraded sites (34.2%). Seedling density (no/pot) was 18 seedlings per pot in soil samples collected on degraded and 10 seedlings per pot on undegraded sites. Seedling mortality (no/pot) was 17 seedlings per pot and 9 seedlings per pot in the soil samples collected on degraded and undegraded sites, respectively. Spearman rank correlation showed strong correlation between seedling density and soil compaction ($r=-0.481$, $P<0.05$). Seedling emergency density (no/pot) reached highest level on degraded sites compared to undegraded sites. Seedling mortality (no/pot) reached highest level on degraded sites. The study suggests that soil compaction and soil bulk density are critical for an efficient resource management in rangelands.

Key words: Soil density, soil compaction, degraded landscapes, semi-arid areas.

INTRODUCTION

The effects of soil compaction on plant productivity have frequently been studied in the context of agriculture, but less information exists about this relationship in the context of rangeland resources. Several studies (Dougill and Cox, 1995a, b; Wilson et al., 1984; Tafangenyasha and Campbell, 1998) have shown that overgrazing leads to increases in runoff and soil erosion. Runoff and soil erosion processes are the result of physical changes in the soil; and vegetation and specific physical change in the soil is related to soil compaction (Morgan, 1980). Soil compaction has been described as simply the increase of soil bulk density and is usually the result of trampling and root thinning (Engels, 2009). Soil compaction and soil strength are also a function of bulk density and water content (Cyzt, 2004; Tokunaga, 2006). The processes by which water enters, moves through and exits soil are essential for sustaining plants and soil organisms, transporting nutrients and recharging surface and ground water. Compaction by animals or natural processes of top soil removal by overland flow can affect soil water movement by increasing bulk density and decreasing porosity and infiltration. These changes can result in less soil water storage, poor nutrient movement, slowed gas exchange and restricted root growth, all of which can cause a reduction in crop yield (Cyzt, 2004).

Soil functions affected by soil compaction include air

*Corresponding author: E-mail: ctafangenyasha@yahoo.com.

Figure 1. Present study area situated about 10 km northeast of St Paul Mission near Nqutu in central KwaZulu-Natal (after Botha and Fedoroff, 1995).

capacity and permeability, available water capacity, water conductivity and root length, penetration resistance and dry bulk density (Taylor and Gardner, 1963; Daddow and Warrington, 1983; Tsimba et al., 1999). Soil susceptibility to erosion depends on soil properties and external factors like climate and soil use (Davidson, 1980; Hudson, 1995). Increased runoff may occur as a result of compaction and increased bulk density. Soil physical properties, including texture, structure, bulk density, soil organic matter (SOM) levels and water content control how well soil resists compaction. Compaction is the process of increasing the bulk density of a soil or aggregate by driving out air. Bulk density is dependent on moisture content. At very high moisture contents, the maximum dry density is achieved when the soil is compacted to nearly saturation, where almost all the air is driven out. Available soil water also declines when the void volume is decreased (Rockich et al., 2001). When this occurs, nutrients and water may become limited because the plant's demands exceed the ability of the roots system to access these resources (Froehlich and McNabb, 1984).

One other useful parameter to consider in soil properties is porosity (Schjonning et al., 1998). Porosity is a measure of the void spaces in a material, and is measured as a fraction, between 0 to 1, or as a percentage between 0 to 100%. To effectively identify harmful changes to the soil through loading, soil properties linked to soil compaction should be investigated. The objective of this paper is to investigate and to explain the importance of soil bulk density to plant establishment on the degraded and undegraded sites in the Umfolozi catchment in central midlands of KwaZulu Natal Province, South Africa. The effects of soil density on the vegetation in the semi-arid areas of the central midlands of South Africa have been little studied thereby prompting an investigation on its nature.

Study area

Covering ten percent of KwaZulu Natal Province, the Umfolozi is the Province's second largest catchment (Figure 1). The study area lies between 28°00'00" S and 28°10'00" S; 30° 37'00" E and 30°55'00"E), 28 km south-west of Vryheid town in central KwaZulu-Natal province (Figure 1). The study area falls within Mondlo Communal Land. The study area and the adjacent areas (Figure 1) have been described by Rein's et al. (2000). The area is known for extreme temperatures with hot summers and cold winter months. The average summer rainfall of the region is 350 mm per annum with range values between 60 to 129 mm (CV ranges from 63 to 217%). The rainfall is concentrated between the months of October and February. Mean annual temperature ranges from 12 to 23°C. Mean annual minimum temperature ranges from 4 to 7°C. The rainfall over the region varies from 700 mm in the south-east to 900 mm in the west, on the Drakensberg range. Frost is a common feature and appears from April to September. The climate of the study area is maintained by the position of the Southern African position in which it lies and its relation to the major atmospheric circulation of the southern hemisphere. The climate, slope and soil conditions combine to determine the potential of a region or site. The area is known for extensive stock farming with goats, sheeps and cattles. The topography of the area consists of mountains, hilly and undulating veld and flat veld. The vegetation consists of typical Karoo veld (Acocks, 1988). The climate, slope and soil conditions combine to determine the potential of the study area to erosion.

According to the Natal Town and Regional Planning Commission (1984) 54% of it has a high natural erosion hazard potential. Ramakgoba (1966) estimates that 17 and 25% of it is already very severely eroded, respectively. Degraded soils have top soil stripped away by overland flow leading to subsoil compaction. The study area is drained by two major tributaries that include the White Umfolozi and Black Umfolozi Rivers. The White Umfolozi and the Black Umfolozi Rivers rise near the study area at altitudes of 1620 and 1524 m a.s.l., respectively, and traverse over 400 km on their course to the sea (Begg, 1988). The Umfolozi catchment contains a wide diversity of physiographic, geological, soil and vegetation. Due to the soils types, the susceptibility to sheet and gully erosion is variable. In their investigation of the 525 km^2 area surrounding Ulundi, Berjak et al. (1986), Liggitt (1988) and Liggitt and Fincham (1989) found that gully erosion was best represented on Dwyka Tillite and to a lesser extent on Pre-Cambrian Rocks, and on slopes less than 9°. The study by Botha (1992) of the 1 360 km^2 area between Dundee and Nqutu included a substantial portion of the Umfolozi catchment, and revea-led the preferential development of gullies in the partly consolidated, bedded sediments of the Masotcheni Formation underlain by Vryheid and Volksrust Forma-tions, Beaufort Group and dolerite bedrock. He also noted that they mostly occur within terrain characterized by concave/ convex topography, highly variable relief and slopes less than 9°.

MATERIALS AND METHODS

An investigation of the influence of soil bulk densities as influenced by agricultural activities was conducted during the dry season of September 2008. The study sites consisted of 10 paired plots in both degraded and undegraded sites. Each plot measured 20 x 20 m and soil samples were collected on the outer edge of the randomly located 1 x 1 m quadrats within the 20 20 x 20 m plots. The plot size of 20 x 20 m plot was determined by using a builder's tape measure. SBD is the ratio of the mass of dry solids to the bulk volume of the soil occupied by those dry solids (Blake and Hartage, 1986). The bulk volume includes the volume of the solids and the pore space. Bulk density of the soil is an important site characterization parameter since it changes for a given soil. It varies with structural condition of the soil, particularly that related to packing. The SBD contains bulk density of the soil based on dry weight and volume using the core method at 0 to 10 cm depth. Samples were collected at randomly identified 20 locations within the study area during the dry season of 2008. Soil bulk density was analysed using the protocols of Blake and Hartage (1986), Doran and Mielke (1984) and Smith and Atkinson (1975). The methods used by Blake and Hartage (1986), Doran and Mielke (1984) and Smith and Atkinson (1975) were found to be repeatable and offering reliable results for the study area.

A shovel, plastic bags to line the excavated hole, water to pour into the plastic bag, and a graduated cylinder were used to determine the volume of water in the bag. Each sample can was identified by the sample site ID code. After bulk density sampling, the soil samples were put in plastic bags for later weighing to obtain the wet weight. Volumetric sampling of collected soil samples were conducted by inserting a plastic bag into the cavity of the ring. The volume of the sample was determined by lining the hole with a plastic bag and measuring the volume of water needed to fill the bag to the level of the surrounding surface. The bag was again removed, soil excavated to a depth of 10 cm, the bag replaced and refilled, and a measure of the volume of the sample between the 0 cm and 10 cm depths calculated. A laboratory dry combustion procedure (loss on ignition (LOI)) was performed on soil samples taken from all 20 sampling points in order to determine the organic matter content (Smith and Atkinson, 1975). Soil samples were dried at 105ºC for 24 h to remove moisture. The dried samples were placed in a muffle furnace at 375ºC for 16 h to incinerate all organic matter. Upon cooling to room temperature, the percent organic matter was calculated by mass after Smith and Atkinson (1975) as follows:

% soil moisture (SOM) weight= [(wet weight-dry weight)/dry weight] × 100

Bulk densities were calculated in the following manner: any dry gravimetric soil samples collected during the dry season of 2009, which weighed more than 65 grams, were assumed to have been collected with the volumetric sampler (i.e., volume = 92.92 cubic centimeters). The bulk density for each of these samples was calculated according to the following Blake and Hartge (1986) formula:

Bulk Density = (Dry weight of sample in grams) / (volume of the sample in milliliters)

Bulk Density = (Dry weight of sample in grams) / 92.92

A mean of all bulk density determinations at a given site-depth was calculated. The number of determinations and the standard deviation of the mean bulk density are also reported. The bulk densities were then calculated using the following Brady and Weil (1996) formula:

Porosity (%) = (1-Bulk density/Particle density)*100

Where 2.65. The mineral grains in many soils are mainly quartz and feldspar, so 2.65 was an adequate average mineral specific gravity for the sand fraction.

Organic matter content can be used as a rough estimate of the total organic carbon content. Soils contain a large variety of organic materials ranging from simple sugars and carbohydrates to the more complex proteins, fats, waxes and organic acids. Important characteristics of the organic matter include their ability to: form water-soluble and water insoluble complexes with metal ions and hydrous oxides; interact with clay minerals and bind particles together; sorb and desorb both naturally-occurring and anthropogenically-introduced organic compounds; absorb and release plant nutrients; and hold water in the soil environment. The loss-on-ignition (LOI) method for the determination of organic matter involves the heated destruction of all organic matter in the soil or sediment. A 10 g was placed in a ceramic crucible which was then heated to between 350 and 44°C overnight (Nelson and Sommers, 1996). The sample was then cooled in a desiccator and weighed. Organic matter content was calculated as the difference between

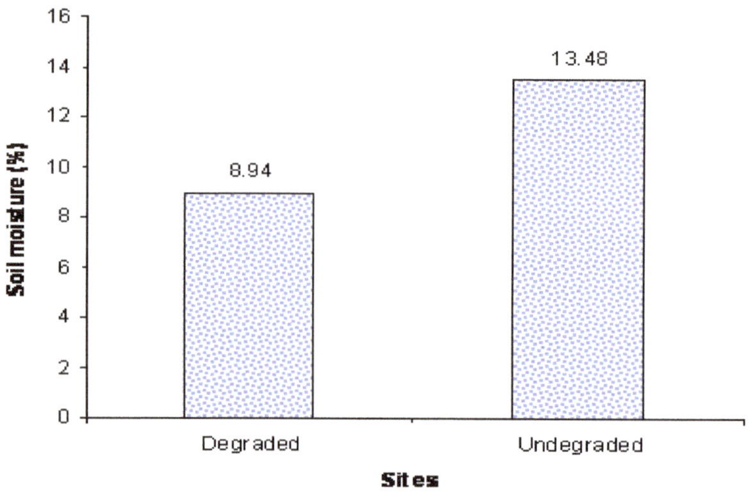

Figure 2. Soil moisture (%) in the soil samples collected from the degraded and undegraded sites.

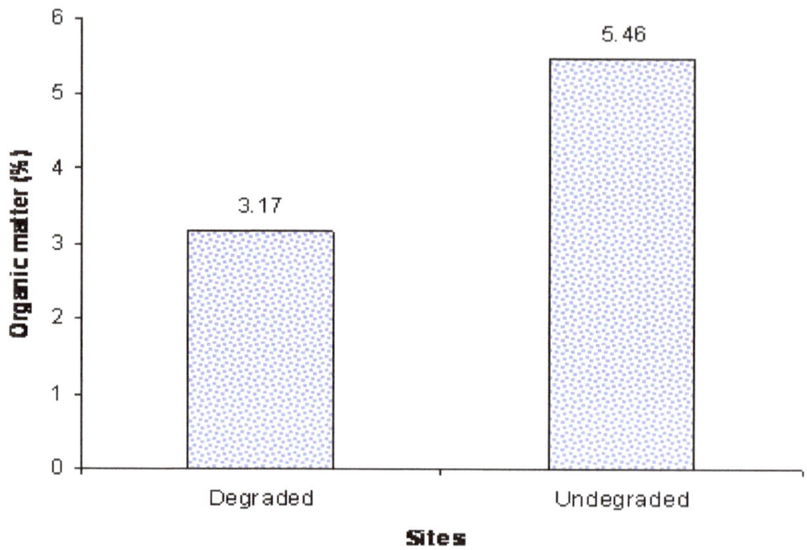

Figure 3. Organic matter (%) in the soil samples collected from the degraded and undegraded sites.

the initial and final sample weights divided by the initial sample weight times 100%. The weight was then corrected for moisture / water content prior to organic matter content calculation.

The factors known to influence seedbank germination included soil compaction, organic matter and soil bulk density. Stepwise multiple regression was carried out using SPSS software (Nie et al., 1975). Stepwise multiple regression identified the order and degree of contribution of the independent variables in terms of prediction of eroded land percentage.

RESULTS

Soil moisture (%) in the soil samples collected from the degraded sites was 8.94 and 13.48% on undegraded sites (Figure 2). Organic matter (%) was 3.17 and 5.46% on degraded and undegraded sites, respectively (Figure 3). Bulk density (g/cm^3) was 0.9 and 0.83 g/cm^3 on degraded and undegraded sites, respectively (Figure 4). The variation in bulk density across the sampled plots is given in Table 1. Porosity was greater on undegraded sites (36.2%) than on degraded sites (34.2%) (Figure 5). Seedling density (no/pot) was 18 seedlings per pot in soil samples collected on degraded and 10 seedlings per pot on undegraded sites (Figure 6). Seedling mortality (no/pot) was 17 seedlings per pot and 9 seedlings per pot in the soil samples collected on degraded and undegraded

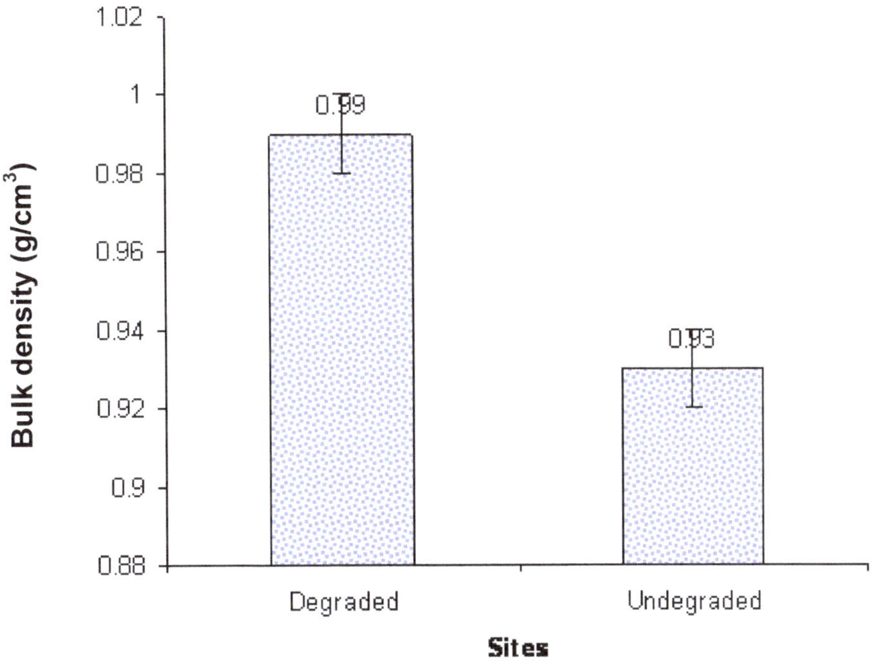

Figure 4. Bulk density (g/cm^3) of the soil samples collected in the degraded and undegraded sites.

Table 1. The variation of bulk density (g/cm^3) across the soil samples collected in the degraded and undegraded sites.

Site no.	Degraded site (g/cm^3)	Undegraded site (g/cm^3)
1	0.96	0.93
2	0.96	0.89
3	0.99	0.97
4	0.96	0.93
5	0.99	0.96
6	1.00	0.95
7	1.00	0.84
8	0.99	0.96
9	1.00	0.88
10	1.00	0.99
11	0.985	0.93
Mean	0.985	0.93
SD	0.0169	0.044
CV (%)	1.81	5.02

sites (Figure 7). Spearman rank correlation showed strong correlation between seedling density and soil compaction ($r=-0.481$, $p<0.05$) (Table 2). Soil moisture and soil organic matter showed significant differences between degraded and undegraded sites (Table 3). Seedling emergency density (no/pot) reached highest level on degraded sites compared to undegraded sites. Seedling mortality (no/pot) reached highest level on degraded sites (Figure 7).

The results of the regression analysis are summarized in Tables 4 to 6. Values of simple r (r values for each variable separately) are given for all the regressor variables, allowing for the fact that these factors do not act in isolation. Given the data set (n=75), correlations were evident for Ivy's erosion class and slope angle, that is the two physical factors. Distance to water, distance to hill and distance to road appeared to play no role in the erosion process in stepwise multiple regression. Stepwise multiple regression identified the order and degree of contribution of the five independent variables in terms of prediction of eroded land percentage. Ivy's erosion class proved to be the most important factor influencing erosion, with a large and highly significant F-value and an r square of 0.142 (Table 4). Slope angle was the second variable to enter the equation, increasing the r square to 0.216 (with a very large F-value associated with the r^2 change).

Significant correlations were evident for plot erosion and soil compaction and bare soil cover. Soil water infiltration, grass cover and litter cover appeared to play no role in erosion process in stepwise multiple regression. Stepwise multiple regression identified the order and degree of contribution of the five independent variables in terms of prediction of eroded land percentage. Soil compaction proved to be the most important factor influencing erosion, with a large and highly significant F-value and an r square of 0.606 (Table 5). Bare soil cover

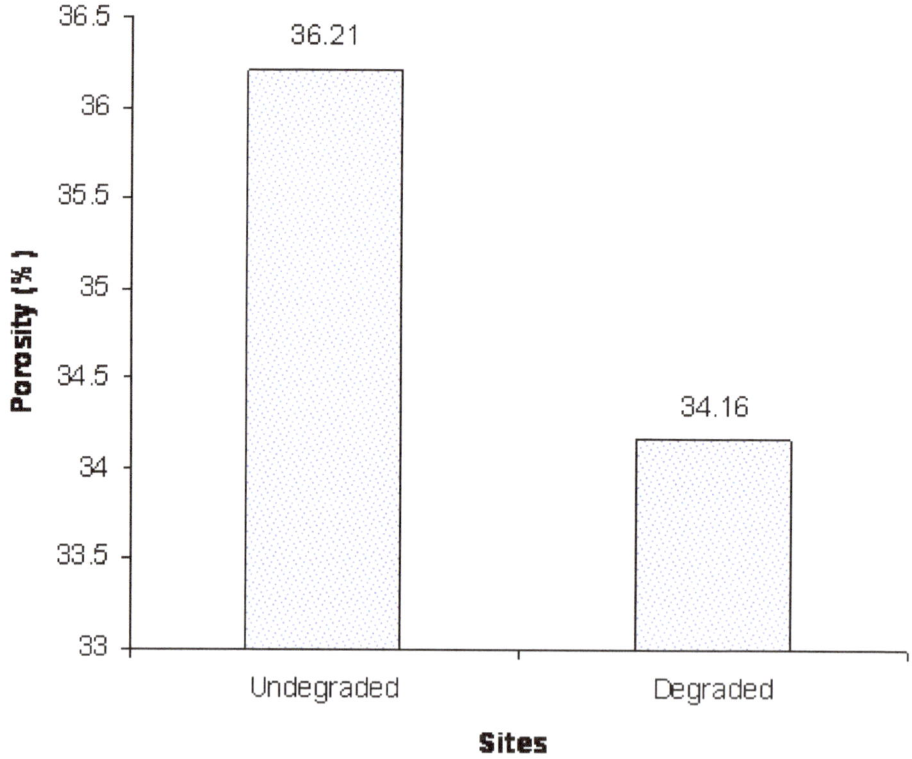

Figure 5. Porosity (%) of the soil samples collected in the degraded and undegraded sites.

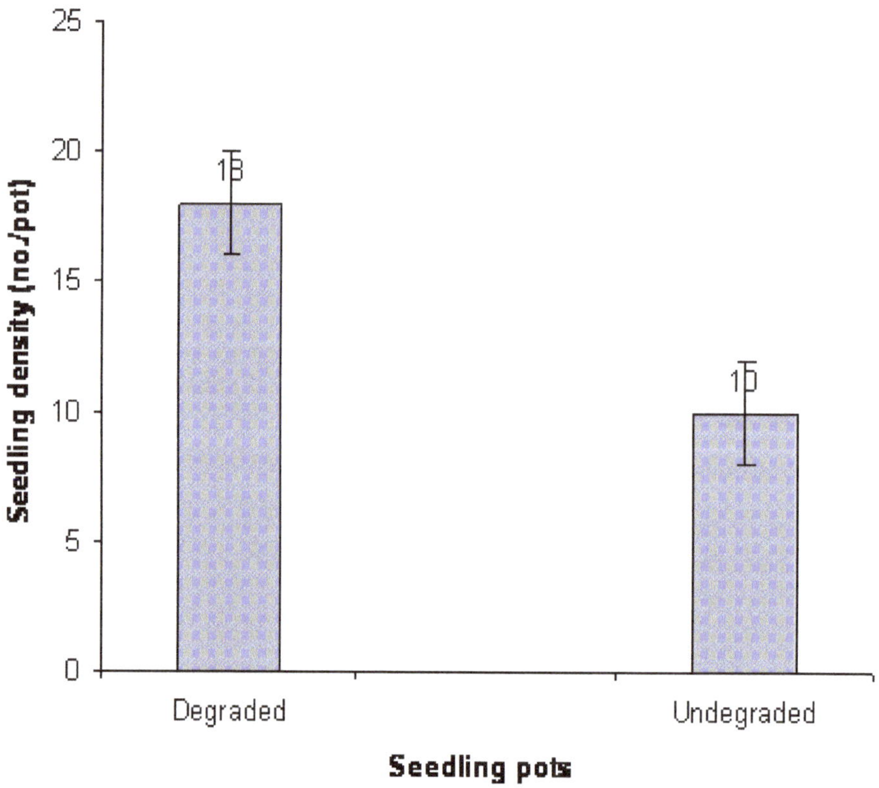

Figure 6. Seedling density (no/pot) in the soil samples collected in the degraded and undegraded sites.

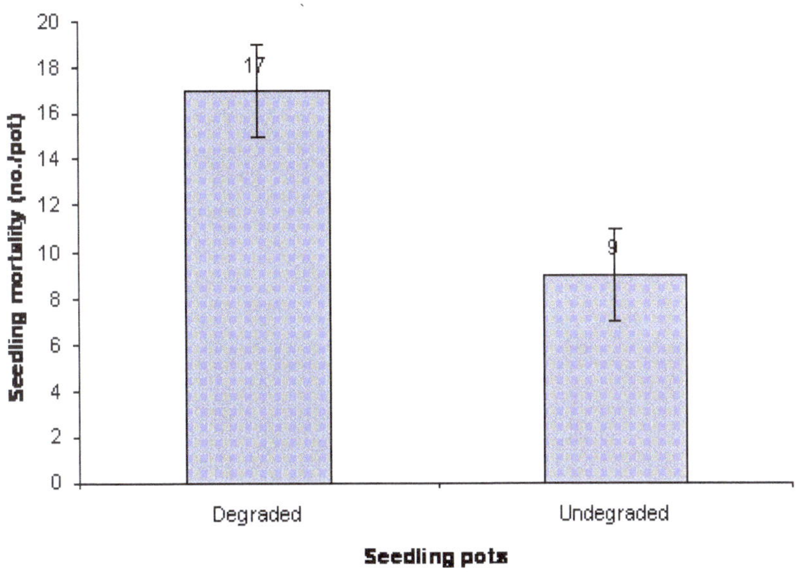

Figure 7. Seedling mortality (no/pot) in the soil samples collected in the degraded and undegraded sites.

Table 2. Spearman rank correlation between seedling density and soil physical properties.

	Soil compaction	TOC	Bulk density
	-0.481	0.192	0.137
Seedling density	0.043	0.446	0.587
	*	ns	ns

ns: not significant; *: significant at 0.05%.

Table 3. Summary of t-test results of SOM and TOC obtained from the degraded and undegraded plots of the Umfolozi catchment area.

Variables (g)	Mean degraded	Undegraded	t-value	Sig.
Soil moisture	8.99	14.22	3.466	*
Soil organic matter	2.91	4.77	1.834	*

*: significant at 0.05%

Table 4. Results of multiple regression with stepwise variable selection using plot level percent of erosion as the dependent variable and Ivy's erosion class, distance to water, distance to hill, distance to road and slope angle as predictor variables.

Predictor variable	Simple r	R square	F-value	Sig.
Ivy's erosion class	0.376	0.142	13.53	**
Distance to water			0.348	ns
Distance to hill			0.009	ns
Distance to road			0.068	ns
Slope angle	0.465	0.216	6.817	*

ns: not significant; *, ** : significant at 0.05 and 0.001%, respectively.

Table 5. Results of multiple regression with stepwise variable selection using plot level percent of erosion as the dependent variable and soil infiltration, soil compaction, grass cover, bare soil cover and litter cover as predictor variables.

Predictor variable	Simple r	r square	F-value	Sig.
Soil infiltration			1.371	ns
Soil compaction	0.779	0.606	4.105	*
Grass cover			0.072	ns
Bare soil cover	0.719	0.517	20.850	**
Litter cover			0.058	ns

ns: not significant; *, ** : significant at 0.05 and 0.001%, respectively.

Table 6. Results of multiple regression with stepwise variable selection using plot seedling density as the dependent variable and soil bulk density, total organic carbon and soil compaction as predictor variable.

Predictor variable	Simple r	r square	F-value	Sig.
Seedling density	0.636	0.404	13.042	**
Bulk density	0.808	0.652	12.924	**
Total organic carbon	0.847	0.718	3.236	*
Compaction			0.027	ns

was the second variable to enter the equation, increasing the r square to 0.517 (with a very large F-value associated with the r^2 change).

Distance to water, distance to hill and distance to road appeared to have no role in the erosion process in stepwise multiple regression. Consequently, with increase in rangeland usage, so the erosion becomes more widespread. Stepwise multiple regression identified the order and degree of contribution of the four independent variables in terms of prediction of seedling density. Seedling density, soil bulk density and total organic carbon were identified by their F-value for their degree of contribution in the stepwise multiple regression. Seedling density proved to be the most important factor influencing erosion, with a large and highly significant F-value and an r square of 0.404. Soil bulk density was the second variable to enter the equation, increasing the r square to 0.652 (with a very large F-value associated with the r^2 change) (Table 6). Total organic carbon was the last variable to enter the equation, increasing the r square to 0.718 (Table 6). Soil compaction appeared to play no role in the establishment of seedlings, possibly due to the importance of water that reduces the stress exerted by compaction on the growing roots. Clearly, the human factors and physical factors stand out as being a greater influence on erosion in this analysis.

DISCUSSION

Human activities in the study area include cultivation, grazing, burning and fuel wood collection. The common tillage treatment used include conventional ploughing (inversion) using a tractor and a single-furrow ox-drawn mould board plough. Using a modified Durham Geo's Pocket soil penetrometer in this study, soil penetration resistance increases to a peak between 20 and 30 cm soil depth. Smith and Dickson (1990) suggested that root penetration was inhibited at soil strength values between 2000 and 3000 kPa and this limits crop emergence and root penetration. The consequence of increased soil compaction is formation of a hardpan near the soil surface. High sheet erosion, rapid breakdown of organic matter and structure are the major factors which contribute to a rapid soil depletion which cannot be sustainably compensated for with inorganic fertilizers. Patches of bare ground are common in the landscape and these changes and fluctuate depending on soil moisture. Bare ground patches are characterized by internal variables such as number of trees, number of tree species, height of trees, percentage soil cover and percentage grass cover. Disturbance through clearing, cultivation, burning and overstocking may be altering the structure of soils through formation of a hardpan.

Degraded sites in the study area have low soil moisture suggesting a more xeric nature that is likely to impede establishment of plants. Organic matter is lower on degraded sites than on undegraded sites because plant cover is sparse. Woody plants are deliberately removed from the landscape to create space for more grass cover that can benefit grazing domestic livestock. The woody cover in the study area is less than 1%. The removal of trees removes the critical protective forces from raindrop impacts and this leaves the soils crusting and with hard-pans. Degraded sites exert pressure on the roots of the few available plants. The study shows a high negative correlation between seedling density and soil compaction. Water supplied to the nursery pots in a seed store experiment (Tafangenyasha et al., in press) was critical in reducing the pressure applied to the roots since seedlings germinated. Gifford et al. (1977) suggests that soil compaction has important hydrologic implications in terms of its contribution to reduced infiltration rates, and increased runoff potentials. In the Mondlo Communal rangeland, soil compaction may occur when weight of livestock or heavy machinery compress soil, causing it to lose pore space. Affected soils absorb less rainfall, thus increasing runoff and erosion. Plants have difficulty in establishing on compacted soil because the mineral grains are pressed together, leaving little space for air and water, which are essential for root growth. The phenomenon of soil-water-compaction has been observed by Engels (2009). Degraded sites in this study have low organic content because plant cover is low due to lack of establishment. The high mortalities among the seedlings in degraded soils in this study may be attributed to poor water holding capacities, compaction and topsoil erosion caused by soil erosion which in turn reduces the water infiltration rate. There were more cotyledons than dicotyledons germinating in the soils in the pots.

The rangeland that is part of the study area is subjected to continuous grazing pressure, there are no partitions that allow seasonal grazing rotations. The increasing and continuous use of degrading landscapes lead to need long time to recover while some degraded landscapes deteriorate to worse conditions. Seedbanks of degraded landscapes in the Umfolozi catchment are viable but once germinated the seedlings are challenged by physical conditions such as compaction and elevated bulk density causing significant (t-test, $p<0.05$) seedling diebacks. The negative emergence and growth conditions of seedlings may be ameliorated by improving the land preparation of the degraded landscapes. Moriuchi et al. (2000) suggest the need for a very long-term data set on plant population dynamics in fragile environments to capture unusual extreme events and explain patterns influenced by weather and edaphic conditions.

Bulk density was 0.99 and 0.83 g/cm^3 on degraded and undegraded sites, respectively. Soil bulk density ranged from 0.84 to 1.00 g/cm^3. The data (Table 1) suggest that degraded sites have higher soil bulk density than undegraded sites. Variation in soil bulk density may be attributable to the relative proportion and specific gravity of solid organic and inorganic particles and to the porosity of the soil (Blake and Hartage, 1986). In this study, porosity was greater on undegraded sites (36.2%) than on degraded sites (34.2%). Soils formed in organic rich materials might have particle densities of 0.9 to 1.4 g/cm^3 while mineral soils have bulk densities between 1.0 and 2.0 g/cm^3 (Blake and Hartage, 1986). Colluvial soils, reaching thicknesses of more than 10 m, mantle the pediment slopes at about 2 to 3° towards the valley bottoms in the study area. In the majority of cases, the percentage composition of the colluvium in the study area comprises sand ranging in proportion from 45 to 65%, silt from 15 to 25% and clay from 10 to 35% (Watson et al., 1984). The hillslopes in the study area are stripped by sheet-wash and the sediment chokes stream channels and in some cases the sheet-wash produces gullies or dongas reported by Botha et al. (1994) and Botha and Fedoroff (1995). Most mineral soils have bulk densities between 1.0 and 2.0 g/cm^3 (Blake and Hartage, 1986).

Blake and Hartage (1986) suggest that root densities are severely impacted at densities greater than 1.6 g/cm^3. If poorly-sorted particles are loosely packed, smaller particles will fill the voids created by the larger particles and again, bulk densities increase. As density increases, pore space decreases and the amount of air and water held in the soil also decrease. Figure 5 shows that Porosity was greater on undegraded sites (36.2%) than on degraded sites (34.2%). Lower soil bulk densities are desirable for plant growth whether the plants are agricultural crops, trees or grasses (Dudley et al. 2002). Low bulk density soils have greater water infiltration rates which minimize runoff, improve water quality and reduce stormwater flow. Soil bulk density measured in this study ranged from 0.84 to 1.00 g/cm^3 and this is close to the 0.88 and 0.92 g/cm^{-3} measured on moderately grazed pastures by Engels (2009). The average bulk density from the extremely grazed pastures is between 1.07 and 1.13 g/cm^3 (Engels, 2009). Soil bulk density ranged from 0.84 to 1.00 g/cm^3. Soil moisture (%) in the soil samples collected from the degraded sites was 8.9 and 13.48% on undegraded sites (Figure 2). Organic matter (%) was 3.17 and 5.46% on degraded and undegraded sites, respectively (Figure 3).

Soil compaction reduces the overall pore volume of soil and change the size and distribution of the pores within the soil (Anderson and Stormont, 2006). The loss of porosity reduces the ability of the soil to conduct water and air. The resulting pore space changes are expected to change the moisture characteristic of the soil. Reduced infiltration capacity results in surface runoff, leading to erosion and transport of nutrients to open water. Organic matter content tends to reduce soil compactibility and to increase its elasticity. The presence of plant tissues, living or dead may influence to some degree the ability of soils to resist compaction. Compaction causes an overall decline in growth, vigour, quality and persistence (Daddow and Warrington, 1983; Taylor and Gardner, 1963). The conditions needed for germination are known to vary considerably between different species and even between different populations of the same species. Growth of the plants in the nursery pots in the greenhouse in this study was made possible after a sustained and managed water supply. Tokunaga (2006) observed that shoot production increased significantly at high water potential and moderate bulk density. Biomass production was greatest when water was readily available and the negative effects of highly compacted soils were often less severe when water was available (Tokunaga, 2006), a situation observed in this study. It follows that when water is available plants tolerate compacted soil. The stepwise regression in its variable selection did not reveal soil compaction as an important variable suggesting that other factors may play a role. In the rangeland, the ability of a soil to recover from compaction has been attributed to climate, mineralogy and fauna (Lortie and Turkington, 2002). The erosion in the study area is increasing and the causal factors have been identified as climate and soil factors. The soils in the study area are expansive when wet and appear to form cracks on the surface when dry. Soils with high shrink-swell capacity, such as vertisols, recover quickly from compaction where moisture conditions are variable (dry spells shrink the soil, causing it to crack). But clays which do not crack as they dry cannot recover from compaction on their own unless they host subterranean animals such as earthworms.

The low survival of seedlings reported in this study suggests that the physical properties of the soils are a constraint to the growth and establishment of seedlings. Other physical conditions regulating the growth of seedlings have implicated fire. Fire is used in the study area to initiate an early growth of grass so that grazing livestock can benefit. The role of fire in seed germination has been recorded as the action of heat in breaking hard seed coats (Kenny, 2000; Egerton-Warburton, 1998). Keeley et al. (1981) suggest that the heat required to fracture a seed coat within the soil underneath the passage of a fire varies with species but is generally in the range 60 to 120°C with an optimum between 80 to 100°C. *Colophospermum mopane* (Benth.) showed high rates of height growth immediately after fire (Mlambo and Mapaure, 2006). Most studies suggest that seedling mortality is high during the germination period (Keith, 2002, Radford et al., 2001), results which tend to agree with the findings in this study. Results of this study support the suggestion that water (Rainfall) is an important determinant of seedling recruitment in the savanna rangeland. The addition of soil treatments and protection from herbivory in early growth stages should increase emergence of seeds from the seedbank and offer opportunities for effective restoration of the degraded rangeland.

The results of the regression analysis are summarized in Table 4 to 6. Values of simple r (r values for each variable separately) are given for all the regressor variables, allowing for the fact that these factors do not act in isolation. Stepwise multiple regression identified the order and degree of contribution of the independent variables in terms of prediction of eroded land percentage and seedling density. When erosion is used as the dependent variable, erosion class, slope angle, soil compaction and bare soil cover proved to be the most important factors influencing erosion. When seedling density is used as the dependant variable, soil bulk density and total organic carbon proved to be the most important variables influencing plant establishment. Soil compaction appeared to play no role in the establishment of seedlings. Clearly, the human factors and physical factors stand out as being a greater influence on erosion in this analysis.

Reduced infiltration capacity results in surface runoff, leading to erosion and transport of nutrients to open water. Organic matter content tends to reduce soil compactibility and to increase its elasticity. The presence of plant tissues, living or dead may influence to some degree the ability of soils to resist compaction. Compaction causes an overall decline in growth, vigour, quality and persistence (Birkeland, 1984).

The soil erosion taking place in the Mondlo Communal Area represents major environmental changes occurring at the landscape (or larger) spatial scales within the savanna biome. The cause-effects of the soil erosion are similar to others reported elsewhere (Dougill and Cox, 1995a, b; Dudley et al., 2002; Tafangenyasha and Campbell, 1998; Cyzt, 2004; Tokunaga, 2006) and include energy forces, protective forces and social forces, but the erosion in Mondlo Communal area is increasing hence the need to elucidate the erosion problems denovo. As this paper shows, the erosion is being influenced largely by poor soil physical properties. Elevated soil bulk density impedes root penetration and seedling establishment.

Conclusions

Soil bulk density has been suggested as a cause of vegetation change in the Umfolozi catchment in this study. The Masotcheni colluvial sediments blanketing the hillslopes in the Umfolozi catchment are sensitive and fragile to land use and their nature in resource management needs to be understood. Further data collection will continue, making it possible for potential interactions between the effects of grazing history, erosion history and topographic position on bulk density.

ACKNOWLEDGEMENTS

We wish to thank the Department of Hydrology at the University of Zululand for granting us the permission to use their equipment without which the work would not have been possible. The Research Office at the University of Zululand assisted with funds for the research expedition.

REFERENCES

Acocks JPH (1988). Veld types of South Africa, Memoirs Botanical Survey of South Africa, 57, Government Printer, Pretoria, p. 316.

Anderson CE, Stormont JC (2006). Changes in the soil moisture characteristic due to porosity variation. Am. Soc. Civil Eng. (ASCE), pp. 1360-1370.
Begg GW (1988). The wetlands of Natal (Part 2) - The distribution, extent and status of wetlands in the Mfolozi catchment. Natal Town and Planning Commission. Pietermaritzburg.
Berjak M, Fincham RJ, Liggitt B, Watson HK (1986). Temporal and spatial dimensions of gully erosion in northern Natal, South Africa, Int. Arch. Photogramm. Rem. Sens., 26: 283-293.
Birkeland PW (1984). Soils and Geomorphology. Oxford Univ. Press, New York, p. 280.
Blake GR, Hartage KH (1986). Bulk density. In: A Klute (ed). Methods of soil analysis. Physical and mineralogical methods. (2nd ed.), Agro. Monogr., 1: 363-375.
Botha GA (1992). The geology and palaeopedology of late quaternary colluvial sediments in northern Natal, South Africa, Unpubl. PhD. thesis, University of Natal, Pietermaritzburg.
Botha GA, Wintle AG, Vogel JC (1994). Episodic late Quartenary palaeogully erosion in northern KwaZulu-Natal, South Africa. Catena, 23: 327-340.
Botha GA, Fedoroff N (1995). Palaeosols in Late Quaternary colluvium, northern KwaZulu-Natal, South Africa. J. Afr. Earth Sci., 21: 291–311.
Brady NC, Weil RR (1996). The nature and properties of soils (11th ed.). Prentice Hall, New York, p. 258.
Cyzt EA (2004). Effects of traffic on soil aeration, bulk density and growth of spring barley. Soil Till. Res., 79: 153-166.
Daddow RL, Warrington G (1983). Growth-limiting soil bulk densities as influenced by soil texture (WSDG-TN-00005). Fort Collins, CO: USDA Forest Service, p. 15.
Davidson DA (1980). Soils and land use planning: Topics in Applied Geography. Longman, Glasgow, p. 129.
Doran JW, Mielke LN (1984). A rapid, low-cost method for determination of soil bulk density. Soil Sci. Soc. Am., 48: 717-719.
Dougill A, Cox J (1995a). Land degradation and grazing in the Kalahari: New analysis and alternative perspectives. ODI Pastoral Dev. Netw., p. 38.
Dougill A, Cox J (1995b). When is ecosystem change land degradation? A reply. ODI Pastoral Develop Netw., p. 38.
Dudley DM, Dudley KW, McDougald NK, George MR (2002). Factors influencing soil-surface bulk density on oak savanna rangeland in the Sierra Nevada Foothills. USDA Forest General Technical Report PSW-GTR-184, p. 190.
Engels CL (2009). The effect of grazing intensity on soil bulk density. North Dakota State University Central Grasslands Research Extension Center. USA, p. 2.
Egerton-Warburton LM (1998). A smoke-induced alteration of the sub-testa cuticle inseeds of the post-fire recruiter, *Emmenanthe penduliflora* Benth. (Hydrophyllaceae). J. Exp. Bot., 49: 1317–1327.
Froehlich HA, McNabb DH (1994). Minimizing soil compaction in Pacific Northwest forests. In: Forest soils and Treatment Impacts. E. L. Stone (Ed). Department of Forestry, University of Tennese. Knoxville, Tennessee, pp. 159-192.
Gifford GF, Hawkins RH (1978). Hydrologic impact of grazing on infiltration: A critical review. Water Resourc. Res., 14: 305–313.
Hudson N (1995). Soil Conservation. (3rd Edition). B.T. Batsford, London,. p. 210.
Keeley SC, Keeley JE, Hutchinson SM, Johnson AW (1981). For electron microscopy. California chaparral Ecol., 62: 1608–1621.
Keeley SC, Keeley JE, Hutchinson SM, Johnson AW (1981). Postfire succession of the herbaceous flora in southern California chaparral. J. Ultrastruct. Res., 26: 31–43.
Keith DA (2002). Population dynamics of an endangered heathland shrub, *Epacris stuartii* (Epacridaceae): Recruitment, establishment and survival. Aust. Ecol., 27: 55–66.
Kenny B (2000). Influence of multiple fire-related germination cues on three Sydney *Grevillea* (Proteaceae) species. Aust. Ecol., 25: 664–669.
Liggitt B (1988). An investigation into soil erosion in the Mfolozi catchment. Unpubl. MSc. thesis, University of Natal, Pietermaritzburg.
Liggitt B, Fincham RJ (1989). Gully erosion - the neglected dimension in soil erosion research. S. Afr. J. Sci., 85: 18-20.
Lortie CJ, Turkington R (2002). The effect of initial seed density on the structure of a desert annual plant community. J. Ecol., 90: 435-445.
Mlambo D, Mapaure I (2006). Post-fire resprouting of *Colophospermum mopane* saplings in a southern African savanna. J. Trop. Ecol., 22: 231-234.
Morgan RPC (1980). Soil erosion: Topics in Applied Geography. Longman, London, p. 113.
Moriuchi KS, Venable DL, Pake CE, Lange T (2000). Direct measurement of the seed bank age structure of a Sonoran Desert annual plant. Ecology, 81: 1166–1168.
Nelson DW, Sommers LE (1996). Total carbon, organic carbon, and organic matter. In: Methods of Soil Analysis, Part 2, 2nd ed., A.L. Page et al., Ed. Inc. Madison, WI. Agronomy. Am. Soc. Agron., 9: 961-1010.
Nie NH, Hull CH, Jenkins JG, Steinbrenner K, Bent DH (1975). Statistical Package for the Social Sciences. McGraw-Hill New York, p. 364.
Radford BJ, Yule DF, McGarry D, Playford C (2001). Crop responses to applied soil compaction and to compaction repair treatments. Soil Tillage Res., 61: 157–166.
Rein SM, Botha GA, Hughes JC (2000). Some physical and chemical properties of sediments exposed in a gully (donga) in northern KwaZulu-Natal, South Africa and their relationship to the erodibility of the colluvial layers. Catena, 39: 11-31.
Rockich DP, Meney KA, Dixon KW, Sivasithamparam K (2001). The impact of soil disturbance on root development in woodland communities in Western Australia. Aust. J. Bot., 49: 169–183.
Schjonning P, Thomsen IK, Christensen BT (1998). Soil porosity as a habitat for microorganisms. In: Børresen, T (Ed.), Proceedings of NJF-Seminar no. 286. Agricultural University of Norway, Ås, Norway, Soil Tillage Biol., pp. 49–56.
Smith RT, Atkinson K (1975). Techniques in Pedology: A Handbook for Environmental and Resource Studies. Elek, London, p. 213.
Smith DL, Dickson JW (1990). Contributions to Vehicle Weight and Ground Pressure to Soil Compaction. J. Agric. Eng. Res., 46: 13-29.
Tafangenyasha C, Campbell BM (1998). Initiation and maintenance of degraded landscape in the Sinamatella area of Hwange National Park, Zimbabwe. J. Environ. Manage., 52: 69-78.
Taylor HM, Gardner HR (1963). Penetration of cotton seedling taproots as influenced by bulk density, moisture content, and strength of soil. Soil Sci., 96: 153-180.
Tokunaga A (2006). Effects of bulk density and soil strength on the growth of blue wildrye (*Elymus glaucus* Buckl) MSc Thesis. Humboldt State University. USA, p. 110.
Tsimba R, Hussein J, Ndhlovu LR (1999). Relationships between depth of tillage and soil physical characteristics of sites formed by smallholders in Mutoko and Chinyika in Zimbabwe. In: Kaumbutho, PG and Simalenga, TE 84-88. Conservation tillage with animal traction. Anim. Traction Network Eastern Southern Afr., p. 173.
Watson A, Williams P, Goudie AS (1984). The palaeoenvironmental interpretation of colluvial sediments and palaeosols of the late Pleistocene hypothermal in southern Africa. Palaeogeogr. Palaeoclimatol. Palaeoecol., 45: 225-249.
Wilson AD, Harrington GN, Beale IF (1984). Grazing management. In Management of Australia's Rangelands In: Harrington G N, Wilson A D, Young M D Eds), Melbourne Mgmt. Austr. Rangel., CSIRO. p. 312.

Characterizing sites for the design and restoration of Kalacha irrigation scheme, Chalbi district, Kenya: Socio-economic and biophysical considerations

E. M. Muya[1]*, K. Gitau[1], S. Obanyi[1], M. Ngutu[2], S. Kuria[2], H. Sharfi[2], D. G. Dub[2], H. Bulle[2], D. M. Miano[3], J. Muriethi[3], G. Keya[3], I. V. Sijali[1], S. Mamo[2], P. M. Maingi[1], M. Okoti[2] and Eng. M. J. Omedi[4] and A. Esilaba[3]

[1]National Agricultural Laboratories, P. O. Box 14733, Nairobi, Kenya.
[2]KARI-Marsabit, P. O. Box 147, Marsabit, Nairobi, Kenya.
[3]Kenya Agricultural Research Institute, P. O. Box 57811, Nairobi, Kenya.
[4]Ministry of Water and Irrigation, P. O. Box 29703, Nairobi, Kenya.

Site characterization was carried out in Kalacha irrigation scheme for the design and restoration of the scheme. Detailed soil survey was carried at the scale of 1:2,000, through systematic observations, where the soils were described in terms of depth, texture, structure and consistence. Based on these characteristics, four soil units were identified and designated as block A, B, C and D, covering 23, 18, 17 and 42% of the scheme, respectively. In each unit, representative soil profiles were identified, characterized in terms of soil physical and chemical properties, and classified according to FAO-UNESCO system of classification. The soils in block A, B, C and D were classified as Petric Calcisols, Salic Fluvisols, Calcic Solonetz and Calcaric Fluvisols, respectively. The aggregate stability of the topsoils for Block A, B, C, and D was found to be 40, 10, 2 and 4%; while the water uptake capacity was 218, 158, 76 and 86 mm, respectively. The highest level of nutrients was found in Block A, followed by Block B, while block C and D had relatively lower levels. The most limiting factors were found to be high salinity, high sodicity, low aggregate stability and high soil pH, hence low nutrient availability.

Key words: Irrigation layout and design, soil quality, irrigation blocks.

INTRODUCTION

Kalacha irrigation scheme was started in 1984 through community initiative with little technical inputs into the design. The result of soil investigation indicated that the soils in the scheme varied considerably in terms of the physical and water-related properties that influence the choice of irrigation methods and practices (Muya et al., 2008). With little technical capacity to face these challenges (of designing the irrigation system), the farmers applied the same irrigation methods and scheduling for different soils and crops. This caused excess water application for the soils and crops which needed relatively less water and under irrigation for those requiring more. Excess water application caused loss of nutrients through leaching and raised the ground water table with salt to the upper soil horizon, hence, increased salt accumulation within the root zones. In order to attain sustainable production of the degraded scheme, the first task is to design the irrigation layout to ensure efficient use of the limited water supply followed by restoration of soil quality and productivity (Muya et al., 2008).

The development of economically viable production system will depend on the interactions between the efficiency of water supply system, the integrated management strategies for restoring soil quality, and the market for envisaged farm products. To improve the efficiency of water supply systems, the necessary physical and

*Corresponding author. E-mail: edwardmuya@yahoo.com.

hydraulic structures must be designed and constructed, based on the conventional design criteria (Muchangi et al., 2005). With the current water deficit, poor water distribution networks and inefficient irrigation practices, agricultural production will not only fail to meet the potential human and market demands, but also lead to further environmental degradation, unless appropriate design and construction of the scheme as well as restoration of the degraded soils are achieved (Muya et al., 2008). The starting point of this important development is to examine the pertinent socio-economic issues that prompted the need for the reconstruction of the scheme, along with the baseline biophysical factors, on the basis of which to delineate soil units (irrigation blocks).

For this purpose, detailed topographical and soil survey were carried out in the area to provide the physical data to assist in planning the layout of the scheme. Apparently the cost of design, construction and restoration of the scheme is frustrating. However, it is a mandatory exercise from the long term standpoint. This is because land degradation has resulted into tremendous loss in soil structure and soil tilth and the available physical and hydraulic structures have failed to distribute water to a level that meets the crop water requirements. A combination of these two factors has lead to the decline in agricultural production to an extent that hardly any farm enterprises are practiced on the ground (Muya et al., 2008). Therefore, this paper examines four important components of irrigation development, namely: socio-economic issues, biophysical factors and the tasks to be addressed in designing the system.

MATERIALS AND METHODS

The community perceptions of the declining production trends and the anthropogenic factors contributing to the decline were captured through structured questionnaires and focus group discussions. Topographical survey was carried out at the scale of 1:2000 using theodolite. Grids were laid down at a 50 m interval. In each grid, detailed observations were made on soils, current irrigation methods and water distribution networks, micro-relief and degree of land degradation. Soil survey, delineation, description and classification were carried out following the standard procedures (FAO, 1977, Kenya Soil Survey Staff, 1987 and FAO-UNESCO, 1997). Observations were made on a rigid grid at 50 m interval, using minipits, in which descriptions were made in terms of soil depth, texture and consistence. Based on these characteristics, the soil units (irrigation blocks) were identified and delineated. The criteria used in differentiating the soils into different units included differences in soil depth, texture, aggregate stability and soil water uptake and retention capacity. For each unit (block), representative soil profile was identified which was characterized in details in terms of physical and chemical characteristics. The selection of the representative soil profiles was achieved by grouping the observation points according to the specified range of soil characteristics. The observation point with soil characteristics similar to other observation points within a given range was identified as a representative profile. The physical characteristics determined included aggregate stability, bulk density, porosity, infiltration rate, and water uptake and storage capacity. Aggregate stability was determined through dry sieving analysis (Hillel, 1982). Porosity, water uptake and storage capacity were determined by subjecting the saturated soil samples to stepwise incremental suctions and calculating the equilibrium moisture content (Hillel, 1982). Infiltration rate was determined using double ring infiltrometers, while hydraulic conductivity was assessed using Inverse Auger Hole methods (Hinga et al., 1980). From each representative soil profile, composite samples were collected at the depth of 20 cm for laboratory determination, using standard procedures provided by Hinga et al. (1980). The soil properties analyzed were plant nutrients such as P, K, Ca, Na, and Mg, using Mehlich Double Acid Method (Anderson and Ingram, 1993), total organic carbon, using Calometeric Method (Nelson and Somers, 1982), total nitrogen, using Micro-Kjeldahl Method, soil pH in 1:1 (W/V) soil-water suspension and micro-nutrients, using Okaleb et al. (2002).

The chemical characteristics (soil quality attributes) of soils were used to determine the productivity of each soil unit delineated, based on their thresholds (Table 1) and response functions (Aune and Lal, 1997; Kamoni and Wanjogu, 2006). These characteristics were soil pH, nitrogen, soil organic carbon (SOC), phosphorous (P) and potassium (K). The productivity index was calculated using the response functions provided by Aune and Lal (1997). The response functions are regressed relationships between each of the soil quality attributes selected and relative yield (RY) of maize, this being the staple food crop in Kenya. Productivity index was calculated using the following soil productivity equation:

$$PI = SQI_1 + SQI_2 + SQI_3 + SQI_4 + SQI_5$$

Where, PI = Productivity index, based on the compounded sufficiency of all the soil quality attributes. $SQ1_1$, SQ_2, SQ_3, SQ_4, and SQ_5 are the soil quality index of pH, soil organic carbon, nitrogen, phosphorous and potassium, respectively. The soil quality index for each soil quality attribute was deduced from the respective response curve on the vertical scale of values ranging from 0.0 - 1.0 (relative yield) against the value of the attribute read on a vertical scale, provided by Aune and Lal (1997). These values were applied into the soil productivity equation to calculate the productivity index.

RESULTS AND DISCUSSION

Socio-economic issues

In Kalacha, irrigation started through the community initiative with little technical inputs into the design. However, the soil productivity was reasonably high in the beginning but after twenty years of continuous irrigation, agricultural production declined to a point where most farmers stopped crop farming. The scheme started in 1984 and by the year 2005, the production had gone down by over 80% for most agricultural products (Muya et al., 2008) (Table 2). The farmers attributed this to low soil quality, which they described as toxic, smeary and slippery when wet and extremely hard when dry. They also claimed that water disappeared in the soil before reaching their crops, thereby causing low crop yields.

From Kalacha perspective, there seems to be less agreement regarding the contentions that sustainable agriculture must be also economically viable and socially acceptable. In this area, the socio-economic sciences and societal expectations are apparently different from

Table 1. Thresholds of soil quality indicators.

Soil quality attributes	Threshold values	Source
K (me%)	0.2 - 1.5	Kamoni and Wanjogu (2006)
K (me%)	0.83	Aune and Lal (1997)
N (%)	0.2	Kamoni and Wanjogu (2006)
C (%)	1.08	Aune and Lala (1997)
P (ppm)	20 - 80	Kamoni and Wanjogu (2006)
Ph	5.5 - 7.0	Kamoni and Wanjogu (2006)
Electrical conductivity (EC)	4.0 mS/cm	Kamoni and Wanjogu (2006)
ESP	6.0	Kamoni and Wanjogu (2006)
Aggregate stability	50%	Kamoni and Wanjogu (2006)

Table 2. Agricultural production in 1984 and 2004 in Kalacha irrigation scheme.

Crop	Yield in Kg/ha	
	1984	2004
Maize	2000	500
Cabbage	2500	480
Kale	3500	350
Tomatoes	3700	300

Source: Muya et al. (2008).

the physical, ecological and agricultural sciences that form important components of environmental sustainability. In this area, urgent human and societal needs for survival require that the envisaged production system be of immediate impacts and economically viable. Anything short of this may be dismissed as socially unacceptable and hence not a viable enterprise. In physical and ecological sense, a production system is said to be sustainable if the compounded sufficiency of soil quality indicators are restored and remains without deterioration over a realistic time span. This means that the soil quality indicators must be sustained or restored within the acceptable environmental thresholds. In Kalacha irrigation scheme, nearly all soil quality indicators have values that have fallen out of the environmental thresholds following severe land degradation in the past. This indicates that the system is in adverse state, thus requiring heavy capital investment for restoration to a functional state. In this case, a healthy and productive system attained through reclamation or amendment may not be profitable and socially acceptable in the short term. Therefore, inherent conflicts exist between the short term interests of individual irrigators and long term interests of the society as a whole. Resolving the conflicts between the short run and long run interests starts by defining the current state and trying to predict what will happen to the environment, individuals and the society as whole if it remains unchanged through intervention. The degraded state defined in Kalacha irrigation scheme consisted of soils whose productive capacity had gone down to an extent that hardly any return would be realized unless restored. Further delay in taking action would result into further degradation until no production would be realized even with heavy capital investments. In this scheme, over 50% of the irrigated area is approaching this undesirable state, which would be a loss to the society in the long run. Design and restoration of the scheme require baseline information on soil followed by understanding the most important challenges associated with the design work.

Biophysical factors

Four soil units were arrived, each with a representative soil profile. They exhibited considerable variations between different soil profiles in terms of soil depth, aggregate stability, infiltration rates and soil moisture holding capacity (Figures 1, 2, 3 and 4). Based on these characteristics, four irrigation blocks (soil units) were identified, namely block A, B, C and D whose representative soil profile were classified as Petric Calcisols, Salic Fluvisols, Calcaric Fluvisols and Salic Solonetz, respectively. Generally, the results showed remarkable variations between different blocks as expressed in the characteristics of representative soil profiles, each profile representing one soil mapping unit (Table 3). The aggregate stability was found to be generally low for all the soil profiles, being less than 50%. Profile 1 had the highest aggregate stability within 20 cm depth, which decreased with the depth. The water holding capacity (WHC) and hydraulic conductivity (HC) were also found to be highest in profile 1. The lowest values were found in profile 3, which had also the lowest soil aggregate stability. The variations in aggregate stability, hydraulic conductivity and soil water holding capacity in different soil profiles could be attributed to differences in sodium concentrations and organic carbon, both being the principle determinants of soil aggregate stability. Since the physical and hydraulic characteristics of soils are the principle determinants of soil moisture

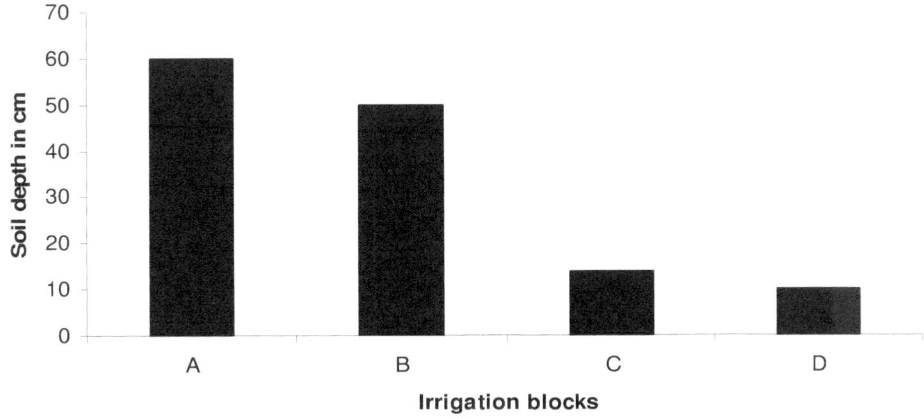

Figure 1. Variation in soil depth.

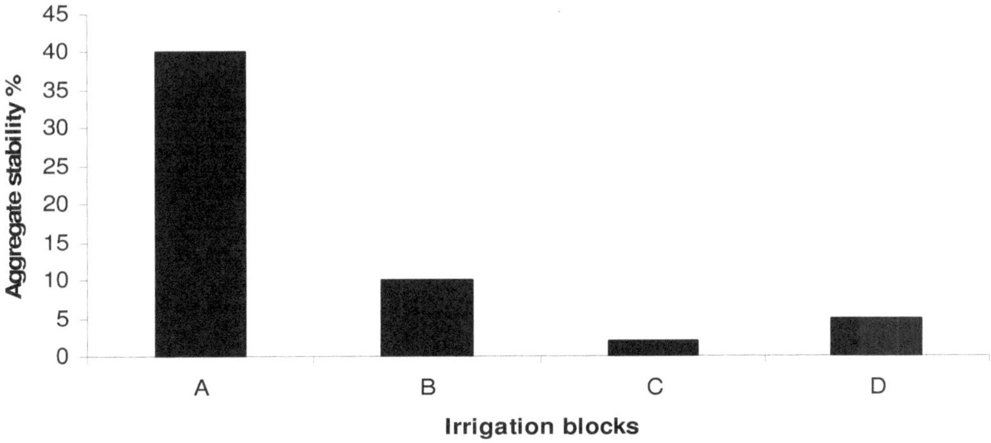

Figure 2. Variation in A. stability.

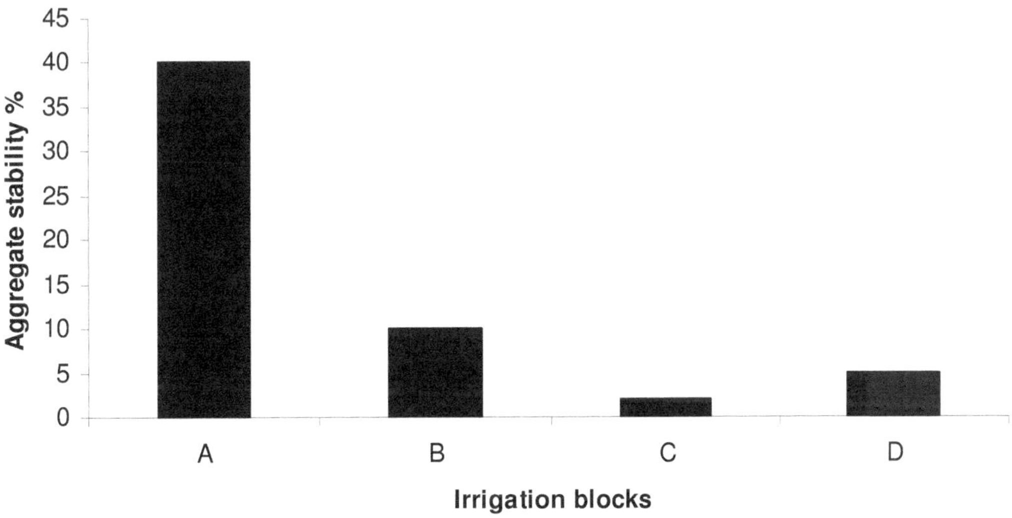

Figure 3. Variation in infiltration.

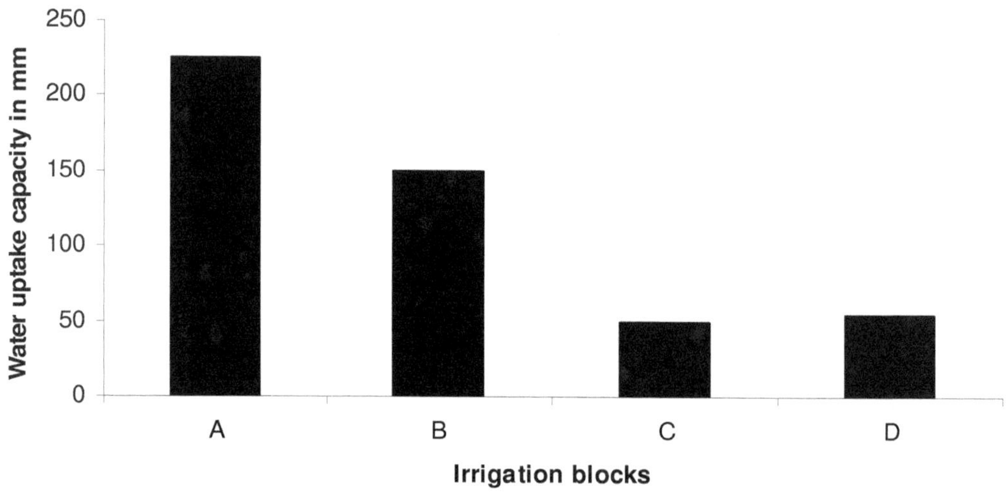

Figure 4. Variation in water uptake.

Table 3. Physical and hydraulic characteristics of soils.

Profile No 1, Block A	**Values at different depths**				
Soil depth (mm)	0 - 10	10 - 20	20-40	40 - 60	60 - 100
Texture	Clay loam	Clay loam	Clay	Clay	Clay
Aggregate stability (%)	40	35	14	15	5
HC (mm/hr)	12.5	11.8	10.6	10.5	3.3
WHC (%)	10	10	22	27	44
Profile No 2, Block B					
Soil depth mm	0 - 10	10 - 15	15 - 30	30 - 60	
Texture	Sandy clay loam	Sandy loam	Clay loam	Clay	Clay loam
Aggregate stability (%)	10	9	9	5	
HC (mm/hr)	10.7	9.6	6.3	4.8	
WHC (%)	8.8	7.6	9.1	33.4	
Profile No 3, Block C					
Soil depth (mm)	0 - 10	`0 - 20	20 - 30		
Aggregate stability (%)	2	2	2		
Texture	Sandy loam	Sandy clay	Loamy sand	Clay	Clay
HC (mm/hr)	1.0	0.9	0.1		
WHC (%)	5	5	11		
Profile No 4, Block D					
Soil depth (mm)	0 - 10	10 - 20	20 - 30	30 - 40	
Texture	Silty clay	Clay	Clay	Clay	
Aggregate stability (%)	4	3	2		
HC (mm/hr)	1.2	0.9	0.1		
WHC (%)	8	10	11		

Table 4. Chemical characteristics of bloc A.

Soil attributes	Values at different depths				
	0 - 10	10 - 20	20 - 40	40 - 60	60 - 100
Soil pH-H_2O (1:2.5)	7.89	7.97	8.1	9.3	9.6
Electrical conductivity (mS/cm)	0.56	1.70	1.88	1.80	1.9
Carbon (%)	1.48	1.22	0.56	0.2	0.2
Nitrogen (%)	0.39	0.22	0.12	0.01	0.01
Phosphorous (ppm)	30	20	25	15	10
Potassium (m.e.%)	1.22	4.8	7.9	6.48	6.22
Calcium (m.e.%)	23.6	17.17	18.8	11.7	18.6
Magnesium (m.e%)	10.4	3.41	4.31	3.21	9.88
Sodium (m.e.%)	1.39	3.5	15.05	17.7	18.8
Sum	36.61	38.88	46.5	39.09	53.5
ESP	3.7	9.02	33.3		

Table 5. Chemical characteristics of bloc B.

Soil attributes	Values at different depths				
	0 - 10	10 - 15	15 - 30	30 - 40	40 - 60
Soil pH-H_2O (1:2.5)	8.5	8.8	9.1	9.3	10.8
Electrical conductivity (mS/cm)	1.02	1.38	1.50	1.5	3.7
Carbon (%)	0.8	0.5	0.1	0.1	0.07
Nitrogen (%)	0.18	0.1	0.09	0.09	0.02
Phosphorous (ppm)	15	20	10	10	5
Potassium (m.e.%)	1.9	2.2	2.2	1.7	2.1
Calcium (m.e. %)	32.8	25.8	36.8	31.7	33.8
Magnesium (m.e %)	9.7	10.2	10.9	11.7	6.7
Sodium (m.e %)	3.8	4.8	7.9	8.7	11.9
Sum	48.2	43.0	57.8	53.8	54.5
ESP	7.88	11.2	13.6	16.17	21.8

uptake and storage capacity, their variation requires different water application time, hence different irrigation scheduling.

There were also variations in the soil chemical characteristics between the four soil profiles (Tables 4 - 7). The variations in the soil macro-nutrients, that is, nitrogen, phosphorous and potassium, indicated that each block require different types and quantity of fertilizer inputs. Similarly, variations in organic carbon indicated that different quantities of organic inputs were required to replenish the level of carbon in different blocks. The levels of salts and sodium also varied with the highest level being recorded in block C. However, the salinity problem being observed only in block C, the overall soil-related constraint is the problem of alkalinity, which requires appropriate management strategies to bring the level of sodium to acceptable limits. The compounded sufficiency of all the soil quality attributes in different irrigation blocks is compared in terms of productivity index (PI) in order to indicate the baseline status of soil quality and productivity (Table 8). As a rule of thumb, the threshold productivity index being 50% (Driessen and Konijn, 1992), all the irrigation blocks had very low productivity. The most limiting factors were found to be high salinity, high sodicity, low aggregate stability and high soil pH, hence low nutrient availability. These are as a result of increased land degradation processes due to unfavorable irrigation and soil management practices, and the situation is bound to be more severe if the current state remains unchanged through appropriate intervention. Block D requires priority intervention because it has the largest acreage and highest number

Table 6. Chemical characteristics of block C.

Soil attributes	Values at different depths				
	`0 - 20	`0 - 15	15 - 20	20 - 25	25 - 30
Soil pH-H_2O (1:2.5)	9.8	9.8	9.9	9.9	9.9
Electrical conductivity (mS/cm)	23.6	23.6	18.8	18.1	19.1
Carbon (%)	0.30	0.30	0.1	0.1	0.1
Nitrogen (%)	0.3	0.3	0.01	0.01	0.01
Phosphorous (ppm)	20	20	17	20	17
Potassium (m.e. %)	1.39	1.39	1.87	1.86	1.76
Calcium (m.e. %)	23.8	23.8	27.9	28.2	29.8
Magnesium (m.e %)	9.8	9.8	9.9	9.7	10.7
Sodium (m.e %)	11.6	11.6	10.8	12.6	13.7
Sum	46.59	46.59	50.47	52.4	55.1
ESP	24.9	24.9	21.3	23.4	24.2

Table 7. Chemical characteristics of bloc D.

Soil depth (mm)	Values at different depths				
	0 - 10	10 - 15	15 - 20	20 - 30	30 - 40
Soil pH-H_2O (1:2.5)	8.40	8.60	9.3	9.3	10.1
Electrical conductivity (mS/cm)	1.03	1.38	1.49	1.61	2.2
Carbon (%)	0.89	0.61	0.2	0.17	0.11
Nitrogen (%)	0.18	0.13	0.05	0.05	0.01
Phosphorous (ppm)	20	20	12	9	7
Potassium (m.e.%)	3.4	3.7	3.2	3.6	3.8
Calcium (m.e. %)	32.7	28.9	22.7	24.6	28.6
Magnesium (m.e %)	11.1	10.1	6.4	6.7	6.5
Sodium (m.e %)	6.4	9.6	12.6	14.7	12.9
Sum	53.6	52.3	44.9	49.6	51.8
ESP	11.9	18.35	28.1	29.6	24.9

Table 8. Soil productivity of different irrigation blocks (soil units).

Block	Extent (%)	PI	Most limiting factors
A	23	60	High soil pH and low soil organic carbon
B	18	48	High soil pH, low aggregate stability, nitrogen and phosphorous
C	17	13	High soil pH, low water holding capacity, low soil organic carbon and high sodium concentration
D	42	5	High soil pH, low water holding capacity, low aggregate stability, low soil organic carbon, low nitrogen and high sodium concentration

of limitations.

The challenges of designing irrigation system

The poor design of the canal system was found to be the main cause of inefficient water distribution. This is because haphazard layout of the canals delayed the flow and increased the opportunity time for water to be lost through seepage and evaporation.

Conclusion and recommendation

The results of topographical survey indicated that the design of the current canal layout was not based on the conventional design criteria and as such was inefficient in distributing water into the irrigated fields. There were considerable variations in soil physical, hydraulic and chemical characteristics that lead to the differentiation of the irrigation fields into four irrigation blocks (soil units), namely: Block A, B, C and D, covering 23, 18, 17 and 42% of the area, respectively. The soils in block A, B, C and C were classified as Petric Calcisols, Salic Fluvisols, Calcic Solonetz and Calcaric Fluvisols respectively. The aggregate stability of the topsoils for Block A, B, C, and D was found to be 40, 10, 2 and 4%; while the water uptake capacity was 218, 158, 76 and 86 mm respectively. The highest level of nutrients was found in Block A, followed by Block B, while block C and D had relatively lower levels. Most of the soils were found to be highly degraded, resulting into low quality and productivity. The most limiting factors were found to be high alkalinity, low aggregate stability and high soil pH, hence low nutrient availability.

ACKLOWDGEMENTS

The authors thank the European Union (EU) for funding the characterization of the site and design of the irrigation layout. They also thank Desert Margin Programmed for funding the initial soil survey and land evaluation that lead to the identification of the critical problem issues in Kalacha. This was the basis of the more detailed site characterization and design of the layout.

REFERENCES

Aune JB, Lal R (1997). Agricultural productivity in the tropics and critical limits of properties of Oxisols, Ultisols and Alfisols. Tropical Agriculture (Trinidad), 74(2): 96-102.

Anderson JM, Ingram JSL (1993).Tropical Soil Biology and Fertility: A handbook of methods. CAB International. Wallingford, UK. p. 35.

FAO (1977). Guidelines for soil profile description, 2nd Edition. Soil Resources Development and Conservation Service, Land and Water Management Division, Rome

FAO-UNESCO (1997). Soil map of the world. Technical Paper No 20, 2nd Edition. ISRIC, Wageningen, The Netherlands.

Hillel D (1982). Introduction to soil physics. Academic Press INC. Orlando, Florida.

Hinga G, Muchena FN, Njihia CM (1980). Physical and chemical methods of soil analysis, Nairobi, Kenya.

Kamoni PT, Wanjogu SN (2006). Rating of land qualities in Kenya. Miscellanous Paper No. 117. Kenya Soil Survey, Nairobi.

Muya EM, Esilaba AO, Lelon JK, Gachini GN (2008). Combating in irrigated agriculture through systematic characterization of irrigated areas for improved irrigation efficiency. Proc. Drylands, Degradation and Desertification Conference, Ben-Gurion University of the Negev, Isreal, 14th – 17th December, 2008.

Muchangi P, Sijali IV, Wendot H, Lemperiere P (2005). Improving the performance of irrigation in Africa. Irrigation systems, their operation and maintenance. Training source book No 1, Nairobi, Kenya.

Nelson DW, Sommers LE (1982). A rapid and accurate method for estimating organic carbon in soil. Proc. Of the Indiana Acad. Sci., 84: 456-462

Okaleb JR, Gathua KW, Woomer P (2002). Laboratory Methods of Soil and Plant Analysis: A working Manual, 2nd Edition. TSBF-CIAT and SACRED Africa, Nairobi, Kenya.

Reducing compaction effort and incorporating air permeability in Proctor testing for design of urban green spaces on cohesive soils

A. M. Ibrahim[1], N. Persaud[2*], R. W. Zobel[3] and A. Hass[4]

[1]Soils and Water Department, Faculty of Agriculture, El-Fayoum University, Fayoum, Postal Code 63514, Egypt.
[2]Crop and Soil Environmental Sciences Department, Virginia Polytechnic Institute and State University, Blacksburg, Virginia 24061-0404, USA.
[3]USDA/ARS Appalachian Farming Systems Research Center, 1224 Airport Road, Beaver, West Virginia, USA.
[4]West Virginia State University, Gus R. Douglass Institute, P. O. Box 1000, Institute, WV 25112-1000, USA.

It is well established that compaction negatively affects agronomic productivity, that air permeability is a sensitive measure of the degree of soil compaction and therefore a good indicator of soil productivity impairment from compaction. Cohesive soils in urban settings are often heavily compacted by the common engineering practice to compact sub-grades of urban construction sites to 95% or more of the optimum density obtained in standardized Proctor tests. The objective of this study was to determine to what extent reducing compaction effort would increase the air permeability of Proctor test specimens. Quantifying this relationship would permit more appropriate Proctor test specifications for the design of urban green spaces on cohesive soils. We designed a portable transient flow apparatus for rapidly measuring air permeability and used it to measure air permeability on Proctor test specimens of three cohesive sub-grade soil materials covering a range of USDA textures (loam, silt loam and silty clay) and Proctor compaction characteristics. We compacted test specimens at their Proctor optimum water content using efforts ranging from 100 to 25% (the lowest practicable value) of that used in the standardized Proctor test. Results confirmed that compaction severely reduces air permeability of the test specimens and indicated that the common practice of compaction to 95% or more of the optimum Proctor density is probably not appropriate for construction of urban green spaces. Reducing compaction effort from 100 to 25% of the standardized Proctor test value increased air permeability 30, 89 and 42 times respectively for the loam, silt loam and silty clay test specimens. More extensive studies are needed to correlate measured air permeability of Proctor test specimens to agronomic productivity of urban green spaces.

Key words: Urban soils, air permeameter, modified Proctor test.

INTRODUCTION

Air permeability is the most commonly-used measure to characterize the ability of air to move through unsaturated soils and is the most direct indicator of the soil's ability to breathe and self-aerate. Most terrestrial plant species cannot transport sufficient oxygen from the leaf area to their roots. Overall plant growth and development is a function of proper root respiration and this in turn would depend on the soil self-aeration capacity (Cheng-He et al., 2006). Air permeability has been long-recognized as an important physical indicator of soil productivity and its improvement is the aim and rationale for many agronomic management practices (Glinski and Stepniewski, 1985).

Measurements of soil air permeability have had a long history and excellent reviews are available (Ball et al., 1981; Corey, 1986; Ghildyal and Tripathi, 1987; Hinsinger and Mettauer, 1989; Scanlon et al., 2002; Ball and Schjonning, 2003; Lal and Shukla, 2004; Switzer and Kosson, 2007). In general, air permeability can be measured on laboratory specimens or in-situ by steady or

*Corresponding author. E-mail: npers@vt.edu.

transient-flow methods (Kirkham, 1947; Grover, 1955; Corey, 1957; Li et al., 2004; Tuli et al., 2005). Steady-flow methods require larger volumes of air, take longer and are more prone to measurement error (Lal and Shukla, 2004). Transient-flow measurements are more rapid, use smaller volumes of air, are less prone to measurement error, but require more sophisticated instrumentation and associated physico-mathematical models and tools for their design and data analysis (Smith et al., 1997a, 1997b, 1998).

Studies have shown that air permeability is a sensitive measure of the degree of soil compaction (Blackwell et al., 1990; Kouno et al., 1992; Stepniewski et al., 1994) and it is well-known that compaction negatively affects agronomic productivity (O'Sullivan, 1992; Assouline and Tessier, 1997; Smith et al., 1997a, 1997b; Lipiec and Hatano, 2003). High compaction levels are especially problematic in urban green spaces where it is common practice to compact sub-grades to 95% or more of optimum Proctor density determined by the standardized ASTM (American Society for Testing and Materials) testing procedures (Baker, 1990; Gilbert, 1991; Harris, 1991; ASTM, 2009). This practice is primarily rooted in customary engineering practices (used to meet grade elevation, strength and load-bearing specifications of soil foundations) that are applied to the construction of urban green spaces without consideration of its intended agronomic uses. For example, it is well-established that high levels of compaction induced during sub-grade construction affects the ability of natural recreational fields to sustain actively-growing turf grass covers (Carrow, 1980; Sills and Carrow, 1983; Kouno et al., 1992); and to provide optimum conditions for player safety and performance (Bramwell et al., 1972; Bonstingl et al., 1975; Baker, 1991; ASTM 1988, 1994, 1995). Indeed, the urban green space maintenance industry has developed costly equipment and practices to mitigate compacted urban green spaces. Clearly, it would be much more cost effective to have in place appropriate ex-ante design and construction practices versus ex-post efforts to mitigate compaction of these soils. Surprisingly, we found no investigations reported in the literature that approached this issue from this ex-ante 'prevention-rather-than-cure' perspective.

Inasmuch as air permeability can serve as an indicator of compaction, it can likewise serve to indicate compaction-induced impairment of the agronomic productivity of soils. Agronomists generally agree that change in porosity and pore size distribution due to compaction has the greatest impact on plant growth and production since these factors control plant root pene-tration and the availability and accessibility of water and nutrients required for optimal biomass production (Ball, 1987; Gupta et al., 1989; Huang and NeSmith, 1999; Stepniewski et al., 1994; Watabe et al., 2000; Richard et al., 2001; Startsev and McNabb, 2001).

These studies make a case for examining whether or not compaction effort should be reduced and whether air permeability measurements should be incorporated in the standardized Proctor test routinely used for the design and construction of urban green spaces. Research considering the appropriateness of the Proctor test in this context is lacking. The objective of this study was to determine what effect reduction of the compaction effort used in the standard Proctor test would have on the air permeability of the test specimens.

MATERIALS AND METHODS

Test specimens

Available air-dried bulk samples of three natural cohesive soil materials were used to prepare the Proctor-compacted test specimens. These materials were selected because they were known to be physically different. The grain size distribution (Figure 1) and ASTM Proctor test for dry bulk density at optimum water content (Table 1) covered a wide range. These three soil materials were designated according to the USDA classification scheme as loam, silt loam and silty clay (Soil Survey Staff, 2004). The percent of sand, silt and clay were 48.4, 35.8 and 15.8 for the loam; 16.9, 57.0 and 26.1 for the silty loam; and 10.7, 46.8 and 42.5 for the silty clay (Figure 1).

Water was added to air dried samples of these three materials to obtain the optimum water content (Table 1). The samples were mixed and left to equilibrate in sealed plastic bags for at least 24 h. The three compaction efforts (designated as blows per layer) represent the ASTM D698 standard effort using a 2.5 kg rammer with three layers compacted at 25 blows per layer (ASTM, 2009) and close to 50 and 75% reduction of this standard effort obtained using 18 and 12 blows respectively per layer (Table 1). The optimum water content and corresponding dry bulk density values were determined following the ASTM D698 standard test (ASTM, 2009) that details how to prepare the sample, the appropriate apparatus, the test conditions and procedures and data recording and analysis. The compaction effort can also be expressed in terms of energy per unit volume. The rammer mass and drop height were 2.5 kg and 30 cm, respectively and the compaction mold volume (10.2 cm ID and 11.6 cm tall) was 940 cm^3. The energy per unit volume was therefore close to 587, 423 and 282 kN-m m^{-3} for the 25, 18 and 12 blows per layer respectively.

AIR PERMEAMETER

Construction and setup

Figures 2a and 2b illustrate the transient flow permeameter used to measure the air permeability of the compacted specimens. The air reservoir chamber consisted of a length of 10 cm ID schedule 40 PVC pipe (Figure 2a). As shown in Figure 2a, a Schrader valve installed near the bottom of the pipe was used to pressurize the air chamber. A type T thermocouple was made of 30-gauge insulated copper-constantan wire and used to monitor air temperature in the chamber (Figure 2a).

The effective length, L_e, (that is measured from the base of the column to the bottom of the sample) of the PVC pipe was designed to permit at least 300 cm^3 of excess volume in the air chamber available for flow through the sample when the chamber was pressurized to a minimum of 20 cm of water column (1962 Pa) above ambient atmospheric pressure. The ideal gas law and observed local barometric pressure (P_{atm}) were used to determine this appropriate effective length. A fixed mass of air (considered ideal) in volume V_o cm^3 at atmospheric pressure (P_{atm}) compressed isothermally to absolute pressure $P = P_{atm} + P'$ (where P' is the excess pressure) would obey Boyle's Law giving $V_o P_{atm} = (P_{atm} +$

Table 1. Dry bulk density (ρ_b) at the optimum water content (θ_g) obtained from ASTM Proctor tests using varying compaction effort on 3 cohesive soils.

Blows per layer	Loam		Silt loam		Silty clay	
	θ_g(g g^{-1})	ρ_b(g cm^{-3})	θ_g(g g^{-1})	ρ_b(g cm^{-3})	θ_g(g g^{-1})	ρ_b(g cm^{-3})
12	15.9	1.770	19.7	1.620	27.2	1.500
18	15.3	1.812	20.0	1.700	25.9	1.525
75	14.4	1.873	17.0	1.775	22.0	1.595

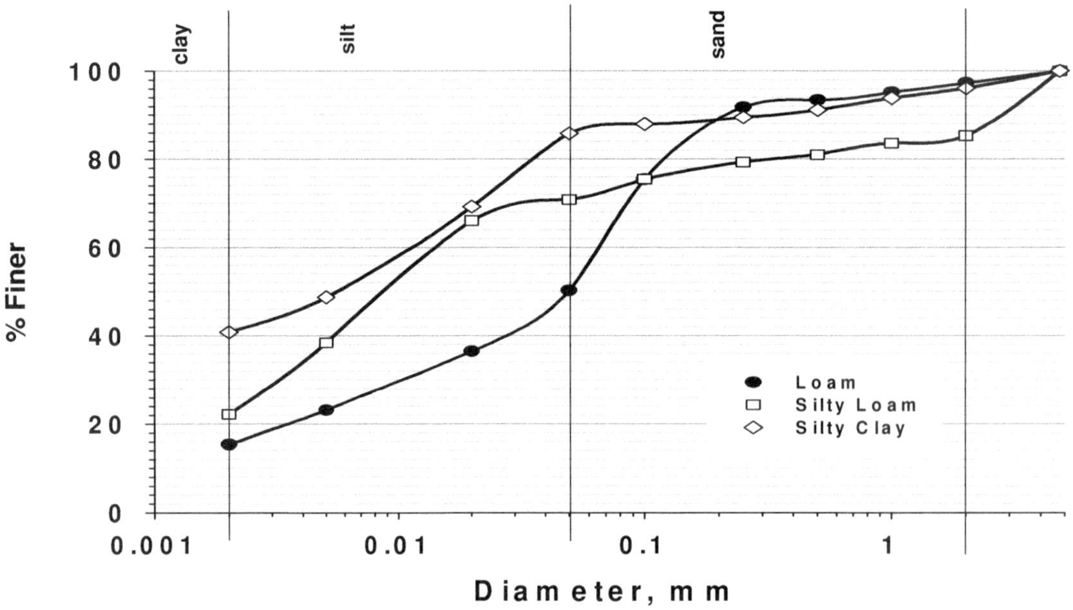

Figure 1. Grain size distribution curves for the three soil materials. Lines represent smooth curves through the data points using natural cubic splines.

P') V where V is the volume after compression. This gives V = (V_o P_{atm}) / (P_{atm} + P'). The change in volume V' after compression is therefore V' = (V_o – V) = V_o - (V_o P_{atm}) / (P_{atm} + P') = V_o (1 – [P_{atm} / (P_{atm} + P')]. The value of P_{atm} was taken from records of hourly barometric pressure measured at the airport in Blacksburg, Virginia (37° 12' 27.5" N and 80° 24' 28.2"W, elevation 650 m above mean sea level, WBAN ID number 53881, WMO ID number 72318) located about 1 km away from the laboratory where the study was conducted. The effective length (L_e in cm) used in this study was 206.4 cm giving an excess volume at 20 cm water column excess pressure = 347.1 cm^3.

The combination pressure regulator and manometer (Figure 2b) was used to precisely control the excess pressure applied and to prevent overloading of the pressure transducer (Figure 2a). As shown, it was constructed with a 1 m long, 15 cm diameter clear extruded Plexiglas pipe chemically welded using methylene chloride to a 20 cm square base of 12 mm thick Plexiglas sheet. Adhesive-backed metric measuring tape graduated in mm (Sargent-Welch, Buffalo, New York 14217, USA) was attached to the outer wall of the clear Plexiglas pipe to measure the difference between level of water in the pressure regulator and manometer and the level in the plastic tubing (Figure 2b). A stopcock installed in the Plexiglas base of the pressure regulator and manometer permitted drainage and adjusting the water level. In order to minimize any radiant heat exchange the air chamber was completely wrapped in high-strength aluminum foil.

Pressure transducer

A general purpose, full-bridge, mV output, 1 psi full scale, gauge pressure transducer (Model PX480A-001GV, Omega Engineering Inc., Stamford, Connecticut 06907, USA) was used to measure excess pressure over time in the air chamber of the permeameter. A datalogger (Model CR-21X, Campbell Scientific Inc., Logan, Utah 84321, USA) was used to log the mV output of the transducer and the temperature output (in °C) of the thermocouple in the permeameter air chamber. The pressure transducer was calibrated by measuring the mV output to step changes in the excess pressure in the permeameter chamber. For this measurement, the chamber was pressurized to 20 cm of water and step decreases in pressure were applied by releasing short burst of air via the Schrader valve. Ten readings at 2 s intervals of the mV output for each pressure step decrease were then taken on the data logger. This process was repeated for starting values of excess pressure equal to 25, 30, 35 and 40 cm of water. The excess pressure values in cm water versus the corresponding average of the ten mV values were plotted and a least squares linear regression line fitted to the plotted points.

Measurement procedure

Prior to the beginning of any measurement, the air-tightness of the

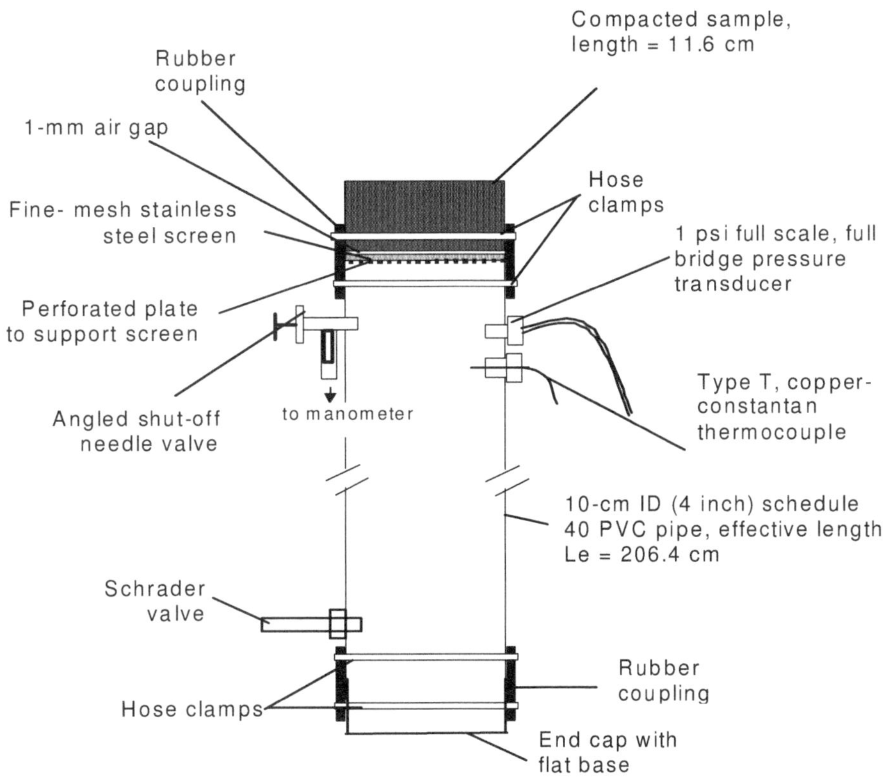

Figure 2a. Air permeameter chamber assembly.

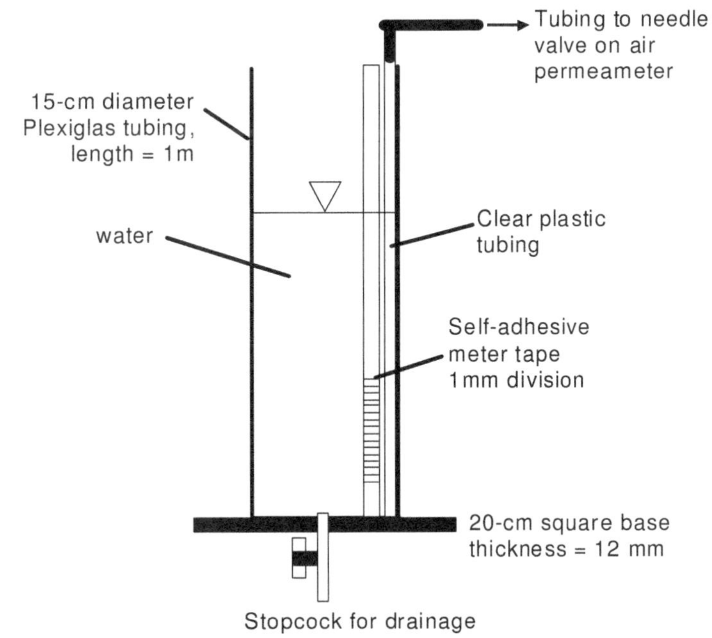

Figure 2b. Pressure regulator and manometer assembly.

permeameter was tested using a soap bubble solution. The Proctor-compacted test specimen was prepared using the soil material at the Proctor optimum moisture content (Table 1) and the standard ASTM D698 apparatus but with a 10 cm ID schedule 40 PVC tube cut to the exact height (11.6 cm) in place of the standard mold. The external and internal diameters of the substitute PVC mold were identical to those of the standard mold.

Compacted test specimens were placed at the sample end of the permeameter and the air chamber over-pressurized to a preset level on the pressure regulator and manometer. Preset excess pressures of 20, 25, 30, 35 and 40 cm of water were used, giving 5 separate tests (numbered as 1 through 5 respectively) on each test specimen for a given compaction effort (an overall total of 45 tests on 9 specimens). The mV output was logged at interval of 3 s and the logged data downloaded manually. The transducer calibration curve equation was used to convert the mV outputs to excess pressure yielding the excess pressure (P') as function of time (t) for each of the 45 tests.

Calculation of air permeability

As proposed by Springer et al. (1995) for transient flow air permeability measurement, the excess pressure (P') in the air chamber decreases with time due to air permeating through the compacted test specimen. The excess volume V' would also decrease as a function of time (t) assuming isothermal conditions in the air chamber. The absolute pressure (P_o) of the air occupying the chamber (volume V_o) is $P_o = P_{atm} + P'_o$ where P'_o is the excess pressure recorded on the datalogger at time t = 0. At some time *t* later, the absolute pressure $P(t) = P_{atm} + P'(t)$. By Boyle's law at any time t, $P(t)V_o = P_o V(t)$ whence $V(t) = P(t)V_o/P_o$ and therefore since $P(t) = P_{atm} + P'(t)$ and $P_o = P_{atm} + P'_o$

$$V'(t) = V_o - V(t) = V_o - \frac{V_o\,[P_{atm} + P'(t)]}{P_o} \quad (1)$$

Assuming quasi-steady state conditions (meaning uniform pressure gradients are rapidly developed along the length of the specimen) and purely viscous flow, the flux (q = volume per unit time per area) through the specimen is:

$$q = \frac{1}{A}\frac{dV'}{dt} = \frac{1}{A}\frac{d}{dt}\left[V_o - \frac{V_o\,(P_{atm} + P')}{P_o}\right] = -\frac{1}{A}\frac{V_o}{P_o}\frac{dP'}{dt} \quad (2)$$

Where; A is the cross-sectional area of the specimen.

Analogously describing steady water flow using Darcy's law, q =

$$K_a \frac{P - P_{atm}}{L} = K_a.\ \frac{P}{L}$$

Here L is the length of the specimen and K_a is a proportionality constant analogous to the saturated hydraulic conductivity in Darcy's law. Therefore,

$$\frac{1}{A}\frac{V_o}{P_o}\frac{dP'}{dt} = -K_a\frac{P'}{L} \quad (3)$$

Separating variables, rearranging algebraically, and integrating between $P' = P_o - P_{atm} = P'_o$ at t = 0 and P' = P(t) at t = t gives ln

$$(P'/P'_o) = .\ \frac{-K_a A P_o}{V_o L} t$$

The implies that plotting the experimental data as P'/P'_o on the y-axis versus time on the x-axis on a semi-logarithmic graph would yield a decreasing straight line with

$$\text{Slope} = \frac{-K_a A P_o}{V_o L}.$$

From this slope the value of K_a can be obtained.

There are two choices for the form of the exponential decay function to be fitted to the $P'(t)/P'_o$ versus *t* data points. Fitting the one-parameter equation $P'(t)/P'_o = e^{-bt}$ to the data points on a semi-logarithmic plot of $P'(t)/P'_o$ versus time (t), forces the plot through $P'(t)/P'_o = 1$ at t = 0. On the other hand fitting the data to the two-parameter equation $P(t)/P'_o = ae^{-bt}$ would relax this condition. Both equations were fitted to the data for each of the 45 air permeability measurements.

A value of P_{atm} = 966.6 cm water was used to calculate $P_o = P_{atm} + P'o$ in all calculations. This value represented the mean of the measured hourly barometric pressure at the airport in Blacksburg, Virginia during the hours when the tests were made. The P_o values were converted to Pa using 1 cm of water = 98.1 Pa. The permeability k was calculated as the extracted K_a value ($cm^2\ Pa^{-1}\ s^{-1}$) times the dynamic viscosity of air (μ_a in Pa s). This value of k in cm^2 was converted to $(\mu m)^2$ by multiplying by $10^8\ (\mu m)^2\ cm^{-2}$.

The value of μ_a is very weakly dependent on pressure but increases with absolute temperature. It's value $\mu_a(T)$ at a given temperature T°K can be calculated using Sutherland's equation (Sutherland, 1893) as;

$$\mu_a(T) = \mu_o\ \frac{T_o + C}{T + C}\left(\frac{T}{T_o}\right)^{\frac{3}{2}}$$

Where; μ_o is the reference value at temperature T_o (°K) and the parameter C (°K) is specific for a given gas. This equation is based on the molecular theory of ideal gases and is valid for a wide range of temperatures. For air, the value $\mu_o = 17.16 \times 10^{-6}$ Pa s at T_o = 273.15°K, and the value of C = 110.4°K. The mean measured air temperature in the chamber during a given permeability test was used to calculate the $\mu_a(T)$ needed to calculate *k* for the given specimen.

RESULTS AND DISCUSSION

The calibration for the general purpose differential (gauge) pressure transducer showed almost perfect linearity. The calibration equation was P' = 1.503 × (mV) - 1.8591 with an r^2 = 0.999. This equation shows the general purpose transducer resolution was quite good, giving 1 mV output for a change of 1.5 cm water column. During the tests, the transducer mV output was read on the data-logger over a firmware-selectable full-scale range of 0 to 50 mV. In this range, the data logger's voltmeter could detect changes of 3.33 μV for a differential voltage measurement. Therefore the pressure measuring system could resolve a change of 0.05 mm of water (approximately = 0.5 Pa) pressure change in the air chamber of the permeameter. The regression line equation showed that a reading of 1.24 mV input into the calibration equation would indicate zero excess pressure in the permeameter air chamber.

The air temperature in the chamber recorded for all

tests never exceeded the mean value by more that 0.35°C. Wrapping the permeameter with aluminium foil was effective in maintaining close to isothermal conditions in the permeameter air chamber during tests although the climate in the laboratory was uncontrolled. Nevertheless, it is recommended that the air chamber should be very well insulated with better materials. With proper insulation the permeameter could be used outdoors at construction sites.

Despite its age, Sutherland's equation was reliable for calculating the air dynamic viscosity at room temperatures. The values of dynamic viscosity (μ_a) calculated by Sutherland's equation at 300°K were very close to the measured value of 1.82 x 10^{-5} Pa s at this temperature reported in the CRC Handbook of Chemistry and Physics (Lide, 1995). Calculated values at 10, 20, 30 and 40°C were respectively (1.763, 1.811, 1.859, 1.906) × 10^{-5} Pa s and practically the same as the corresponding values of (1.77, 1.81, 1.85, 1.89) × 10^{-5} Pa s reported by Scanlon et al. (2002).

The experimental data on $P'(t)$ / P'_o versus time were plotted on semi-logarithmic graphs for all 45 tests. In all cases these plots were linear until an inflexion point that occurred at some point after $P'(t) / P'_o < 0.2$. This could only mean that after this point the assumptions made in developing the foregoing physico-mathematical model discussed above were no longer applicable. These assumptions would imply linearity as long as the air flow through the voids in the test specimen was purely viscous. Gaseous viscous advective flux generally predominates over viscous slip and diffusive fluxes even under low pressure gradients when pore radii are much greater than the mean free path of the air molecules. But if the pore radii are small (as would be expected for the partially saturated compacted test specimens), viscous slip (when mean free path ≈ pore radii), Knudsen diffusive (mean free path > pore radii), or molecular diffusive flow (mean free path >> pore radii) would be expected to predominate over viscous advective fluxes as the pressure decreases (Alzayadi, 1975; Alzayadi and Moore, 1978; Thorstenson and Pollock, 1989). Also, the mean free path is inversely proportional to pressure at a given temperature implying it is greater at lower pressure. Therefore, depending on the pore-sizes and their distribution, the predominance of advective viscous flow would decrease relative to viscous slip, Knudsen diffusive, or molecular diffusive flows at some value of the excess pressure in the chamber. In this case, the quasi-steady state conditions for purely viscous flow would no longer be valid. This changeover value would depend on the permeability of the test specimen. For the test specimens in this study this value occurred at some point after $P'(t) /P'_o < 0.2$.

Since the primary interest and the assumptions made in this study were to quantify the air permeability under predominantly viscous flow, the ($P'(t) /P'_o$) versus time data points for all 45 test cases were truncated and the portion for $P'(t) /P'_o > 0.2$ used to estimate *k*. The two-parameter fitting equation $P(t) /P'_o = ae^{-bt}$ tended to give better and more consistent fits to the data for all 45 tests compared to the one-parameter equation $P'(t) /P'_o = e^{-bt}$. The coefficient of determination (r^2) for all fits with the two-parameter equation was never < 0.98 and never < 0.94 for the one-parameter equation. Two-tailed t-tests (assuming equal variances) showed that the mean of the k-values over the 5 tests for the one-parameter and two-parameter fitting equations were not statistically different at the 5% level of significance. Nevertheless, the k-values obtained using the two-parameter equations were considered more valid.

A typical case (for the silty loam specimen at 18 blows per layer) for the two-parameter fits is illustrated in Figure 3. The tests numbered as 1 through 5 are for targeted excess pressures of 20, 25, 30, 35 and 40 cm of water, respectively. The mean and standard deviation of the k-values obtained using the two-parameter fitting equation for the 5 tests for each of the 9 treatment combinations were calculated and showed that the 5 k-values for a given treatment combination fell within the mean ± < 2 standard deviations.

Since there was negligible differences in the slope of the fitted lines for different values of P'_o, the 5 k-values obtained using the two-parameter fitting equation were treated as repeated measurements for a given test specimen. A two-way analysis of variance was conducted using the soil material and compaction effort as factors. The results showed highly significant effects for soil material, compaction effort and their interaction. Interestingly, the results (Table 2) showed in order of decreasing overall mean permeability silt loam > loam > silty clay. Figure 1 shows the order of decreasing (clay + silt) % was silty clay > silt loam > loam, the order of decreasing sand was loam > silt loam > silty clay and the order of decreasing gravel (> 2mm) was silt loam > loam = silty clay. It would appear that the higher % of gravel in the silt loam had a disproportional effect on the air permeability.

In general, decreasing the compaction effort increased the permeability (Table 2). In most cases the means were significantly different (or very close to being significant) at the 5% level using the Fisher's LSD values in Table 2. However the pattern of these increases were different for the different soil materials (Figure 4) and explained the significant interaction obtained in the analysis of variance. Overall, the mean permeability estimates (Table 2) were very low as would be expected for Proctor-compacted specimens at the optimum water content. More surprising was the rather insignificant (from an agronomic perspective) increase in permeability with decreasing compaction effort (Table 2 and Figure 4). For natural un-compacted soil, low air permeability is taken as $k < 10$ $(\mu m)^2$ and high permeability as $k > 250$ $(\mu m)^2$ (Ball and Schjonning, 2003).

Decrease in air permeability from compaction is due

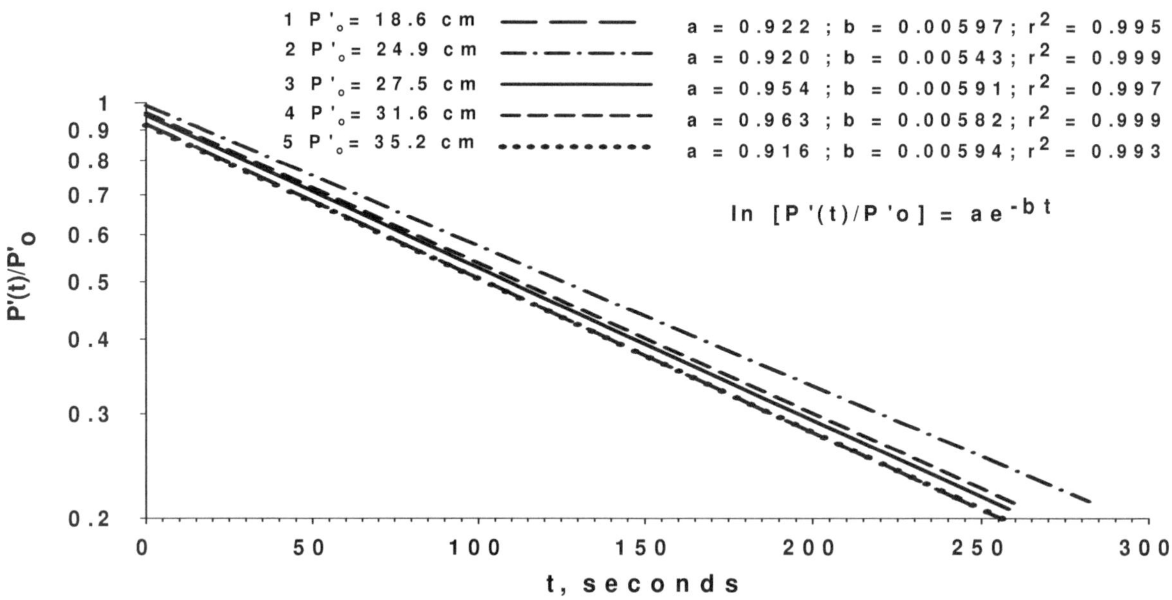

Figure 3. Semi-logarithmic plots for fitting the two-parameter equation $P(t)/P'_o = ae^{-bt}$ to a truncated set of data points for $P'(t)/P'_o$ versus time (t) measured for the 5 values of P'_o.

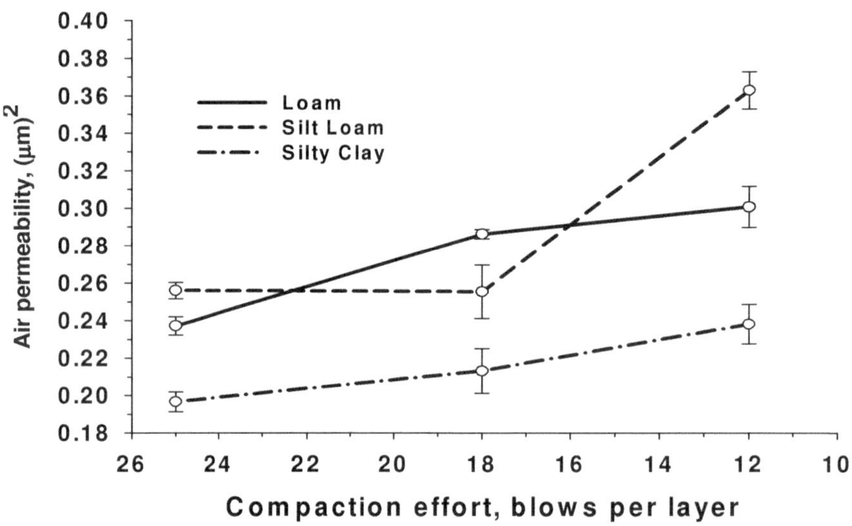

Figure 4. Air permeability (estimated using the two-parameter fitting equation $P(t)/P'_o = ae^{-bt}$) with decreasing compaction effort. Error bars represent the standard error of the mean of 5 repetitions on a given test specimen.

primarily to its effect on total porosity and pore size distribution as a result of re-arrangement of the soil particles (Sridharan et al., 1971; O'Sullivan, 1992; Assouline and Tessier, 1997; Smith et al., 1997; Lipiec and Hatano, 2003).

If θ_a is the volume fraction of air and $V \sim L^3$ is the total bulk volume of the specimen then, assuming a specific gravity of water = 1, $V(1 - \theta_a) = V_s(1 + \theta_g \rho_p)$ where V_s is the volume of dry solids, θ_g is the gravimetric water content and ρ_p is the mean particle density. Therefore, $V_s/V = (1 - \theta_a)/(1 + \theta_g \theta_p)$ and since by definition $\rho_b = W_s/V$ and $W_s = V_s\rho_p$, then

$$\rho_b = \frac{W_s}{V} = \frac{V_s \rho_p}{V} = \frac{(1-\theta_a)\rho_p}{1+\theta_g\rho_p} \tag{4}$$

Using a reasonable value for $\rho_p = 2.65$ g cm^{-3} with the values of θ_g and ρ_b in Table 1 shows that the volumetric air content of the test specimens was ≤ 5% of the sample

Table 2. Air permeability measured using the two-parameter fitting equation for three sub-grade soil materials under varying compaction effort. Values represent the mean of 5 repetitions on a given test specimen.

Soil material	Compaction effort, blows per layer			Overall mean
	25	18	12	
Loam	0.237	0.286	0.301	0.275
Silt Loam	0.256	0.253	0.363	0.291
Silty Clay	0.197	0.213	0.238	0.216
Overall mean	0.230	0.252	0.301	0.261

LSD = 0.026 $(\mu m)^2$ and 0.035 $(\mu m)^2$ for comparison of means at the 5 and 1% significance levels, respectively

volume.

Yet this question remained short of no compaction: what level of reduction in compaction of the specimens would produce substantial increase in the permeability? Although reducing from 25 to 12 blows per layer produced statistically significant differences (Table 2) the actual differences appeared insignificant from the agronomic perspective. Reducing the compaction effort to 6 blows per layer would be equivalent to 141 kN-m m^{-3} and would be the lowest possible reduction practicable to still achieve uniform compaction in the Proctor test. The Proctor density (g cm^{-3}) at optimum water content (%) were determined as 1.681 at 18.3; 1.607 at 19; and 1.365 at 26.6 for the loam, silt loam and silty clay, respectively.

Using Equation (4) with these values showed that the volumetric air content would be appreciably higher at 6 blows per layer and therefore an increase in permeability was to be expected. When the air permeability measurements were initiated on test specimens compacted using 6 blows per layer, it was immediately apparent that an appreciable change had indeed occurred since 40 cm water excess pressure (P'_o) was dissipated in less than 10 to 20 s. It was not possible to make reliable $P'(t)/P'_o$ versus time measurements at $P'_o < 40$ cm water.

The k-values using $P'_o = 40$ cm water estimated using the one-parameter fitting equation was 7.21, 22.46 and 8.53 $(\mu m)^2$ for the loam, silt loam and silty clay, respectively. Corresponding values using the two-parameter fitting equation was 7.12, 22.78 and 8.34 $(\mu m)^2$. These latter air permeability values, albeit still considered low, were 30, 89 and 42 times the values obtained using the standardized Proctor compaction effort (25 blows per layer) for the loam, silt loam and silty clay test specimens, respectively (Table 2).

CONCLUSION

We compacted test specimens at their Proctor optimum water content using efforts ranging from 100 to 25% (the lowest practicable value) of that used in the standardized Proctor test. Results confirmed that compaction severely reduces air permeability of the test specimens and indicated that the common practice of compaction to 95% or more of the optimum Proctor density is probably not appropriate for construction of urban green spaces. Reducing compaction effort from 100 to 25% of the standardized Proctor test value increased air permeability 30, 89 and 42 times, respectively for the loam, silt loam and silty clay test specimens. More extensive studies are needed to correlate measured air permeability of Proctor test specimens to agronomic productivity of urban green spaces.

The transient-flow permeameter we designed to measure air permeability of Proctor test specimens is rapid, was quite reliable and had good resolution. The most important parameter in the design is the volume (V_o) of the air chamber. Solving the equation $\ln (P'/P'_o) = (-K_a A P_o t) / V_o L$ for V_o, gives V_o in $cm^3 = -(K_a A P_o t)/[L \ln (P'/P_o')]$ with $K_a = [k \text{ in } (\mu m)^2 \times 10^{-8}] / \mu_a$. This permits specifying V_o based on expected values of the air permeability and the time (*t*) for the excess pressure in the air chamber to dissipate.

ACKNOWLEDGEMENTS

Funding for this study was provided by the United States Department of Agriculture, Agricultural Research Service (USDA/ARS) under Specific Collaborative Agreement 58-1932-6-636 between the USDA/ARS Appalachian Farming Systems Research Laboratory, Beaver, West Virginia and Virginia Polytechnic Institute and State University, Blacksburg, Virginia; and Cooperative Agreement #58-1932-5-539 between the USDA/ARS Appalachian Farming Systems Research Laboratory, Beaver, West Virginia and West Virginia State University, Institute, WV.

REFERENCES

Alzayadi AA (1975). Flow of gases in porous media. Ohio State Univ. Press, Columbus, Ohio, USA. 172 p.

Alzayadi AA, Moore CA (1978). Combined pressure and diffusional transition region flow of gases in porous media. Amer. Institute of Chem. Engineers J. 24:35–43,

Assouline S, Tessier D (1997). Effect of compaction on soil physical and hydraulic properties: experimental results and modeling. Soil Sci.

Soc. Am. J. 61: 390 –398.
ASTM (1988). Natural and artificial playing fields: characteristics and safety features. ASTM Special Technical Publications 1073. ASTM, Philadelphia.
ASTM (1994). Safety in American football. ASTM Special Technical Publications 1305. ASTM, Philadelphia.
ASTM (1995). Safety in baseball/softball. ASTM Special Technical Publications 1313. ASTM, Philadelphia.
ASTM (2009). Standard test methods for laboratory compaction characteristics of soil using standard effort 12 400 ft-lbf/ft^3 (600 kN-m/m^3) ASTM D698. In: ASTM Annual Book of Standards Volume 4.08, ASTM, Philadelphia.
Baker SW (1990). Sands for sports turf construction and maintenance. Sports Turf Research Institute, Bingley, West Yorkshire, United Kingdom. 58p.
Baker SW (1991). Temporal variation of selected mechanical properties of natural turf football pitches. J. Sports Turf Res. Inst. 67: 53 – 65.
Ball BC, Harris W, Burford JR (1981). A laboratory method to measure gas diffusion and flow in soil and other porous materials. J. Soil Sci. 32: 323–333.
Ball BC (1987). Air permeability and gas diffusion measurements to quantify soil compaction. In Monnier, G. and M. J. Goss (ed), Soil Compaction and Regeneration: pp.15–23.
Ball BC, Schjonning P (2003). Air permeability. In Methods of Soil Analysis, Part 4, Physical Methods. Soil Sci. Soc. Amer. Inc., Madison, Wisconsin, USA: pp.1141 – 1158.
Blackwell PS, Ringrose-voase AJ, Jayawardane NS, Olsson KA, Mckenzie DC, Mason WK (1990). The use of air-filled porosity and intrinsic permeability to air to characterize structure of macropore space and saturated hydraulic conductivity of clay soils. Eur. J. Soil Sci. 41: 15 – 228.
Bonstingl R, Morehouse C, Niebe IB (1975). Torques developed by different types of shoes on various playing surfaces. Med. Sci. Sports Exerc. 7: 127 –131.
Bramwell S, Requa R, Garrick J (1972). High school football injuries: a pilot comparison of playing surfaces. Med. Sci. Sports Exerc. 4: 166 – 169.
Carrow RN (1980). Influence of soil compaction on three turfgrass species. Agron. J. 72: 1038 - 1042
Cheng-He H, Ouyang Y, Zhang JE (2006). Effects of a compact layer on soil O_2 diffusion. Geoderma 135: 224 –232.
Corey AT (1957). Measurement of water and air permeability in unsaturated soil. Soil Sci. Soc. Am. Proc. 21: 7 – 10.
Corey AT (1986). Air permeability. In Methods of Soil Analysis Part1 Physical and Mineralogical Methods (2nd edition). Am. Soc. Agron. Madison, Wisconsin, USA: pp.1121–1136.
Ghildyal BP, Tripathi RP (1987). Soil physics. New York, Wiley, 656p.
Gilbert OL (1991). The ecology of urban habitats. Chapman and Hall, New York.
Glinski J, Stepniewski W (1985). Soil aeration and its role for plants. CRC Press, Boca Raton, Florida, USA. 229 p.
Grover BL (1955). Simplified air permeameters for soil in place. Soil Sci.
Soc. Am. Proc. 19: 414 – 418.
Gupta SC, Sharma PP, De Franchi SA (1989). Compaction effects on soil structure. Adv. Agron. 42: 311 – 338.
Harris JA (1991). The biology of soils in urban areas: In: Bullock P, Gregory PJ (ed.) Soils in the urban environment. Blackwell Scientific Publications, Oxford, UK. pp139–152.
Hinsinger P, Mettauer H (1989). Soil air permeability and its application. 1. Air permeability: theory and methodology. Agronomie 9: 3 –12.
Huang B, NeSmith SD (1999). Soil aeration effects on root growth and activity. Acta Hortic. 504: 41–52
Kirkham D (1947). Field method for determination of air permeabilty of the soil in its undisturbed state. Soil Sci. Soc. Am. Proc.11: 93 –99.
Kouno K, Ogata S, Takiyama S (1992). Soil compaction effects on root development of grasses and some soil physical properties. Jpn. J. Soil Sci. Plant Nutr. 63:154–160.
Lal R, Shukla M (2004). Principles of soil physics. New York, Marcel Dekker. 716p.
Li H, Jiu J, Luk M (2004). A falling-pressure method for measuring air permeability of asphalt in laboratory. J. Hydrol. (Amsterdam) 286: 69 – 77.
Lide DR (Editor-in-Chief). (1995). CRC handbook of chemistry and physics: a ready-reference book of chemical and physical data, 76th edition. CRC Press, Boca Raton, Florida, USA.
Lipiec J, Hatano R (2003). Quantification of compaction effects on soil physical properties and crop growth. Geoderma 116: 107 – 136.
O'Sullivan MF (1992). Uniaxial compaction effects on soil physical properties in relation to soil type and cultivation. Soil Tillage Res. 24: 257 – 269.
Richard G, Cousin I, Sillon JF, Bruand A, Guérif J (2001). Effect of compaction on the porosity of a silty soil: influence on unsaturated hydraulic properties. Eur. J. Soil Sci. 52: 49 – 58.
Scanlon BR, Nicot JP, Wassman JW (2002). Soil gas movement in unsaturated systems. In. Soil physics companion, Warrick AW (ed.) CRC Press, Boca Raton, Florida, USA. pp.297 –341.
Sills MJ, Carrow RN (1983). Turfgrass growth, N use, and water use under soil compaction and N fertilization. Agron. J. 75: 488 – 492.
Smith CW, Johnston MA, Lorentz S (1997a). The effect of soil compaction and soil physical properties on the mechanical resistance of South African forestry soils. Geoderma 78: 93 –11.
Smith JE, Robin MJL, Elrick DE (1997b). A source of systematic error in transient-flow air permeameter measurements. Soil Sci. Soc. Am. J. 61: 1563–1568.
Smith JE, Robin MJL, Elrick DE (1998). Improved transient-flow air permeameter design: dampening the temperature effects. Soil Sci. Soc. Am. J. 62: 1220–1227.
Soil Survey Staff. (2004). Soil survey laboratory methods manual. Version No. 2.0. USDA-NRCS
Soil Survey Investigations Report No. 42. Washington, D.C.: U.S. Government Printing Office.
Springer DS, Cullen SJ, Everett LG (1995). Laboratory studies on air permeability. p. 217-248. In LG Everett, SJ Cullen (ed.) handbook of vadose zone characterization and monitoring, Lewis Publishers, Boca Raton, Florida.
Sridharan A, Altschaeffl AG, Diamond S (1971). Pore-size distribution studies. J. Soil Mech. Found. Div. ASCE, 97: 771 – 787.
Startsev AD, McNabb DH (2001). Skidder traffic effects on water retention, pore-size distribution, and van Genuchten parameters of boreal forest soils. Soil Sci. Soc. Am. J. 65: 224 – 231.
Stepniewski W, Glinski J, Ball BC (1994). Effects of compaction on soil aeration properties. In Soane BD, van Ouwerkerk C (ed) Developments in Agricultural Engineering 11, Soil compaction in crop production. pp.167–189.
Sutherland W (1893). The viscosity of gases and molecular force. Philos. Mag. Series 36(5): 507 – 531.
Switzer C, Kosson DS (2007). Evaluation of air permeability in layered unsaturated materials. J. Contam. Hydrol. 90: 125–145.
Thorstenson DC, Pollock DW (1989). Gas transport in unsaturated porous media, the adequacy of Fick's Law. Rev. Geophys. 27: 61–78.
Tuli A, Hopmans JW, Rolston DE, Moldrup P (2005). Comparison of air and water permeability between disturbed and undisturbed soils. Soil Sci. Soc. Am. J. 69: 1361 – 1371.
Watabe Y, Leroueil S, Le Bihan JP (2000). Influence of compaction conditions on pore-size distribution and saturated hydraulic conductivity of a glacial till. Can. Geotech. J. 37: 1184 – 11

Resolving the gully erosion problem in Southeastern Nigeria: Innovation through public awareness and community-based approaches

Obidimma C. Ezezika[1,2]* and Olorunfemi Adetona[1,3]

[1]Yale School of Forestry and Environmental Studies, 195 Prospect Street, New Haven, CT 06511 USA.
[2]McLaughlin-Rotman Centre for Global Health 101 College Street, Suite 406, Toronto, Ontario, Canada M5G 1L7.
[3]Department of Environmental Health Science, College of Public Health, University of Georgia, Athens, GA 30602, USA.

The formation of gullies is one of the greatest environmental disasters in Southeastern Nigeria. Large areas of agricultural lands are lost or have become unsuitable for cultivation due to gully erosion. There have been numerous attempts to curb gully erosion in the region; especially through large-scale engineering projects, however, little has been discussed about ways to prevent their onset or the use of community-based low-technology approaches to mitigate their development. We reviewed pertinent literature, interviewed key stakeholders, observed gully erosion sites and had informal discussions with key experts in the field on the causes of gully erosion and community-based approaches in tackling the problem. Based on these discussions, interviews and observations, we showed that many eroding gullies in the region are human-induced and therefore can be prevented. The results also reveal that many of the causes of gully erosion can be traced to poor land management practices and to a lack of innovation and awareness measures. We conclude that community-based, low-technology land management practices and public awareness programs through workshops could halt the development of many gullies in the Southeastern region of Nigeria.

Key words: Erosion, gully, environmental management, soil, Southeastern Nigeria, workshop, public awareness.

INTRODUCTION

The formation of gullies has become one of the greatest environmental disasters facing many towns and villages in Southeastern Nigeria (Adekalu et al., 2007; Okpala, 1990). This region is fast becoming hazardous for human habitation. Hundreds of people are directly affected every year and have to be re-located. Large areas of agricultural lands are becoming unsuitable for cultivation as erosion destroys farmlands and lowers agricultural productivity (Egboka et al., 1990).

Erosivity and erodibility are the factors that contribute to erosion and gully formation. Erosivity is a function of rainfall, a natural phenomenon which is outside human control and manipulation. Rainfall intensities can be high in Southeast Nigeria. Obi and Salako (1995) reported that rainfalls with intensities between the range of 100 to 125 mm h^{-1} are likely to occur more than five times a year. Storms with 25 mm/h intensity have been reported by Hudson (1981) to be erosive. Erodibility, on the other hand, is dependent on soil properties, topography, and land management.

Appropriate land management is very important in an area like Southeast Nigeria where the geotectonic, geologic, and geohydrologic characteristics of the region make many areas within it susceptible to gully erosion. For example, cuestas, fractures and joints are common features in the gully-erosion-prone areas of Southeastern Nigeria and have been identified as significant factors in the formation of gully erosion (Gobin et al., 1999; Okagbue and Uma, 1987).

The loss of soil degrades arable land and eventually renders it unproductive. As a result, there are significant

*Corresponding author. E-mail: ezezika@gmail.com, obidimma.ezezika@mrcglobal.org.

per capita shortages of arable land. The effects of this erosive action are made more severe by recent and rapid population growth in the Southeastern region of Nigeria. Loss of agricultural output is one of the greatest economic costs of gully erosion (Pimentel et al., 1995). Large portions of land have been destroyed in recent years in towns such as Ekwulobia and Nanka[1]. In addition, highways are damaged, leading to numerous vehicle accidents and large displacement of residential houses.

A number of studies have been conducted on the causes of gully erosion in Southeastern Nigeria and ways to control them. These studies primarily revolve around geological properties of the region and proffer solutions that require large-scale engineering efforts with minimal regard to indigenous knowledge. They also deal more with combating gully erosion rather than their prevention and pay little attention to ways to control gully erosion from the perspective of members of the community that are affected by the erosion.

To bridge this gap, we conducted this short study intended to explore the causes of gully erosion and methods of preventing and controlling gully erosion in the region. This study was informed by the perspectives of those affected by gully erosion and further explored the human dimensions of the problem and potential solutions, with particular attention to low-cost, community-based approaches. To our knowledge, this study represents the first in which community based research has been used to study soil erosion problem in the region, and solutions proposed through a community-based framework.

Based on the results of this study, we posit that community-based land management practices and public awareness on the causes and solutions could prevent the onset of gully erosion within the Southeastern region of Nigeria.

METHODS

This study started at the Yale School of Forestry and Environmental Studies when a team of students developed a tentative proposal on sustainable and practical approaches to tackle gully erosion in Oko and Nanka, two towns in Anambra State in Southeastern Nigeria. A number of specialists in Nigeria were contacted and the team reviewed literature on the topic. The preparation of this study relied on multiple sources of evidence including direct observations, short informal interviews with stakeholders and documentation. However, the bulk of this study arose from direct observations and informal discussions among the team and experts.

A literature review of academic articles, news articles and publicly available documents were used in the study. We also interviewed six individuals, who represented public sector views within South-eastern region of Nigeria. The roles played by these interviewees included scientist, policymaker, farmer, surveyor, village head and nurse. The interview was designed as a loosely semi-structured format, and lasted approximately 20 min in length. It included ques-tions on understanding the gully erosion problem as well as specific questions investigating best practices for preventing and controlling erosion.

The study also included visits to gully erosion sites in the towns of Oko, Nanka and Ekwulobia and interaction with stakeholders affected by the erosion. In addition, informal discussions with key experts were held on the issue.

RESULTS

Many eroding gullies in Southeastern Nigeria are human-induced

The literature review suggests that although some gullies are initiated by natural cracks in the earth, much of the gully erosion in the Southeast region of Nigeria can be linked to human activities, which can be managed and controlled through proper land use, forest vegetation, etc. (Gobin et al., 1999; Olife et al., 2007; Osadebe and Akpokodje, 2007; Egboka et al., 1990; Igwe,1999). Although the interviewees acknowledged that the nature of the soil is a strong contributing factor to the rapid development of gully erosion in the region, they noted that most of the causes of erosion are linked to poor land management practices. Observations made in Nanka and Oko and other studies also identified the prime contribution of poor land management practices towards the development of gullies as shown in Table 1 (Akpan et al., 2009; Egboka et al., 1990; Okagbue and Uma, 1987; Idike, 1992).

Some gully erosions can start very small. For example, in 2000, small sheet erosion on the side of a major road in the town of Ekwulobia in Anambra State developed into a deep gully within three years. This displaced many residents in the neighborhood, making the major road impassable. There are other incipient sites, which are preventable and can be tackled before they get too severe. According to a land surveyor who was interviewed, there has been an increase in houses without proper drainage systems, causing increased surface runoff and floods during heavy rainfall. For example, from observations made in this study, the ground surface was paved in many residential properties, with rainwater flowing into the surrounding areas. In addition, vegetation, which can retain fluid and retard water flow were absent in these residences due to the paved surfaces. This inability or unwillingness of many households to employ proper drainage systems is a major cause of gully erosion.

Gully erosion can be mitigated through low-cost approaches

Although some gullies have become too severe to remedy, and will require huge engineering efforts, others can be prevented and incipient ones tackled through a variety of best land management practices and low-cost approaches. Control measures to stem gully erosion that are incipient are most effective when erosion is still at an early stage – when sheet or rill erosion is still the erosion form

[1] These towns are located within the Awka-Orlu Cuesta and are prone to fractures and landslides (Egboka et al.,1990).

Table 1. Primary causes of gully erosion in Southeastern Nigeria.

Cause of gully erosion
Roads without proper drainage or catchments pits
Unguided cultivations that cause flooding
Indiscriminate channeling of flood water on sloped terrain, especially in loose soil structure area
Intense rainfall on metal roofs without drains, which congregate to form large, unguided water channels
Large cracks in hills that form flood channels especially on loose soil ecological terrain
Grading of roads, thereby reducing road levels and creating waterways
Poor drainage systems

– and erosion is still amenable to low technology intervention. Unlike soil properties and topography, which can only be manipulated by human action, land management is often very much under human control. Proper land use and watershed management can be used to reduce surface water runoff and control infiltration in order to dampen erosive forces and reduce the erodibility of soils. The importance of proper land use management for erosion control had also been noted in at least one study conducted in the area (Igwe, 1999). Some of the immediate control measures that were either alluded to by interviewees or corroborated by previous studies included: reduction of surface runoff from impervious surface; provision of collection system for roof rain water runoff and its subsequent reuse; drainage of surface runoff, including the establishment of interception, division and primary (trunk) drains and the construction of interception ponds; planting of grasses on available favorable surfaces to reduce the amount of bare soils exposed to the erosive force of the rains, and to control infiltration; forestation in the areas more susceptible to gully formation and the planting of local cover crops such as indigenous leguminous plants. However, lack of awareness about the cause of the problem was very evident within the community. Many households in the community either do not know the cumulative effect of the lack of proper drainage systems or do not care because there are no direct and immediate repercussions for their poor land management. There is a lack of legislative frameworks to ensure that households refrain from practices that cause gully erosion, and enable these communities enact enforcement mechanisms. In addition, information on the causes of gully erosion and how it can be prevented are scarce. Many of these communities are not aware of the major causes of gully erosion and how it can be prevented, or how their actions are contributing to the problem.

Community-based land management practices coupled with public awareness is key

Gully erosion can be prevented when communities engage in more public awareness initiatives and put in place structures that can govern and implement best land practices. Effective land management practices coupled with pragmatic public awareness measures within the community is crucial to mitigate further onset of gully erosion in the region (Figure 1).

Interviewees recommended that communities take part in initiatives that adopt and engage in small-scale, low technology input and best land management practices. By ensuring that the public is more aware of the gully erosion problem and the best mitigating practices, as well as by putting in place proper legislative and administrative measures for land management practices, the initiation and reoccurrence of gully erosion may be prevented (Okagbue and Uma, 1987).

For example, residents who pave their lands may not be aware that their actions could exacerbate gully erosion in the community because rainwater flowing out of their land to the surrounding regions causes surface run-offs and could initiate erosion. However, a strong public awareness program could help resolve the ignorance surrounding this issue. On the other hand, if a neighbor, knowing the consequences, still decides to forego a drainage system, the existence of a community regulatory framework to ensure that each landlord is responsible for all rainwater that falls within his property can help mitigate against poor land management practices. The framework may include provisions wherein an individual can be fined or face disciplinary actions from the community for his or her poor land management practices.

Finding a channel of gully erosion control through workshops

One of the most effective avenues for combining a public awareness program and effective land management policy is through workshops in various towns and villages that accesses local knowledge. As a strategy, these workshops should include the following elements: awareness – sensitize the community on issues concerning erosion; education – relate to the community the causes of erosion, suggest control measures that can be undertaken on the individual and communal levels, and initiate best management practices for the control of surface water and infiltration; recruitment – recruit students and youths in the area to disseminate necessary information, and equip them to communicate erosion control measures

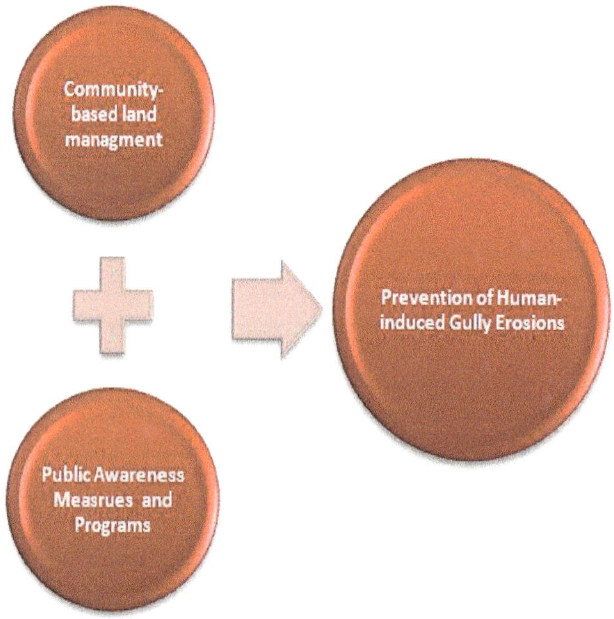

Figure 1. Improved gully erosion control through a community-based approach and public awareness program.

control measures and best land management practices. They will also help in the evaluation and monitoring process; learning – gather local knowledge that could be useful and/or impact on the overall objectives of the project as they relate to the community. In addition, find out what the local attitudes to gully erosion are, and how land management practices have changed over time: demonstration – show to the community the practice of some of the control measures during the workshop; evaluation and monitoring – put a system in place to monitor and evaluate the effectiveness of the project, and develop indices that tie improvements in erosion control to the project; working example – put a system in place that will show case project as a working example of a community based erosion control effort; use it as an educational tool, where possible, for other communities that may be interested in starting community based erosion control projects; future small and large scale interventions – use the proposed project as a reference for future attempts to assist communities in Southeastern Nigeria to control erosion.

These workshops should include mitigation measures such as identification of the causes of gully erosion by all stakeholders and review of the different measures of mitigation, targeted to each stakeholder group. It should also involve steps to implementing new policy and policy suggestions, leadership training and ways to secure funds and investments.

DISCUSSION

A number of studies have shown that the Southeastern region of Nigeria is susceptible to gully erosion due to the nature of the soil, topography and geology (George et al., 2008; Okagbue and Uma, 1987; Onu, 2005; Osadebe and Akpokodje, 2007; Stamp, 1938; Teme and Youdeowei, 2004). For example, Nanka and Oko, the towns for this study, are located within the Awka-Orlu Cuesta, which is an area susceptible to ground surface cracks and landslides during the rainy season. In 1988, there was a landslide initiated by fractures in which many properties, including sixty homes, were reported to have been lost (Egboka et al., 1990). The two cuestas in the area – the Udi-Nsukka and the Awka-Orlu Cuestas – are thought to have been formed due to compression and consequent uplift of the landscape. These cuestas make the region susceptible to gully erosion. Cuestas are ridges that have gentle and steep slopes on their opposite faces. The red lateric soils of the area overlay white sands referred to as the Nanka Sands (Egboka et al., 1990; Okagbue and Uma, 1987). The soils have low silt/clay content which decreases with depth. The white sands are cohesionless, very permeable and have high infiltration rates, making control of gullies difficult after they have cut through the red soils at the surface. In addition, the rise in the water table due to heavy rain falls in the rainy season contributes to an increase in hydraulic head, high subterranean flow rate, and the enhancement of gully formation (Egboka et al., 1990).

However, this study has also shown that despite the susceptibility of the region to gully erosion, it can be prevented through enhanced public awareness programs and better land management practices. Although these practices are insufficient to resolve the huge gully erosion sites, these practices can prevent the occurrence of future

Table 2. Mitigating strategies for erosion control in Southeastern Nigeria.

Strategy for erosion control
1. Local government officials and legislative representatives should be more aware of the problem of gully erosion, along with concrete suggestions to combat and prevent erosion in the local context.
2. The Federal Ministry of Environment together with local communities should design public awareness programs on the causes and mitigation measures of gully erosion.
3. Community leaders should ensure that households account for surface runoff from impervious surfaces within their residential property and other best land management practices.
4. Organizations and churches should be more involved in public awareness measures and capacity building programs such as planting of grasses on available favorable surfaces, afforestation projects in the areas more susceptible to gully formation, and planting of local cover crops.
5. The State and local government should organize workshops for community leaders, households and youths on measures of gully erosion mitigation.

future ones. It is important for community leaders to organize awareness programs that provide reliable information to the community on gully erosion mitigation strategies. These programs can be an incipient process for providing the community with methods of tackling erosion, sharing of local knowledge on these issues, and raising awareness. These workshops could serve to introduce strong policies to bring about better management and monitoring of effective land and farming practices and to mitigate the formation of gully erosion in Southeastern Nigeria.

Farming is still an important occupation in the area and much of this is done on a subsistence level. It has been opined that "the farmer [in the region] can only be convinced through measures and approaches that are economically feasible, non complex and can accommodate his values and ethics" (Okagbue and Uma, 1987). It is thus important to access local knowledge to ensure the success of any erosion mitigation and soil project in the Southeastern region of Nigeria, while stressing the need to concentrate on low technology, low capital control measures. It is also important to note that control measures to stem incipient gully erosion are most effective when erosion is still at an early stage and amenable to low technology intervention.

The constitutional framework for the management of the environment in Nigeria remains very centralized. This results in a system where government at the state and federal levels holds the primary responsibility for developing and applying the legislative framework, including supervisory and regulatory roles, for environmental management (ARD, 2002). This framework can be improved by increasing the capacity of local governments to integrate environmental planning and management into their activities.

Gully erosion occurs locally, usually starts small and is best mitigated at the community and local levels. Therefore, a system where the local communities most affected by erosion drive the mitigation strategy is both empowering and most effective. We propose the action plans in Table 2 which are geared towards engagement at the level of the local community.

The workshops will allow for rich interaction among community residents, community leaders and village chiefs, the academic community, legislative and policy representatives in order to get a full representation of those affected by and influencing erosion control in South eastern Nigeria. They will also provide the opportunity to determine the major causes of soil erosion in the region through the consultation of all of the stakeholder groups, creating a comprehensive body of information for use by the local organizations, and residents.

Additionally, by interfacing with the community leaders and chiefs of the various villages, the present institutional and legal frameworks available for addressing gully erosion can be assessed. This would facilitate possible legal and regulatory changes that will enable these communities, towns, and villages at the local level to enhance their inherent capacity to protect against gully erosion. This creates a hub for developing innovative ideas and erosion control programs. It also moves the responsibility to the affected community and creates structures through these workshops or other discussion fora that emphasize the exchange of ideas and the demonstration and application of techniques and skills to manage erosion in the region.

Conclusion

This study focused on developing erosion mitigation measures for Southeastern Nigeria where soil erosion is continuing to be a huge environmental problem, and local communities are currently ill equipped to combat the problem. Through the use of direct observations, short informal interviews, and informal discussions among the study team and soil erosion experts, it was discovered that there was a general lack of awareness of the human factors that contribute to the development of gullies in the communities where the study took place. It was determined that there is great opportunity to adapt small scale, low technology land management practices in order to prevent erosion and the development of gullies through community based approaches and public awareness programs. Finally, it is believed that equipping and shifting responsibility towards the community would enhance erosion control efforts.

ACKNOWLEDGEMENTS

The author is grateful to the Soil Erosion Team at the Yale School of Forestry and Environmental Studies for their useful discussions and contribution towards this study. The team included: Olorunfemi Adetona, Obinna Aduba, Obidimma Ezezika, Anne Hamilton, Ngozi Ndulue, Lisa Leombruni, Peyton Smith and Ekene Udoye, Gordon Geballe, Roy Lee, Florencia Montagnini, and Thomas Siccama.

REFERENCES

Adekalu KO, Olorunfemi IA, Osunbitan JA (2007). Grass mulching effect on infiltration, surface runoff and soil loss of three agricultural soils in Nigeria. Bioresour. Technol., 98(4): 912-917.

Akpan A, George N, George A (2009). Geophysical investigation of some prominent gully erosion sites in calabar, southeastern Nigeria and its implications to hazard prevention. Disaster Adv., 2(3): 46-50.

ARD I (2002). Nigeria Environmental Analysis Final Report. Retrieved May/02, 2011, from http://pdf.usaid.gov/pdf_docs/PNACP627.pdf.

Egboka BCE, Nwakwor GI, Orajaka IP (1990). Implications of palæo- and neotectonics in gully erosion-prone areas of southeastern nigeria. Natural Hazards, 3: 219-220-231.

George N, Akpan A, Obot I, Akpanetuk N (2008). Geoelectrical investigation of erosion and flooding using the lithologic compositions of erosion and flood-stricken road in ukanafun local government area, Akwa Ibom State, Southern Nigeria. Disaster Adv., 1(4): 46-51.

Gobin AM, Campling P, Deckers JA, Feyen J, Poesen J (1999). Soil erosion assessment at the Udi-Nsukka cuesta (Southeastern Nigeria). Land Degrad. Dev., 10(2): 141-160.

Hudson NW (1981). Soil conservation. New York: Cornell University Press.

Idike FI (1992). On appraising soil erosion menace and control measures in southeastern Nigeria. Soil Technol., 5: 57-58-65.

Igwe CA (1999). Land use and soil conservation strategies for potentially highly erodible soils of Central-Eastern Nigeria. Land Degrad. Dev., 10: 425-426-434.

Obi ME, Salako FK (1995). Rainfall parameters influencing erosivity in South-eastern Nigeria. Catena, 24: 275-275-287.

Okagbue CO, Uma KO (1987). Performance of gully erosion control measures in southeastern nigeria. Forest Hydrol. Watershed Manag., 167: 163 -172.

Okpala AO (1990). Nigerian population growth and its implications for economic development. Scandinavian. J. Dev. Altern., 9(4): 63-77.

Olife IC, Okaka AN, Dioka CE, Meludu SC, Orisakwe OE (2007). Iodine status and the effect of soil erosion on trace elements in Nanka and Oba towns of Anambra state, Nigeria. Annali Di Chimica, 97(9): 895-903.

Onu DO (2005). Determinants of farm-level soil conservation and erosion control adoption and utilisation behaviours in the ecologically vulnerable areas of Imo state, Nigeria. J. Rural Dev., 24(4): 521-544.

Osadebe CC, Akpokodje EG (2007). Statistical analysis of variability in properties of soils in gully erosion sites of Agulu-Nanka-Oko area, southeastern Nigeria. J. Mining. Geol., 43(2): 197-202.

Pimentel D, Harvey C, Resosudarmo P, Sinclair K, Kurz D, McNair M (1995). Environmental and economic costs of soil erosion and conservation benefits. Science (New York, N.Y.), 267(5201): 1117-1123.

Stamp LD (1938). Land utilization and soil erosion in Nigeria. Geogr. Rev., 28(1): 32-33-45.

Teme SC, Youdeowei PO (2004). Geotechnical investigations for design of foundations for erosion and flood control structures at Unwana Beach, Afikpo, Ebonyi state, South-eastern Nigeria. Fifth International Conference on Case Histories in Geotechnical Engineering, April 13-17, 2004. 1: 9. New York, N.Y.

The utilization of wood ash as manure to reduce the use of mineral fertilizer for improved performance of maize (*Zea mays* L.) as measured in the chlorophyll content and grain yield

F. O. Adekayode[1]* and M. R. Olojugba[2]

[1]Department of Crop, Soil and Pest Management, Federal University of Technology, Akure, Ondo State, Nigeria.
[2]Department of Soil Science and Management, Joseph Ayo Babalola University, Ikeji Arakeji, Osun State, Nigeria.

The increasingly high cost of mineral fertilizers and a preferred economic disposal of wood ash had necessitated a research in the use of wood ash to reduce the rate of fertilizer application for maize production. The comparative effects of wood ash and NPK 15-15-15 on maize were investigated in Akure (7°15'N, 5°15'E), Nigeria in 2008. The four treatments investigated were 4 t/ha wood ash, 300 kg/ha NPK 15-15-15, 200 kg/ha NPK 15-15-15 plus 2t/ha wood ash mixture and a control plot without manure; while the land preparation was ploughed and harrowed once. Downy mildew resistant (DMR) open pollinated maize variety was planted at 60 × 30 cm in a randomized complete block experiment replicated four times, and the manure was applied in a ring at 10 cm from the maize plant. Pre-treatment soil analysis that entailed taking soil samples before the application of manure was carried out. Two maize crops were investigated and the means of data were computed for comparisons across treatments. The total leaf area per plant, leaf area index and the chlorophyll content were correlated to the grain yield. The profitable response was computed to confirm the economic viability. The chlorophyll content and the grain yield were significantly high in 200kg/ha NPK 15-15-15 + 2t/ha wood ash mixture and in 300 kg/ha NPK 15-15-15 compared to other treatments. Corresponding higher net revenues were also obtained which confirmed a mixture of wood ash and inorganic mineral fertilizers to be a suitable agronomic practice for profitable maize production.

Key words: Wood ash, mineral fertilizers, chlorophyll content, maize grain yield.

INTRODUCTION

The improved performance of maize as a result of added manure is due to the supply of nutrient elements to the plants, and maize being a crop that has exhausting effects on the soil needs the supply of the necessary nutrients in the correct proportions to produce a satisfactory yield (Ristanovic, 2001). Nitrogen, phosphorus and potassium and other nutrient elements play great physiological importance in formation of chlorophyll, nucleotides, phosphotides, alkaloids as well as in many enzymes, hormones and vitamins for optimum grain yield (Mohamed et al., 2008). Nitrogen deficiency could exert a particularly marked effect on maize crop yield as the plant would remain small and rapidly turned yellow if sufficient nitrogen was not available for the construction of protein and chlorophyll (Kogbe and Adediran, 2003). The lack of chlorophyll had been observed to diminish carbohydrate production and a restriction in the assimilating power of the plant (Amany et al., 2006).

The involvement of other macro and micro nutrients in the metabolic processes of plants including maize had been reported in previous research (Kogbe and Adediran, 2003). The processes of anabolism and catabolism of the carbohydrate metabolism would only proceed normally when the organic compounds had been esterised with phosphoric acid. Potassium had been found to

*Corresponding author. E-mail: adekay98@yahoo.com.

accumulate in the parts of the plant in which cell division and growth processes were actively proceeding; calcium influenced the water economy of the plant and the protein carbohydrates in many physiological processes. Magnesium was confirmed a constituent of the chlorophyll, protochlorophyll, pectin and phytin while the micro nutrients, though required in very small quantities, were also involved in the plant metabolic processes (Modi and Asanzi, 2008).

The source of the nutrients essential for plant's growth could either be organic or inorganic or from the two combined. The inorganic sources involved the use of fertilizers and the usefulness of fertilizers in increasing plant growth to a designated optimum had been reported in previous research (Niehues et al., 2004; Stewart et al., 2005; Mohamed et al., 2008). The use of organic manure in form of poultry and goat manures and wood ash had been found to complement the use of inorganic fertilizer as both manures improved crop yield (Brady and Weil, 1999; Meludu, 2005; Odiete et al. 2005). Previous research conducted by Roger and Sharland, (1997), Nottidge et al.(2005), Mbah and Nkpaji (2009) had shown that when synthetic fertilizers were not applied, the use of wood ash produced significant effects on the growth and yield of many crops, particularly maize. Saarsalmi et al. (2001), Bougnom et al. (2009), Rodriguez et al. (2009) in the investigation of the effects of wood ash fertilization on soil chemical properties observed that wood ash significantly increased the effective cation exchange capacity and base saturation and decreased the concentration of exchangeable aluminum in the soil.

Wood ash which is the inorganic and organic residue remaining after the combustion of wood had been confirmed to be a good source of potassium, phosphorus, magnesium, calcium and the micronutrients (Kakier and Summer, 1996; Demeyer et al., 2001; Saarsalmi et al., 2001). The burning of sesbenia wood and incorporation of the ash into soil increased grain yield of maize markedly ICRAF (1996) while the application of ash to young maize plants had significantly increased the yield of maize (Odiete et al., 2005). Iremiren (1989) observed increase in yield of maize resulting from ash through trash burning to be consistent to observed increase in the amount of phosphorus in the soil. Previous studies had indicated wood ash to have the same liming effects as commercial lime and in several studies where traditional limestone and wood ash had been compared, the result had confirmed wood ash to give better plant growth responses than limestone because of the additional nutrients that the ash contained (Roger and Sharland, 1997; Awodun et al., 2007). The effects of organic and inorganic manure in increasing the chlorophyll content in maize had also been documented (Amujiyegbe et al., 2007; Mohamed et al., 2008).

The objectives of this research were to investigate the effectiveness of wood ash as manure for a sustainable improved maize production.

MATERIALS AND METHODS

Physical setting

The experiments were conducted in 2008 in the Teaching and Research Farm of the Federal University of Technology, Akure ($7^0$15'N, $5^0$15'E). The experimental site had been continuously used for arable maize and cassava crop production for five years without any manure application. The experimental design was a randomized complete block with four treatments replicated four times. Each plot size was 3 × 3 m and discard of 1 m between plots giving experimental area of 18 × 18 m (324 m^2). Maize seeds resistant to Downy Mildew and Streak (DMRESR) obtained from the Ondo State Agricultural Development Project (OSADEP), Akure were planted at 60 x 30 cm to give a plant population of 55,000 stands per hectare. The four treatments were 4 t/ha wood ash, 300 kg/ha NPK 15-15-15, 200 kg/ha NPK 15-15-15 plus 2 t/ha wood ash mixture and a control plot without manure; and the manure was applied three weeks after planting in a ring at 10 cm from the maize plant while the land preparation was ploughed and harrowed once.

Soil sample collection and laboratory analysis

Soil samples were taken with a soil auger to a depth of 30 cm both before and after manure application. After the manure application, soil samples were collected to a depth of 30 cm ninety days after planting at harvest for the first maize crop planted in April, and to observe the residual effects of the applied manure, soil samples were taken at ninety days after planting for the second maize crop planted in August 2008. The soil samples collected were air dried and sieved through a 2 mm mesh and analyzed for soil chemical properties (Carter, 1993). The soil pH was determined in water using a glass electrode pH meter. Organic carbon was determined by oxidising soil sample with dichromate solution and later titrated with ferrous sulphate solution. The total nitrogen was determined using micro-kjeldahl method and the available phosphorus extracted colorimetrically by the molybdenum blue method. The exchangeable cations were extracted by leaching 5 g of soil with 50 ml of ammonium acetate at pH 7. The potassium and sodium in the leachate were determined with a column model 21 flame spectrophotometer while the calcium and magnesium were determined with atomic absorption spectrophotometer.

Measurement of the total leaf area per plant, leaf area index and determination of chlorophyll content

The total leaf area of each randomly selected fifteen maize stands per plot was taken and the corresponding leaf area index was computed. The leaf area was measured following the procedure of Stewart and Dwyer (1999) and Elings (2000) by multiplying the length of leaf by the widest width by alpha, where alpha is 0.743 (L × W × 0.743). The leaf area index was computed by dividing the total leaf area of a maize plant stand by the total land area occupied by the single stand (Mauro et al., 2001). The chlorophyll was determined by extraction in 80% acetone and reading the absorbance of the solution at 645, 663 and 652 nm (Ibitoye, 2005).

Maize yield parameters

Harvesting was carried out at ninety days after planting and the 14% grain moisture content confirmed with the use of grain moisture tester and the grain yield per hectare was taken.

Table 1. The means of soil chemical properties at harvest in the first and second cropping season.

Treatment	pH	Organic matter (%)	% N	Avail P (ppm)	Na	K	Ca	Mg
					(cmol/kg)			
Wood ash at 4t/Ha	7.9	2.92	1.84	9.4	0.19	0.23	3.42	0.43
NPK 15-15-15 at 300kg/Ha	6.9	3.14	2.12	12.4	0.17	0.26	1.98	0.22
NPK 15-15-15 at 200kg/Ha plus wood ash at 2t/Ha	7.3	3.28	2.06	10.2	0.16	0.24	3.26	0.48
Control without manure	5.4	1.6	0.11	8.4	0.15	0.17	1.32	0.26
LSD (0.05)	0.3	0.36	0.02	6.9	0.03	0.02	0.49	0.02

Table 2. Effects of manure application on the total leaf area, leaf area index, chlorophyll content and maize grain yield in tons per hectare.

Treatment	Total leaf area/plant (cm^2)	Leaf area index	Chlorophyll content (mg/100g)	Grain yield/Ha (mg/Ha)
Wood ash at 4 t/Ha	3874.2	2.15	1.25	1.62
NPK 15-15-15 at 300 kg/Ha	3954.1	2.19	1.46	1.65
NPK 15-15-15 at 200 kg/Ha plus Wood Ash at 2 t/Ha	4039.1	2.24	1.50	1.97
Control without manure	3715.2	2.06	0.88	0.82
LSD (0.05)	NS	NS	0.03	0.02

Statistical analysis

The average values of the soil and grain yield parameters of the first and second maize crops were determined for statistical analysis of means comparison, using the least significant difference (lsd at 5% probability).

RESULTS

The pretreatment soil analysis of the experimental plot indicated the soil to be sandy clay loam with a composition of 67% sand, 12% silt and 21% clay. The soil chemical properties indicated pH 5.3, 1.25% organic matter, 9.5 ppm available phosphorous, 0.12% nitrogen, 0.15 cmol/kg sodium, 0.19 cmol/kg potassium, 1.32 cmol/kg calcium and 0.24 cmol/kg magnesium while the analysis of the wood ash used showed composition of pH 11.8, 1.19% nitrogen, 8.7 ppm phosphorus, 0.87% potassium, 0.16% sodium, 1.84% calcium, 0.63% magnesium.

Table 1 showed the mean values of soil chemical properties. The pH values were significantly higher in the plots treated with wood ash compared with the other plots while the organic matter values were 3.45 and 3.30% in 200 kg/ha NPK 15-15-15 plus 2 t/ha wood ash mixture and 300 kg/ha NPK 15-15-15, respectively compared to other treatment plots. Significantly higher nitrogen and potassium values were obtained in plots treated with 300 kg/ha NPK 15-15-15 and in 200 kg/ha NPK 15-15-15 plus 2 t/ha wood ash mixture compared to other treatment plots. Calcium and magnesium were significantly higher in plots treated with wood ash compared with other treatments.

Table 2 showed the mean values of total leaf area, leaf area index, chlorophyll content and the maize grain yield. There were no significant differences in total leaf area and leaf area index among treatments, though a higher value of 4039.1 cm^2 was obtained in 200 kg/ha NPK 15-15-15 plus 2 t/ha wood ash mixture and in a decreasing order of magnitude of 3954.1, 3874.2 and 3715.2 cm^2 in 300 kg/ha NPK 15-15-15, 4 t/ha wood ash and control plot without manure respectively for the total leaf area. A corresponding trend of 2.24, 2.19, 2.15 and 2.06, respectively were obtained for the leaf area index. Significantly higher chlorophyll content values of 1.50 and 1.46 mg/g were obtained in 200 kg/ha NPK 15-15-15 plus 2 t/ha wood ash mixture and 300 kg/ha NPK 15-15-15 with correspondingly higher nitrogen content compared with other treatment plots. Significantly higher yield value of 1.93 t/ha was obtained in 200 kg/ha NPK 15-15-15 plus 2 t/ha wood ash mixture and followed in a decreasing order of magnitude by 1.89, 1.67 and 0.82 t/ha in 300 kg/ha NPK 15-15-15, 4 t/ha wood ash and control plot without manure, respectively while the corresponding percentage increase in yield over the control plot without manure were 115.8, 101.2 and 97.6%, respectively. The grain yield values were correlated with the total leaf area, the leaf area index and the chlorophyll content with correlation values 0.96, 0.97 and 0.95, respectively. Table 3 which showed the profitable analysis indicated net revenue of 77,630.00, 66,780.00, 45,255.00 and ₦10,255.00 in 200 kg/ha NPK 15-15-15 plus 2 t/ha wood ash mixture, 300 kg/ha NPK 15-15-15, 4 t/ha wood ash and control plot without manure, respectively.

Table 3. The profitable analysis of maize production with the application of wood ash and NPK 15-15-15 at various rates.

Farm operations	Wood ash at 4t/Ha (₦)*	NPK 15-15-15 at 300Kg/Ha (₦)*	Wood ash at 2t/Ha plus 100Kg/Ha NPK 15-15-15 (₦)*	Control without manure application (₦)*
Land preparation (Ploughing and Harrowing)	10,000.00	10,000.00	10,000.00	10,000.00
Manure procurement /transportation	20,000.00	1,500.00	12,500.00	Nil
Purchase of maize seeds	3,900.00	3,900.00	3,900.00	3,900.00
Planting (8 pd at ₦ 500.00/pd)**	4,000.00	4,000.00	4,000.00	4,000.00
Application of manure (8 pd at ₦ 500.00/pd)**	4,000.00	4,000.00	4,000.00	4,000.00
Cost of one supplementary weeding (20 pd at ₦ 500.00/pd)**	10,000.00	10,000.00	10,000.00	10,000.00
Cost of harvesting (20 pd at ₦ 500.00/pd)**	10,000.00	10,000.00	10,000.00	10,000.00
Cost of shelling and bagging	3,000.00	3,000.00	3,000.00	3,000.00
Sub-total cost	64,900.00	46,400.00	57,400.00	44,900.00
Miscellaneous expenses	3,245.00	2,320.00	2,870.00	2,245.00
Total cost	68,145.00	48,720.00	60,270.00	47,145.00
Yield (T/Ha)	1.62	1.65	1.97	0.82
Gross revenue (Sales at ₦ 70,000.00 per ton)	113,400.00	115,500.00	137,900.00	57,400.00
Net revenue	45,255.00	66,780.00	77,630.00	10,255.00
% Increase in yield over plot without manure	97.6	101.2	115.8	

*Exchange rates: 1US Dollar = 146 Nigerian Naira (₦ 146.00); ** pd = person days (Farm labour wage).

DISCUSSION

The pH 5.3 of the soil in the experimental site before the application of manure indicated the soil to have medium acidity level. Soil with a pH range of 5.2 - 5.6 had been reported to be of medium acidity (Brady and Weil, 1999). This justified the investigation with the use of wood ash as soil amendment. This agreed with the principle on the use of wood ash only on soils that were not alkaline since such essential elements as iron, manganese and zinc would be locked up and made unavailable to plants on highly alkaline soils (Demeyer et al., 2001). Also the generally low level of plant nutrients in the experimental plot was as a result of a continuous cultivation of the experimental site for arable maize and cassava crops without application of manure to replenish the soil of nutrients. Tisdale et al. (2003) had explained with Mitscherlich's principle of plant's positive response to applied nutrients that were previously limiting in the soil.

The significantly higher pH values of 7.7 and7.5 in 4 t/ha wood ash and 200 kg/ha NPK 15-15-15 plus 2 t/ha wood ash mixture plots was due to the liming effects of the wood ash added. Demeyer et al. (2001) and Lickaaz (2002) had described wood ash to be similar to burned or hydrated lime as it contained oxides and hydroxides of potassium, sodium, calcium and magnesium. This also could be responsible for the significantly higher sodium, calcium and magnesium levels in plots treated with wood ash compared with the plot treated with the 300 kg/ha NPK 15-15-15 and control plot without manure. Nitrogen and potassium, being major components of NPK 15-15-15 inorganic compound fertilizer caused their significantly high levels in 300 kg/ha NPK 15-15-15 and 200 kg/ha NPK 15-15-15 plus 2 t/ha wood ash mixture. The

The phosphorus content did not show significant differences among treatments because phosphorus formed a major component in both NPK fertilizers and wood ash (Ayeni et al., 2008), while the pretreatment soil analysis indicated the soil not to be originally deficient in phosphorus. The significantly high organic matter values in 200 kg/ha NPK 15-15-15 plus 2 t/ha wood ash mixture and 300 kg/ha NPK 15-15-15 signified high soil fertility which reflected in significantly high maize grain yield which corroborated earlier observation of Roger and Sharland (1997) and Bruno et al. (2002).

The order of magnitude of total leaf area and the leaf area index values which were higher in 200 kg/ha NPK 15-15-15 plus 2 t/ha wood ash mixture and which also reflected in the yield of maize grain corroborated with the earlier assertion of leaf area to be essential for simulation of light interception and photosynthate production (Stewart and Dwyer, 1999; Subedi and Ma, 2005). Mauro et al. (2001) had also explained leaf area index and leaf angle distribution as important parameters for estimating the exchange of energy and gases in vegetative canopies. The total leaf area, leaf area index and chlorophyll content which had positive correlation with the maize grain yield (r = 0.96, 0.97 and 0.95 respectively) confirmed the assertion of Mohamed et al. (2008) that chlorophyll content and photosynthesis were biochemical processes which increased the grain yield of maize. The highest grain yield value of 1.97 t/ha in 200 kg/ha NPK 15-15-15 plus 2 t/ha wood ash mixture compared with other treatments showed a high yield obtainable when wood ash alone or in combination with inorganic manure was used to improve soil fertility for increased crop production (Amujityegbe et al., 2007; Roger and Sharland, 1997; Awodun et al., 2007).

The higher net revenue of seventy seven thousand six hundred and thirty naira obtained in 200 kg/ha NPK 15-15-15 plus 2 t/ha wood ash mixture compared to the sixty six thousand seven hundred and eighty naira obtained in 300 kg/ha NPK 15-15-15 confirmed that the rate of inorganic fertilizer can be reduced by replacement with wood ash to obtain a highly profitable maize production for an economic emancipation of maize farmers (Oladeebo and Ezekiel, 2006).

Conclusion

The significantly higher maize grain yield obtained in 200 kg/ha NPK 15-15-15 plus 2 t/ha wood ash mixture compared with when 300 kg/ha NPK 15-15-15 or 4t/ha wood ash, confirmed that a combination of both organic and inorganic nutrient sources gave higher maize grain yield than when each was applied separately and the reduction in the rate of inorganic fertilizers would increase the profit margin of maize production.

REFERENCES

Amany AB, Zeidan MS, Hozayn M (2006). Yield and Quality of Maize (*Zea mays* L.) as affected by slow release nitrogen in newly reclaimed sandy soil. Am. Eurasian J. Agric. Environ. Sci. 1(3): 239- 242.

Amujiyegbe BJ, Opabode JT, Olayinka A (2007). Effect of organic and inorganic fertilizer on yield and chlorophyll content of maize (*Zea mays* L.) and sorghum (*Sorghum bicolor* (L.) Moench). Afr. J. Biotechno. 6(16): 1869 - 1873.

Awodun MA, Ojeniyi SO, Adeboye A, Odedina SA (2007). Effect of Oil palm Bunch Refuse Ash on Soil and Plant Nutrient Composition and Yield of Maize. Am Eurasian. J. Sustain. Agric. 1(1): 50 - 54.

Ayeni LS, Oso OP, Ojeniyi SO (2008). Effect of sawdust and wood ash application in improving soil chemical properties and growth of cocoa (*Theobroma cacao*) seedlings in the Nurseries. Medwell. Agric. J. 3(5): 323 - 326.

Bougnom BP, Mair J, Etoa FX, Insm H (2009). Compost with woodash addition: A risk or a chance for ameliorating acid tropical soils. Geoderma 153: 402 - 407.

Brady NC, Weil RR (1999). The Nature and Properties of Soils. (12th Editon). Prentice Hall, New Jersey pp. 881.

Bruno G, Johannes L, Wolfgang Z (2002). Ameliorating physical and chemical properties of highly weathered soils in the tropics with charcoal. Biology and Fertility of Soils 35(4): 219 - 230.

Carter MR (1993). Soil Sampling and Methods of Soil Analysis. Canadian Society of Soil Science. Lewis Publishers, London p. 823.

Demeyer A, Voundi NJC, Verloo MG (2001). Characteristics of wood ash and influence on soil properties and nutrient uptake: An Overview. Bioresource Technology 77(3): 287 - 295.

Elings A (2000). Estimation of leaf area in tropical maize. Agron. J. 92: 436 - 444.

Ibitoye AA (2005). Laboratory Manual on Basic Methods in Analytical Chemistry. Concepts IT and Educational Consults, Nigeria p. 67

International Center for Research in Agroforestry (ICRAF) (1996). Annual Report of International Center for Research in Agroforestry (ICRAF), Kenya p. 330.

Iremiren GO (1989). Response of maize to trash burning and nitrogen fertilizer in a newly opened secondary forest. J. Agric. Sci. (Cambridge) 113(02): 207 - 210.

Kakier U, Summer ME (1996). Boron availability to plant from coal combustion by-products. Water, Air and Soil Pollution 89: 93 - 110.

Kogbe JOS, Adediran JA (2003). Influence of nitrogen, phosphorus and potassium application on the yield of maize in the savannah zone of Nigeria Afr. J. Biotechnol. 2: 345 - 349.

Lickaez J (2002). Wood Ash: An Alternative Liming Material for Agricultural Soils. Agdex 534-2. htpp//www/.agric.gov.ab.ca/dept.docs.nsf/all/agdex/534-2.pdf

Mauro A, Homem A, Elizabeth A, Walter-Shea, Mark AM (2001). Test of an extended mathematical approach to calculate maize leaf area index and leaf angle distribution. Agricultural and Forest Meteorology 108(1): 45 - 53.

Mbah CN, Nkpaji D (2009). Response of Maize (*Zea mays* L.) to Different Rates of wood ash Application in acid Ultisol in Southeast Nigeria. J. Am. Sci. 5(7): 53 - 57.

Meludu NT (2005). Use of human waste in sustainable crop production in Niger. J. Environ. Extension 5: 65 - 70.

Modi AT, Asanzi NM (2008). Seed Performance of Maize in Response to Phosphorus Application and Growth as related to Phytate-Phosphorus occurrence. Crop Science 48: 286 - 297.

Mohamed SA, Ewees SA, Sawsan A, Seaf EY, Dalia MS (2008). Improving maize grain yield and its quality grown on a newly reclaimed sandy soil by applying micronutrients, organic manure and biological inoculation. Res. J. Agric. Biol. Sci. 4:537 - 544.

Niehues BJ, Lamond RE, Godsey CB, Olsen CJ (2004). Starter Nitrogen Fertilizer Management for continuous N0-Till Corn Production. Agron. J. 96: 1412 - 1418.

Nottidge DO, Ojeniyi SO, Asawalam DO (2005). Comparative Effect of Plant Residue and NPK Fertilizer on Nutrient Status and Yield of Maize (Zea mays L.) in a Humid Ultisol Niger. J. Soil Sci. 15: 1 - 8.

Odiete I, Chude VO, Ojeniyi SO, Okozi AA, Hussaini GM (2005). Response of Maize to Nitrogen and Phosphorus sources in Guinea Savanna Zone of Nigeria. Niger. J. Soil Sci. 15: 90 - 101.

Oladeebo JO, Ezekiel AA (2006). Economic efficiency of maize farmers in Oyo West Local Government Area of Oyo State, Nigeria. Proceedings of the Second Annual Conference of the School of

Agriculture and Agricultural Technology, Federal University of Technology, Akure, Nigeria. 2006, 186 - 191.
Ristanovic D (2001). Maize. In: R.H. Raemaekers (Editor) Crop Production in Tropical Africa. Goekint Graphics nv. Belgium. pp. 23 - 45.
Rodriguez L, Salazar P, Preston TR (2009). Effect of biochar and biodigester effluent on growth of maize in acid soils. Livestock Research for Rural Development 21(7). http://www.lrrd.org/lrrd21/7/rodr21110.htm
Roger J, Sharland RW (1997). Understanding traditional perceptions of wood ash; a means of communicating soil fertility. Centre for Information on Low External Input and Sustainable Agriculture (ILEA) Newsletter 13(3): 23 - 29.
Saarsalmi A, Malkonen E, Piirainen S (2001). Effects of wood ash fertilization on forest soil chemical properties. Silva Fennica 35(3): 355 - 368.
Stewart, DW, Dwyer LM (1999). Mathematical characterization of leaf shape and area of maize hybrids. Crop Science 39: 422 - 427.
Stewart WM, Dibb DW, Johnston AE, Smyth TJ (2005). The Contribution of Commercial Fertilizer Nutrients to Food Production. Agron. J. 97: 1 - 6.
Subedi KD, Ma BL (2005). Ear Position, Leaf Area and contribution of individual leaves to grain yield in conventional and leafy maize hybrids. Crop Science 45:2246 - 2257.
Tisdale SL, Nelson WL, Beaton JD, Havlin JL. (2003). Soil Fertility and Fertilizers, 5th Edition, Pearson Education, Inc. New Jersey. p. 634.

Effect of quantity and placement distances of inorganic 15-15-15 fertilizer in improving soil fertility status and the performance and yield of maize in a tropical rain forest zone of Nigeria

F. O. Adekayode[1] and M. O. Ogunkoya[2]

[1]Department of Crop, Soil and Pest Management, Federal University of Technology, Akure, Nigeria.
[2]Federal College of Agriculture, Akure, Ondo State, Nigeria.

The high cost of mineral fertilizer in crop production had necessitated an investigation into the optimum fertilizer rate for profitable maize production. The study was conducted in Akure (7°15'N, 5°12'E) in Nigeria in 2007 and revalidated in 2008. The experimental design was a split plot method in a randomized complete block with fertilizer quantity as the main plot and fertilizer placement distance of maize plant as the sub plots. The NPK 15-15-15 fertilizer rates in the main plot were 0,150,200,250 and 300 kg NPK/ha, while distances of application of the maize plants that formed the subplots were 10, 15 and 20 cm. Improved maize growth parameters were observed in plots treated with fertilizers at 300 and 250 kg per hectare and at placement distances of 10 and 15 cm with higher yield and corresponding higher net revenue. The economic analysis confirmed the efficient use of fertilizer to lift small holder maize farming from subsistence levels to a more business-oriented agricultural economy.

Key words: Optimum fertilizer use, inorganic fertilizer, soil fertility, maize performance and yield.

INTRODUCTION

Fertilizer had been defined as any organic and inorganic material added to a soil to supply certain elements essential to the growth of plants (Brady and Weil, 1999). Fertilizers provide typically in varying proportions, the three major plant nutrients (nitrogen, phosphorus, potassium), the secondary plant nutrients (calcium, magnesium, sulphur) and when required, trace elements (boron, manganese, iron, zinc, copper, molybdenum and chlorine) (Tisdale et al., 2003).

The use of fertilizers in improving crop production had been discussed in previous study as Stewart et al. (2005) evaluated the percentage contribution of fertilizer to the increased yield of agricultural crops that ranged from 40 to 60% in the U.S. and England while Niehues et al. (2004) reported starter nitrogen fertilizer as an effective efficient way of stimulating early growth and improving yields of corn in Kansas.

The effects of fertilizer subsidy in Brazil which caused an increase in the production of wheat, soybean and other agricultural commodities with positive effects on the Brazilian economy were reported (Hall et al., 1983). Bationo et al. (1993) reported an increased yield in pearl millet in Niger, West Africa on the addition of crop residue with fertilizer. The economics of fertilizer application in such crops as cacao was discussed in Opeke (1997) who reported a likely average of 30% yield increase from the application of combined nitrogen and phosphorus fertilizer.

The use of fertilizer was emphasized in the African fertilizer summit held in Abuja, Nigeria in June 2006 by New Partnership for Africa Development (NEPAD), African Union and International Fertilizer Development Centre (IFDC). The summit was held to increase the awareness of the role that fertilizer could play in stimulating sustainable productivity growth in African agriculture and to discuss approaches for rapidly increasing efficient fertilizer use by African smallholder farmers. The beneficial effects of fertilizer application on soils for a sustainable food crop production had made the need for information on fertilizer supply and use for

*Corresponding author. E-mail:adekay98@yahoo.com.

increased food production desirable.

Maize being one of the most widely grown crops in the world and a staple that provided half of the calorie consumed in many countries in Sub-Saharan Africa had been reported to have a high requirement for nutrients which justified its being a good indicator of the nutrient status of the soil as it responded readily to the application of fertilizer (Iken and Amusa, 2004; Ayoola and Makinde, 2009; Yu Helong et al., 2009). The common practice of fertilizer application among farmers in Akure where the research was conducted and in fact in other agrarian towns around had been through broadcasting and localized placement. The broadcasting method had several disadvantages of higher rates to be used and the fertilizer being easily washed away with runoff during heavy rains would make such application method being uneconomical. The localized placement was being practiced haphazardly. There was therefore the need to identify a very appropriate method to guarantee optimum fertilizer usage for maize and other food crops for profit maximization in a business-oriented agricultural economy.

The objective of this study was to investigate the optimum fertilizer rate and the placement distance from the plant to effectively improve soil nutrient status and the performance and yield of maize.

MATERIALS AND METHODS

Site description

Two field experiments were conducted in Akure (7°15'N, 5°12'E) in Nigeria in 2007 and revalidated in 2008. The latitude and longitude data were taken with the use of a 12 channel Garmin 72 Global Positioning System (GPS) receiver. The site before the study in 2007 was characterized as a gentle terrain which had been cropped continuously for five years without fertilizer application. The climate was humid tropical with the rainy and dry seasons distinctly defined. The climatic data were obtained from Ondo state agroclimatological unit, Ministry of Agriculture and Natural Resources, Akure. The experiment was carried out in the rainforest agro-ecological zone of Ondo State.

Experimental design, land preparation and planting of maize.

The experimental design was a split plot method in a randomized complete block with fertilizer quantity as the main plot and fertilizer placement distance from maize plant as the sub plots. The NPK 15-15-15 fertilizer rates in the main plot were 0,150,200,250 and 300 kg NPK/ha, while distance of application of the maize plants that formed the subplots were 10, 15 and 20 cm. The land was ploughed and harrowed and late maturing Downy Mildew Resistant (DMR) open pollinated yellow maize planted in April at a spacing of 75 x 25 cm to give a plant population of 53,000 stands per hectare (Phillips, 1977).

Fertilizer treatment

NPK 15-15-15 application that formed the main plot was applied at 0, 150, 200, 250 and 300kg/ha which gave a corresponding weight of 0, 2.8, 3.8, 4.7 and 5.6 g per maize plant, respectively at three weeks after planting.

Soil sampling and analysis

Pre-planting soil samples were taken for analysis before tillage operations of ploughing plus harrowing and the fertilizer application while post planting soil sampling was carried out at the maize flowering stage of 65 days after planting. Soil samples from three points in each plot were bulked air-dried and sieved through a 2 mm sieve and analysed for the chemical properties following the laboratory procedures described by Carter (1993). The particle size distribution was determined using 50 g of soil in 0.1 M NaOH as dispersing agent using Hydrometer (ASTM, 1524) methods. The soil pH was determined in water using a glass electrode pH meter. Organic carbon was determined by oxidising soil sample with dichromate solution and later titrated with ferrous sulphate solution. The total nitrogen was determined using micro-Kjeldahl method and the available phosphorus determined by the Bray P-1 method. The exchangeable cations were extracted by leaching 5 g of soil with 50 ml ammonium acetate at pH 7. The potassium and sodium in the leachate were determined with a flame spectrophotometer while the calcium and magnesium were determined with atomic absorption spectrophotometer.

Agronomic parameters

The mean values of plant height, stem girth, total leaf area, leaf area index, ear weight, number of grains/ear and yield of four maize plants with in a 1 m × 1 m quadrat in each subplot were taken for statistical analysis.

Measurement of the total leaf area per plant and leaf area index

The total leaf area of randomly selected four maize plants per subplot was taken and the corresponding leaf area index computed. The leaf area was measured following the procedure of Stewart and Dwyer (1999), Elings (2000) by multiplying the length of leaf by the widest width by alpha where alpha is 0.743 (L x W x 0.743). The leaf area index was computed by dividing the total leaf area of a maize plant stand by the total land area occupied by the single stand (Mauro et al., 2001).Harvesting was carried out 120 days after planting and the 14% grain moisture content confirmed with the use of grain moisture tester, while the grain yield per hectare was taken.

Statistical analysis

The plant height, stem girth, total leaf area, leaf area index and yield values were subjected to statistical analysis using Statistical Package for the Social Sciences (SPSS), while mean comparisons were made using Duncan Multiple Range Test (DMRT) at 5% probability. The fertilizer rate and the distances of application were each correlated to the growth parameters of plant height, stem girth, total leaf area and leaf area index and the yield parameters of ear weight, number of grains per ear and the yield per hectare. The economic analysis based on expenditure and income on maize production at the varied NPK 15-15-15 fertilizer rates and placement distances were computed.

RESULTS

Figures 1 and 2 show rainfall distribution pattern, relative humidity and temperature regime of Akure in 2007 and

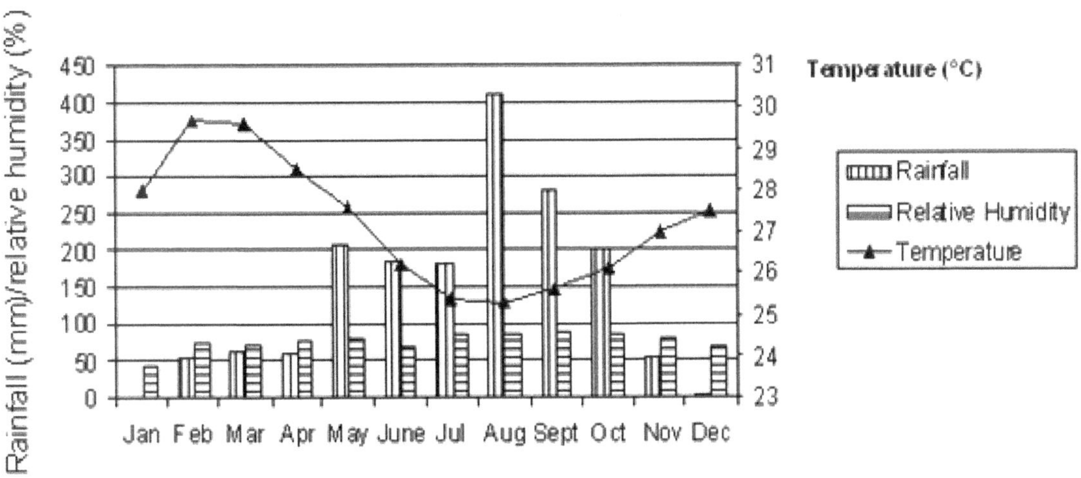

Figure 1. The chart of rainfall distribution, relative humidity and temperature regime of Akure in 2007. Source: Ondo State Agroclimatological Unit, Ministry of Agriculture and Natural Resources Akure.

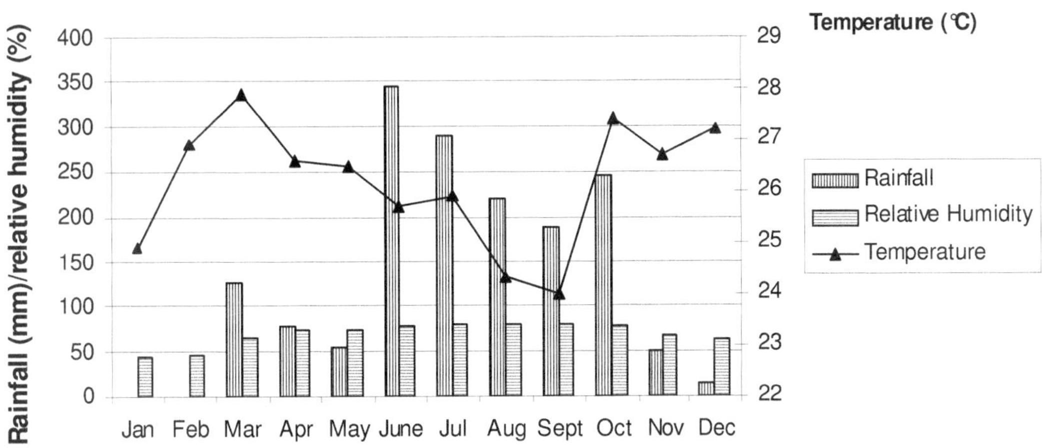

Figure 1b. The chart of rainfall distribution, relative humidity and temperature regime of Akure in 2008. Source: Ondo State Agroclimatological Unit, Ministry of Agriculture and Natural Resources Akure.

2008 respectively. In the two years, the rainy and dry seasons were distinctly identified. There were higher rainfall values between the months of May and October in 2007 while higher rainfall values were observed between March and October in 2008.

Even though temperature did not follow a corresponding trend with rainfall, the months of June to October had lower values when compared to higher range between November and April in 2007 while in 2008 lower temperature was obtained in January and between the months of June and September (Figure 1b).

Soil analysis

The pre-treatment soil analysis showed the soils to be slightly acidic with low organic matter, total nitrogen, calcium, magnesium, potassium and phosphorus (Table 1).

Effects of NPK fertilizer application on soil chemical properties at post planting

Table 2 shows the effects of fertilizer application on soil chemical properties at post planting. The application of the NPK fertilizer increased the organic matter, percentage nitrogen, phosphorus and potassium content compared to the values observed in the pre-treatment soil analysis. This apparently showed in the mean values observed in plots without fertilizer application (0 kg NPK/ha) compared to plots treated with 150, 200, 250 300 kg NPK/ha respectively. The pH did not however follow any sequence at post planting as soils ranged from

Table 1. Pre-planting soil analysis.

Soil properties	Value
pH	5.6
Organic matter (%)	1.57
Available Phosphorus (ppm)	7.10
Total Nitrogen (%)	0.16
Sodium (cmol/kg)	0.17
Potassium (cmol/kg)	0.17
Calcium (cmol/kg)	2.80
Magnesium (cmol/kg)	1.63

Table 2. The effects of NPK 15 – 15 – 15 on soil chemical properties at post planting.

Fertilizer (Kg/Ha)	Distance of application (cm)	pH	Organic matter (%)	Nitrogen (%)	Available K (ppm)	Na (cmol/kg)	K (cmol/kg)	Ca (cmol/kg)	Mg (cmol/kg)
0	10	6.0	1.57	0.16	7.3	0.18	0.17	3.03	1.90
	15	5.6	1.60	0.16	7.1	0.17	0.17	2.97	1.63
	20	5.7	1.64	0.17	7.2	0.17	0.17	2.80	1.80
150	10	5.9	3.23	0.19	8.3	0.19	0.27	3.40	2.10
	15	6.2	3.05	0.22	9.2	0.17	0.24	3.53	2.53
	20	6.1	3.04	0.18	8.1	0.16	0.21	3.60	2.17
200	10	6.3	3.69	0.25	11.4	0.20	0.33	3.80	2.40
	15	6.2	3.26	0.24	11.9	0.18	0.33	4.03	2.10
	20	6.3	3.07	0.20	10.9	0.19	0.26	3.67	2.00
250	10	6.3	3.83	0.29	12.2	0.20	0.32	3.60	2.30
	15	6.8	3.64	0.29	12.4	0.20	0.34	4.13	2.40
	20	6.3	3.15	0.23	11.8	0.21	0.25	3.83	2.07
300	10	6.8	4.23	0.32	12.6	0.18	0.36	4.10	2.440
	15	6.7	4.00	0.32	13.1	0.20	0.38	4.03	2.30

slightly acidic to medium acidic level.

Effects of fertilizer quantity on maize growth parameters

Table 3 shows the effects of fertilizer application on maize plant height, stem girth, total leaf area and leaf area index. There was observed significantly high maize plant height in plots treated with 200, 250 and 300 kg/ha NPK compared with 0 and 150 kg/ha NPK fertilizer with the highest value of 197.6 cm in 300 kg NPK/ha and the least value of 167.9 cm in plot without fertilizer (0 kg NPK/ha). Similar trends of significantly high values were observed for stem girth, total leaf area and the leaf area index. The fertilizer rate was positively correlated to the growth parameters with correlation coefficient values of 0.90, 0.96, 0.99 and 0.98 in plant height; stem girth, total leaf area and leaf area index, respectively (Table 9).

Effects of fertilizer on maize yield parameters

Table 4 shows the effects of fertilizer quantity on maize ear weight, number of seeds per ear and yield per hectare. The ear weight, number of grains per ear and the grains yield per hectare showed significantly high values in plots treated with 300 kg/ha NPK fertilizer compared to other plots. The values decreased in a decreasing order of magnitude in 250, 200, 150 kg NPK/ha and plot without fertilizer (0 kg NPK/ha). The fertilizer rate was positively correlated with the yield parameters with correlation coefficient values of 0.99, 0.94 and 0.97 in ear weight, number of grains per ear and grains yield per

Table 3. Effects of quantity of fertilizers on the maize growth parameters of plant height, stem girth, total leaf area and leaf area index.

Fertilizer (kg/ha)	Plant height (cm)	Stem girth (cm)	Total leaf area (cm^2)	Leaf area index
0	167.90^c	3.80^d	3521.00^d	1.88^d
150	175.50^b	4.50^c	3806.60^c	2.03^c
200	198.10^a	5.40^b	4021.80^b	2.15^b
250	197.00^a	5.60^a	4030.40^b	2.15^b
300	197.60^a	5.60^a	4161.50^a	2.22^a

Figures followed by the same letters were not significant using Duncan multiple range text (DMRT) at 5% probability.

Table 4. Effects of quantity of fertilizers on the maize yield parameters of cob weight, number of seeds per cob and yield per hectare.

Fertilizer (kg/ha)	Ear weight (gm)	Number of grains per ear	Grains yield (kg/ha)
0	217.60^c	386^d	894.60^e
150	234.10^d	455^c	1123.50^d
200	241.20^c	565^b	1465.20^c
250	247.10^b	567^b	1561.80^b
300	256.90^a	570^a	1661.70^a

Figures followed by the same letters were not significant using Duncan multiple range text (DMRT) at 5% probability.

Table 5. Effects of fertilizer placement distance from maize plant on the maize growth parameters of plant height, stem girth, total leaf area and leaf area index.

Distance (cm)	Plant height (cm)	Stem girth (cm)	Total leaf area	Leaf area index
10	187.60^a	5.00^b	3929.80^a	2.09^a
15	186.60^a	5.10^a	3940.50^a	2.00^a
20	186.50^b	4.80^c	3854.50^b	2.06^b

Figures followed by the same letters were not significant using Duncan multiple range text (DMRT) at 5% probability.

hectare, respectively (Table 9).

Effects of distance of fertilizer placement on the maize growth and yield parameters

Tables 5 and 6 show the effects of distance of fertilizer placement on maize growth and yield parameters, respectively. There were significantly higher mean values in plant height, stem girth, total leaf area and the leaf area index when fertilizer was placed at a distance of 10 and 15 cm compared to fertilizer placed at a distance of 20 cm from the plant. The ear weight, number of seeds per ear and the grains yield per hectare had lower significant values when fertilizer was placed at a distance of 20 cm when compared with when it was placed at 10 and 15 cm.

The interaction effects of fertilizer and distance on maize growth and yield parameters

Tables 7 and 8 show the interaction effects of fertilizer and placement distance on maize growth and yield parameters, respectively. Higher values were obtained by application of fertilizers at a rate of 200, 250 and 300 kg/ha and at distances of 10 and 15 cm compared with the lowest values observed in 0 and 150 kg fertilizer per hectare and at a distance of 20 cm of maize plants. At each rate of fertilizer application, the placement distance of 20 cm consistently had lower values.

Table 5 shows the correlation between the fertilizer rate and the distances of application respectively to the growth parameters of plant height, stem girth, total leaf area and leaf area index and also to the yield parameters of ear weight, number of grains per ear and

Table 6. Effects of fertilizer placement distance from maize plant on the maize yield parameters of cob weight, number of seeds per cob yield per hectare.

Distance (cm)	Ear weight (gm)	Number of grains per ear	Grains yield (kg/ha)
10	240.80[a]	510 [a]	1351.50[a]
15	241.90[a]	510[a]	1352.70[a]
20	235.40[b]	505[b]	1317.70[b]

Figures followed by the same letters were not significant using Duncan multiple range text (DMRT) at 5% probability.

Table 7. The interaction effects of fertilizer and distance on maize growth parameters.

Fertilizer (kg/ha)	Distance (cm)	Plant height (cm)	Stem girth (cm)	Total leaf area	Leaf area index
0	10	166.90	3.80[c]	3505.90	1.87
	15	167.30	3.80 [c]	3549.10	1.89
	20	169.40	3.80 [c]	3508.10	1.87
150	10	176.50	4.60[b]	3837.10	2.05
	15	175.60	4.70 [b]	3886.40	2.07
	20	174.30	4.30 [b]	3696.30	1.97
200	10	198.60	5.40[a]	4037.70	2.15
	15	199.00	5.50[a]	4034.70	2.15
	20	196.70	5.20[a]	3993.10	2.13
250	10	197.70	5.70[a]	4059.00	2.16
	15	197.60	5.80[a]	4052.70	2.16
	20	195.70	5.30[a]	3979.30	2.12
300	10	198.10	5.60[a]	4209.10	2.24
	15	198.50	5.70[a]	4179.60	2.23
	20	196.30	5.40[a]	4095.90	2.18

the yield per hectare. Increasing fertilizer had a positive correlation with maize growth and yield parameters while the increasing distances of fertilizer application had a negative correlation with the maize growth and yield parameters. Increasing the distance from 15 to 20 cm caused reduction in maize growth and yield values.

Effects of fertilizer rates and placement distance on maize profitable response

Table 10 shows the profitable response of maize to fertilizer rates and placement distance. The highest maize yield of 2,622.3 and 1,876.7 kg/ha were obtained in plots treated with 300 and 250 kg NPK fertilizer per hectare and at placement distances of 15 and 10 cm of maize plants and gave correspondingly higher net revenues of ₦134,284.00 and ₦74,636.00 respectively. The lowest yield value of 871.3 kg/ha was obtained in plot without fertilizer treatment (0 kg fertilizer per hectare) and the lowest net revenue of ₦27, 900.00 observed in plot treated with 150 kg fertilizer per hectare and at a distance placement of 20 cm from the maize plant.

DISCUSSION

The tropical climatic pattern had been validly confirmed from the rainfall distribution, relative humidity and temperature values shown in Figures 1 and 2. Akintola (1986) in previous study of rainfall distribution in Nigeria for a period of 1892 to 1983 (91 years) described the tropical climate in relation to rainfall distribution and temperature pattern.

Table 8. The interaction effects of fertilizer quantity and distance of fertilizer placement on maize yield parameters.

Fertilizer (Kg/ha)	Distance (cm)	Ear weight (g)	Number of Grains per ear	Grains yield (kg/ha)
0	10	219.70	387	915.30^{g}
	15	216.00	385	895.30^{g}
	20	217.00	385	871.30^{g}
150	10	232.70	455	1124.30^{f}
	15	238.00	460	1140.30^{f}
	20	231.70	448	1105.00^{f}
200	10	243.00	565	1464.00^{e}
	15	242.30	569	1481.30^{e}
	20	238.30	559	1450.00^{e}
250	10	248.70	568	1572.70^{d}
	15	252.70	569	1570.00^{d}
	20	240.00	563	1539.70^{d}
300	10	260.00	574	1876.70b
	15	260.70	568	2622.30a
	20	250.00	562	1681.30c

Table 9. Correlation between fertilizer rate and the distances of application and maize growth and yield parameters.

	Plant height	Stem girth	Total leaf area	Leaf area index	Ear weight	Number of grains per ear	Grains yield (kg/ha)
Fertilizer rate	0.90	0.96	0.99	0.98	0.99	0.94	0.97
Distances of application	-0.90	-0.65	-0.80	-0.33	-0.77	-0.87	-0.85

The lowest fertility status as observed in the pre-treatment soil analysis could be due to the previous continuous cropping of the land without fertilizer application. This made the response of maize to the NPK fertilizer treatment apparent. Tisdale et al. (2003) had previously explained Mitscherlich's principle on the positive response of crops to the supply of the limiting elements in the soil. The critical limits of nutrients in the soil had previously been discussed in the study of productivity limitation of soils in North Western Nigeria (Kparmwang and Malgwi, 1997).

The significantly higher plant height, stem girth, total leaf area and leaf area index observed in plots treated with 300, 250 and 200 kg NPK fertilizer over 150 and 0 kg fertilizer showed the improvement of maize performance resulting from increased rate of NPK fertilizer (Odiete et al., 2005). The NPK 15-15-15 fertilizer supplied the elements nitrogen, phosphorus and potassium in equal proportions and therefore higher rates resulted in higher amounts of the nutrient elements. The best luxuriant growth expressed through higher total leaf area in plots treated with 300, 250 and 200 kg NPK fertilizer per hectare and at placement distances of 10 and 15 cm could be adduced to higher availability of potash which had been described to stimulate the synthesis of carbohydrates for the development of the framework substance of maize plants and this was explained from previous research to be accelerated with sufficient quantities of nitrogen (Hershey, 2002; Kiran, 2004). The leaf area index; a factor influencing crop growth due to the influence on photosynthesis had a significantly higher value of 2.22 in plot treated with 300 kg NPK fertilizer per hectare and this showed higher availability of the nutrient element nitrogen, phosphorus and potassium for the plants improved growth (Gobron, 2009). The significantly higher ear weight, number of grains per ear and the yield per plot treated with 300 kg NPK/ha showed improved yield with an optimal NPK fertilizer application (Whitbread et al., 2004; Odiete et al., 2005). Previous study had reported phosphate to principally affect the development and the set of the ear grains (Babalola and Salako, 2006). The placement distance

Table 10. Income and expenditure showing the profitable response of maize production to NPK 15 -15 -15 at varying rates and distances.

S/N	Farm operations	Cost (₦)* of farm operations with fertilizer rate and at the distance of application														
		0 kg/Ha NPK			150 kg/ha NPK			200 Kg/ha NPK			250 kg/ha NPK			300 Kg/ha NPK		
		10 cm	15 cm	20 cm	10 cm	15 cm	20 cm	10 cm	15 cm	20 cm	10 cm	15 cm	20 cm	10 cm	15 cm	20 cm
1	Land preparation (ploughing and harrowing)	10000	10000	10000	10000	10000	10000	10000	10000	10000	10000	10000	10000	10000	10000	10000
2	Purchase of maize seeds	2500	2500	2500	2500	2500	2500	2500	2500	2500	2500	2500	2500	2500	2500	2500
3	Planting (8 pd** at ₦ 500.00 /pd)	4000	4000	4000	4000	4000	4000	4000	4000	4000	4000	4000	4000	4000	4000	4000
4	Purchase of NPK 15-15-15	Nil	Nil	Nil	15000	15000	15000	20000	20000	20000	25000	25000	25000	30000	30000	30000
5	Application of NPK 15-15-15	Nil	4000	4000	4000	4000	4000	4000	4000	4000	4000	4000	4000	4000	4000	4000
6	Purchase of Primextra (Pre-emergence herbicide	5000	5000	5000	5000	5000	5000	5000	5000	5000	5000	5000	5000	5000	5000	5000
7	Application of Primextra	3000	3000	3000	3000	3000	3000	3000	3000	3000	3000	3000	3000	3000	3000	3000
8	Cost of one supplementary weeding (20 pd** at ₦ 500.00 /pd)	10000	10000	10000	10000	10000	10000	10000	10000	10000	10000	10000	10000	10000	10000	10000
9	Cost of harvesting	10000	10000	10000	10000	10000	10000	10000	10000	10000	10000	10000	10000	10000	10000	10000
10	Cost of storage in the crib	3000	3000	3000	3000	3000	3000	3000	3000	3000	3000	3000	3000	3000	3000	3000
11	Cost of shelling and bagging	3000	3000	3000	3000	3000	3000	3000	3000	3000	3000	3000	3000	3000	3000	3000
12	Total cost	41500	41500	41500	60500	60500	60500	65500	65500	65500	70500	70500	70500	75500	75500	75500
13	Maize yield (kg/ha)	915.3	895.3	871.3	1124.3	1140.3	1105	1464	1481.3	1450	1572.7	1570	1539.7	1681.3	1876.7	2622.3
14	Gross revenue (1 kg sold at ₦80,000.00)	73224	71624	69704	89944	91224	88400	117120	118504	116000	125816	125600	123176	134504	150136	209784
15	Net Revenue	31724	30124	28204	29444	30724	27900	51620	53004	50500	55316	55100	52676	59004	74636	134284
16	% Increase in Yield over 871.3 kg/ha	5.1	2.7	Nil	29.0	30.9	26.8	68.0	70.0	66.4	80.5	80.2	76.7	115.4	201.0	93.0
17	% Increase in net revenue over ₦27900	13.7	8.0	1.1	5.5	10.1	Nil	85.0	90.0	81.0	98.3	97.5	88.8	167.5	381.3	111.5

*Exchange rates: 1US Dollar = 156 Nigerian Naira (₦156.00), **pd = person days (Farm labour wage).

of 10 and 15 cm indicated optimum distances for efficient uptake of nutrients by the plants while the fertilizer application distance of 20 cm resulted in less fertilizer elements to be available to maize plants which caused the negative correlation between fertilizer placement distances and maize growth and yield parameters. Previous study had confirmed placement to influence the effectiveness of fertilizer on plant's performance. (Yu Helong et al., 2009).

The highest maize grain yield resulting from higher fertilizer application at the appropriate distances of 10 and 15 cm showed the efficient application method that could result in higher economic returns to maize farmers. Ojo (2000) and Zhiying Xu et al. (2009) observed that application of fertilizer to maize farms gave significantly higher yield with correspondingly improved financial status to producers while Bifarin et al. (2008) observed an economic empowerment for maize farmers when there was an increased production resulting from increased use of fertilizers.

CONCLUSION

Increasing the rate of NPK fertilizer from 150 kg/ha to increasing magnitude of 200, 250 and 300 kg/ha improved soil nutrient status and at the NPK fertilizer appropriate placement distances of 10 and 15 cm from the maize plant correspondingly improved maize yield which also resulted in higher economic returns to the maize farmers.

REFERENCES

Akintola JO (1986). Rainfall Distribution in Nigeria, 1892 to 1983. Impact Publishers Nigeria Ltd. Ibadan, Nigeria. p.380.

Ayoola OT, Makinde EA (2009). Maize Growth, Yield and Soil Nutrient Change with N-Enriched Organic Fertilizers. Afri. J. Food, Agric. Nutr. Dev., 9(1): 580–592.

Babalola OA, Salako FA (2006). Comparative study of Ogun

Phosphate rock and single superphosphate on three varieties of cowpea and maize and their effects on soil. Nig. J. Soil Sci., 16: 84–89.

Bationo A, Christianson CB, Klaij MC (1993). The effect of crop residue and fertilizer use on pearl millet. Nutrient Cycling in Agroecosystems, 34: 251–258.

Bifarin JO, MoyinJesu EI, Essiet EE (2008). Economic Analysis of post harvest losses of maize in Ondo State. In: Agele SO, Adekunle VAJ, Olufayo MO, Osho, IB (eds) Agriculture in Nigeria Wetlands: Proceedings of the Fourth Annual Conference of the School of Agriculture and Agricultural Technology, Federal University of Technology, Akure, Nigeria, pp: 237-240.

Brady NC, Weil RR (1999). The Nature and Properties of Soils. (12th Editon). Prentice Hall Inc, New Jersey, p.881.

Carter MR (1993). Soil Sampling and Methods of Soil Analysis. Canadian Society of Soil Science. Lewis Publishers, London. p.823.

Elings A (2000). Estimation of leaf area in tropical maize. Agron. J. 92: 436–444.

Gobron N (2009). Leaf Area Index. ftp://ftp.fao.org/docrep/fao/011/0197e/i0197e15.pdf

Hall NH, Stulp VJ, Reed MR (1983). Effects of a Fertilizer Subsidy on Brazilian Crop Production and Exports. Am. J. Agric. Econ., 65:571-575.

Hershey D (2002). How do the amounts of fertilizer affect plant growth. MadSci. Network. http://www.madsci.org/posts/archives/2002-04/1019088560.Bt.r.html

Iken JE, Amusa NA (2004) Maize Research and Production in Nigeria. Afri. J. Biotechnol. 3: 302–307.

Kparmwang T, Malgwi WB (1997). The Genesis, Classification and Productivity Limitations on Sandstone Soils in North Western Nigeria. In: Singh BR (ed) Management of Marginal Lands in Nigeria Proceedings of the 23rd Annual Conference of Soil Science Society of Nigeria held at Usmanu Danfodiyo University, Sokoto, Nigeria. pp: 43–50.

Kiran RJ (2004). Effect of Time of Weeding and Levels of N and P_2O_5 Fertilizers on the Grain Yield of Maize. Nepal. Agric. Res. J., 5:69–70

Mauro A, Homem A, Elizabeth A Walter-Shea, Mesarch MA (2001). Test of an extended mathematical approach to calculate maize leaf area index and leaf angle distribution. Agric. For. Meteorol. 108(1): 45–53.

Niehues BJ, Lamond RE, Godsey CB, Olsen CJ (2004). Starter Nitrogen Fertilizer Management for continuous N0-Till Corn Production. Agron. J., 96: 1412–1418.

Odiete I, Chude VO, Ojeniyi SO, Okozi AA, Hussaini GM (2005) Response of Maize to Nitrogen and Phosphorus sources in Guinea Savanna Zone of Nigeria. Nig. J. Soil Sci., 15:90–101.

Ojo SO (2000) Factor Productivity in Maize Production in Ondo State, Nigeria. Appl. Trop. Agric., 5: 57–63.

Opeke LK (1997). Tropical Tree Crops. Spectrum Books Limited, Ibadan. p.326.

Phillips TA (1977). An Agricultural Notebook. (With special reference to Nigeria) Lowe and Brydone Printers Limited, Great Britain. p.310.

Stewart DW, Dwyer LM (1999). Mathematical characterization of leaf shape and area of maize hybrids. Crop Sci.. 39: 422–427.

Stewart WM, Dibb DW, Johnston AE, Smyth TJ (2005). The Contribution of Commercial Fertilizer Nutrients to Food Production. Agron. J., 97: 1–6.

Tisdale SL, Nelson WL, Beaton JD, Havlin, JL (2003). Soil Fertility and Fertilizers, 5th Edition, Pearson Education, Inc. New Jersey. p.634.

Whitbread A, Mushayi P, Waddington S (2004). Modeling the Effects of Phosphorus on Maize Production and Nitrogen Use Efficiency on Small Holder Farms in Sub-Humid Zimbabwe. Fourth International Crop Science Congress in Brisbane, Australia 2004.

Yu Helong, Chen Guifen, Liu Dayou (2009). Research and Application of Precision Fertilization on Maize. Int. Federation Information Processing Book Series. 293: 477–486.
http://www.springerlink.com/content/027rt/h143342721/

Zhiying X, Zhengfei G, Jayne TS, Black R. (2009). Factors influencing the profitability of fertilizer use on maize in Zambia. Agric. Econ., 40(4):437–446.

Comparative study of different organic manures and NPK fertilizer for improvement of soil chemical properties and dry matter yield of maize in two different soils

Adeniyan O. N.*, Ojo A. O., Akinbode, O. A. and Adediran J. A.

Institute of Agricultural Research and Training, Obafemi Awolowo University (OAU), P. M. B. 5029 Ibadan, Nigeria.

A pot experiment was conducted to compare different organic manures with NPK fertilizer for improvement of chemical properties of acid soil from farmer's field in coastal area of Epe and nutrient depleted soil from research field of Institute of Agricultural Research and Training, Moor Plantation, Ibadan. Maize was planted for 12 weeks and dry matter yield was determined. Results showed that application of 5 ton/ha of each of the evaluated organic manures and 100 kg/ha NPK 15-15-15 fertilizer improved chemical properties of both acid and nutrient depleted soils compared with unfertilized soil. Application of different types of organic manures reduced the acidic levels of both the soils. Cow dung application resulted in the highest pH levels of 6.37 and 6.50 in acid soil and nutrient depleted soil respectively while NPK fertilizer gave lowest pH levels of 5.28 and 5.74 for both soils. Also, application of different types of organic manures enhanced soil organic C, total N, available P, exchangeable K and CEC better than NPK fertilizer in both soils. The study indicated that among the organic manures evaluated, cane rat droppings improved soil chemical properties best. Plant dry matter yield increased with application of NPK fertilizer compared with compost, poultry manure and cane rat droppings in both soils. In acid soil, application of NPK fertilizer gave the highest dry matter yield of 4.77 g/plant while in nutrient depleted soil; application of NPK fertilizer gave the highest dry matter yield of 5.58 g/plant.

Key words: Manure, fertilizer, dry matter, pH levels.

INTRODUCTION

Acid soils are highly weathered and contain large quantities of Al and Fe hydrous oxides that have the ability to adsorb major elements onto their surfaces such that much of added nutrients are fixed instead of being made available for crop use (Akinrinade et al., 2006; Enwezor et al., 1981). Vast areas of tropical lands that were once fertile have been rendered unproductive due to continuous cultivation and erosion which caused physical degradation, loss of soils organic matter and decreased cation exchange capacity (CEC) and as well as increased Al and Mn toxicity (Mba, 2006). As these soils suffered multi-nutrient deficiencies, application of mineral fertilizers has become mandatory to increase crop yields. However, mineral fertilizers are commonly scarce, costly, having imbalanced nutrition and their use could exacerbate the problem of soil acidity (Oguike et al., 2006; Nottidge et al., 2006). The problem is worsened by the continuous removal of government subsidies on fertilizers, and poor distribution systems. This is why Aduayi (1985), and Agbim and Adeoye (1994) recommended the use of crop residue and other organic wastes as supplements to inorganic fertilizers.

The expansions in animal enterprises and cassava processing industries have been followed by the

*Corresponding author. Email: adeniyantayo@yahoo.com.

production of a large amount of wastes in urban and peri-urban areas, which becomes potential environmental hazard. However, there is a renewed interest in proper and effective use of organic manures derivable from these wastes to get rid of such wastes and to improve agricultural soil fertility. These manures are known to decompose in soils to form humus and humid substances, which play a dominant role along with clay micelles in the complex soil reactions that, enhance the CEC of soils. The chemical transformations involving humus in soil colloidal ion exchange complexes are summarized by Brady (1990), and Agbim and Adeoye (1994) to increase the net positive or negative charge, depending on the pH prevailing. It has been stated by Akpegi et al. (2006) that this is a crucial factor to the soil buffer capacity. According to them, at low pH the hydrogen ions are said to dissociate from the carboxyl (-CoOH), hydroxyl, (-OH), and the phonetic (R-OH) sites, where R is a benzene ring, leaving an increased net negative surface charge on the colloid. An important factor in soil reclamation using organic matter derivable from organic residues is its role in improving the soil physical properties, by forming stable aggregates to improve soil structure, buffer the soil, improve aggregate stability, the population of soil microorganisms, and enhance water-retention capacity and the applied manures raise the soil pH in acid soils (Sanchez and Miller, 1986; Spaccini et al., 2002; Ogunlade et al., 2006; Olatunji et al., 2006). Repeated application of organic residues on soils has long-lasting positive effects on soils, (Adeniyan and Ojeniyi, 2003, 2005).

According to Olatunji et al. (2006), the application of organic manure haD been found to have higher comparative economic advantage over the use of inorganic fertilizer. A study conducted by Nwajiuba and Akinsanmi (2002) in Southeastern Nigeria, showed that returns per ha were higher in organic farms though outputs were slightly less in inorganic farms. Therefore, replenishment of nutrients and improvement in quality of nutrient depleted and acid soils could be achieved through the application of inorganic fertilizers, organic manures or a combination of both organic and inorganic fertilizers (Adeniyan and Ojeniyi, 2005). The aim of this work therefore was to investigate the effects of different organic manures and NPK 15-15-15 fertilizer on soil chemical properties and maize dry matter yield in acid soil and nutrient depleted soil.

MATERIALS AND METHODS

Collection and preparation of soil samples

The soils used for the pot experiment were collected from one of the experimental sites of Institute of Agricultural Research and Training, Moor Plantation Ibadan (Latitude 7° 22′ N, 3° 50′ E) and a farmer's field at Epe (Latitude 6° 27′ N, 3° 57′ E). Ibadan soil was classified as nutrient depleted soil after it has been cropped for many years (over 35 years) and therefore was unable to support plant growth without heavy fertilization or a long time fallowing. The soil at Epe was coastal acidic. The dominant soils in Ibadan and Epe are Alfisol and Entisol respectively. According to the USDA (1975) classification, both soils are oxic paleustalf. This contains predominantly kaolinite minerals and is abundant in iron oxides. The soils were air-dried and sieved through 2 mm sieve and thereafter 10 kg soil sample was weighed into a perforated polythene pot.

Treatments

Five different organic manures (cow dung, cassava peelings compost, poultry manure, rabbit droppings and cane rat droppings) and NPK 15: 15: 15 inorganic fertilizer were applied to the potted soils at the rate equivalent to 5 ton/ha of the organic manure and 60 kg/ha of NPK fertilizer for the pots to be planted with maize. The rates amounted to 5 kg and 3.59 g per pot of organic manure and NPK fertilizer respectively. The pots were arranged in a completely randomized design with three replicates. The manures were incurporated into the soils two weeks before planting while the NPK fertilizer was applied at split-dose first at three weeks after planting and second at maize tasselling. The control pots were planted without any treatment application. Two seeds of maize (variety DMR-ESR Yellow) were planted and were later thinned to one healthy seedling at germination. The soil was watered daily to field capacity throughout the growing period. The pot experiment was terminated at maize maturity (12 weeks after planting).

Soil and plant samples analyses

Pre-planting soil samples from acid soil and nutrient depleted soil were collected from ten points at Epe and Ibadan sites respectively and then bulked separately, while post-planting soil samples were collected from the three replicates according to the treatments applied and bulked accordingly. Analyses of acid and nutrient depleted soil samples were then carried out. The air-dried soil samples were allowed to pass through 2 mm sieve. Soil pH was determined in 1:2 soil:water ratio. Soil organic matter was determined by Walkey-Black dichromate method. Total N was determined by steam distillation technique (Bremner and Keeney, 1966), soil available P was determined by Bray and Kurtz (1945) extraction method. Exchangeable bases were determined by the ammonium acetate method, while Ca and Mg were determined using EDTA titration. Organic manures were also analyzed for total N, P and exchangeable bases.

At 12 weeks after planting, the shoot was harvested and washed with mild concentration (2%) of detergent, rinsed with deionized water, put in an envelope and oven dried at 65°C for 48 h. The plant samples were ground with Willey mill to pass through 0.5 mm sieve. The ground samples were digested with nitric-perchloric-sulphuric acid mixture (AOAC, 1970). Total N was determined by the micro-Kjeldahl procedure (Jackson, 1967). Phosphorus was determined colorimetrically by the vanado-molybdate method, K and Ca were determined on flame photometer, while Mg was determined using atomic absorption spectrophotometer. Concentration of nutrients was expressed as the percentage of dry plant material. Dry matter yield amount was determined. Nutrient uptake was calculated using the formula: Dry matter × nutrient content of plant (%) (Ombo, 1994).

Statistical analysis

The data collected were statistically analyzed using analysis of variance procedure. Means were compared using Duncan multiple range test at 5% level of probability.

Table 1. Chemical and physical analysis of soils used for the pot experiment.

Parameter	Acid soil	Nutrient depleted soil
pH	5.10	6.20
Organic carbon (%)	0.80	0.67
Total nitrogen (%)	0.08	0.06
Available phosphorus ($mgkg^{-1}$)	2.74	4.14
Exchangeable K ($cmol\ kg^{-1}$)	0.19	0.25
Exchangeable Ca ($cmol\ kg^{-1}$)	2.38	1.12
Exchangeable Mg ($cmol\ kg^{-1}$)	0.78	1.42
Exchangeable Na ($cmol\ kg^{-1}$)	0.56	0.36
CEC ($cmol\ kg^{-1}$)	3.91	3.15
Sand (%)	86.00	87.00
Silt (%)	10.00	8.00
Clay (%)	4.00	5.00
Soil textural class	Sandy loam	Sandy loam

Table 2. Chemical analysis of poultry manure used for the pot experiment.

Organic manure	Organic C (%)	N (%)	P (%)	Parameter K (%)	Ca (%)	Mg (%)	Zn (ppm)	Cu (ppm)
Cow dung	13.5	1.30	0.58	2.15	0.99	0.52	129	128
Poultry manure	15.1	2.21	2.98	2.05	3.28	1.06	700	180
Cassava peelings compost	12.3	1.70	0.86	1.50	1.06	0.63	208	140
Rabbit droppings	16.7	1.04	0.99	2.05	1.13	0.54	493	168
Cane rat droppings	20.9	1.95	2.06	3.30	2.82	1.99	838	196

RESULTS AND DISCUSSION

Table 1 shows the initial chemical properties of the Epe soil (coastal acid soil) and Ibadan soil (nutrient depleted soil) used for the experiment. The chemical properties of both soils indicated that they have low nutrients (organic C, total N and CEC). The pH of the soil sample from Epe was slightly more acidic with the recorded pH value of 5.10 compared to the soil sample from Ibadan with the recorded pH value of 6.20. Also, the Epe soil was higher in organic C, total N and CEC with the recorded values of 0.80%, 0.08% and 3.91 $cmolkg^{-1}$, respectively, compared to the soil sample from Ibadan with the recorded values of 0.67%, 0.06% and 3.15 $cmolkg^{-1}$. Table 2 shows the chemical contents of the organic manures applied. Nitrogen, phosphorus and calcium contents of the poultry manure were the highest compared to other organic manures used for the experiment.

It was observed from Tables 3 and 4 that application of organic manures to both the acid soil and nutrient depleted soil increased considerably the soil organic C, total N, available P, exchangeable K and CEC compared to using NPK fertilizer. The increase in the levels of soil organic C and organic matter was expected, since, organic manures have the ability of increasing soil organic matter content (Ojeniyi, 2000). It has also been reported that the contents of some major nutrients in the soil were slightly dependable on the level of organic matter (Adeniyan and Ojeniyi, 2005). Organic matter shows a greater capacity to retain nutrients in forms that can easily be taken up by plants over a longer period of time. The nutrients in the (N-P-K) fertilizer (inorganic fertilizer) were already in the mineralized form and it provides a ready source of nutrients to the soils. By implication, the nutrients released from NPK fertilizer were for a short period of time because leaching of nutrients may be higher in the soil treated with NPK fertilizer than the soil treated with organic manures. Several workers have reported longer residual effect of organic manures when applied to the soil (Adeniyan and Ojeniyi, 2003' Adetunji, 1997).

The acid soil and nutrient depleted soil treated with organic manures had significant pH values ranging from 6.20 to 6.35 and 6.21 to 6.38, respectively compared to soil treated with NPK fertilizer with pH values ranging from 5.28 to 5.75, respectively. Under both situations, the values in NPK fertilizer treated plots (5.28, 5.75) were not significantly different from their respective control values (5.08 and 5.53). These results indicated that organic manures have greater potential of raising soil pH

Table 3. Effects of applied treatments on acid soil chemical properties.

Treatment	pH H_2O	Organic C (%)	Total N (%)	Avail. P ($mgkg^{-1}$)	Exch. K ($cmolkg^{-1}$)	CEC ($cmolkg^{-1}$)
Poultry manure	6.30^{a}	1.83^{ab}	0.38^{ab}	4.80^{a}	0.83^{a}	2.91^{a}
Rabbit droppings	6.28^{ab}	1.55^{c}	0.20^{c}	3.90^{c}	0.71^{c}	2.59^{b}
Cane rat droppings	6.20^{b}	1.96^{a}	0.47^{a}	4.27^{abc}	0.91^{a}	3.10^{a}
Cow dung	6.37^{a}	1.70^{b}	0.28^{b}	4.13^{abc}	0.78^{b}	2.64^{ab}
Cassava peelings compost	6.25^{ab}	1.89^{ab}	0.40^{ab}	4.30^{abc}	0.89^{a}	2.94^{a}
NPK	5.28^{cd}	0.44^{d}	0.26^{bc}	4.03^{bc}	0.66^{d}	2.39^{c}
Control	5.08^{b}	0.06^{e}	0.09^{d}	2.20^{d}	0.11^{e}	0.90^{d}

Means with different letter(s) are statistically different by DMRT.

Table 4. Effects of applied treatments on nutrient depleted soil chemical properties.

	pH H_2O	Organic C (%)	Total N (%)	Avail. P ($mgkg^{-1}$)	Exch. K ($cmolkg^{-1}$)	CEC ($cmolkg^{-1}$)
Poultry manure	6.38^{a}	2.94^{a}	0.68^{ab}	6.37^{ab}	1.02^{ab}	2.97^{ab}
Rabbit droppings	6.21^{b}	2.57^{de}	0.57^{c}	6.33^{ab}	0.92^{b}	3.12^{a}
Cane rat droppings	6.32^{a}	2.98^{a}	0.71^{a}	5.93^{ab}	1.12^{a}	3.39^{a}
Cow dung	6.50^{a}	2.49^{e}	0.59^{b}	6.13^{ab}	0.80^{d}	2.84^{b}
Cassava peelings compost	6.38^{a}	2.92^{ab}	0.69^{a}	6.60^{a}	1.07^{a}	3.20^{a}
NPK	5.74^{cd}	2.85^{b}	0.56^{c}	5.73^{ab}	0.85^{c}	2.89^{b}
Control	5.53^{d}	0.49^{e}	0.04^{d}	3.25^{c}	0.20^{e}	1.13^{c}

Means with different letter(s) are statistically different by DMRT.

compared to NPK fertilizer. This implies that organic manures could serve as good amendment materials in ameliorating acid soils. Soil acidity has been reported as one of the major constraints to crop growth in tropical region (Busari et al., 2005). This is due to the usual accompanying effect of Al and Mn toxicity, nutrients imbalance and deficiencies and their consequential detrimental effects on crop growth and yield (Oguntoyinbo et al., 1991). While comparing the nutrients enhancement between the acid soil and nutrient depleted soil with the application of poultry manures, it was observed to be higher in nutrient depleted soil.

Specifically, the increase in soil nutrients with the application of organic manures to the acid soil and nutrient depleted soil was more obvious with the application of cane rat droppings with respect to organic C, total N, exchangeable K and CEC compared to other organic manures. This might be due to the fact that the nitrogen, K, Ca and Mg contents of the cane rat manure applied were relatively high compared with values in other organic manures. However, these values were not significantly different from that of poultry manure and cassava peelings for soil N, P, K, Ca and Mg.

The dry matter yields of the maize plant shoot from acid soil and nutrient depleted soil that were treated with organic manures and NPK fertilizer are shown in Table 5. The organic manures and NPK fertilizer significantly increased dry matter yield over control (unfertilized pot). The application of NPK fertilizer gave the highest dry matter yield under acid soil and nutrient depleted soil. However, this was not significantly different from that of compost, poultry manure and cane rat droppings.

Conclusion

Application of different types of organic manures and NPK fertilizer enhanced availability of soil nutrients and cation exchange capacity considerably in both acid soil and nutrient depleted soil. The applications of organic manures performed favorably the same or better than NPK fertilizer, and cane rat manure seemed to be the best among the organic manures applied. To derive maximum benefits, large quantity of organic manure will be required for large scale farming which the cane rat droppings may not provide. The poultry manure and cassava peelings performed favorably almost well like cane rat droppings and NPK fertilizer when applied to the acid soil and nutrient depleted soil. Therefore, the present study clearly indicates that the marginal and unproductive soils can be reclaimed by the application of cassava peelings and poultry manure for the optimization of crop production.

Table 5. Effects of applied treatments on dry matter yield of maize plant.

	Dry matter	Yield (g/plant)
Treatment	Acid soil	Nutrient depleted soil
Compost	3.17[ab]	2.49[c]
Poultry manure	2.88[ab]	2.90[b]
Rabbit droppings	2.60[b]	3.80[ab]
Cane rat droppings	3.38[ab]	3.74[ab]
Cow dung	2.27[b]	1.81[cd]
Cassava peelings	2.59[b]	2.60[bc]
NPK	4.77[a]	5.58[a]
Control	1.90[c]	1.12[d]

Means with different letter(s) are statistically different by DMRT.

REFERENCES

Adetunji MT (1997). Organic residue management, soil nutrient changes and maize yield in humid Utisol. Nutrient Recycling Agro Ecosyst., 47: 189-195.

Adeniyan ON, Ojeniyi SO (2003). Comparative effectiveness of different levels of poultry manure with NPK fertilizer on residual soil fertility, nutrient uptake and yield of maize. Moor J. Agric. Res., 4(2): 191-197.

Adeniyan ON, Ojeniyi SO (2005). Effect of poultry manure and NPK 15-15-15. and combination of their reduced levels on maize growth and soil chemical properties. Nig. J. Soil Sci., 15: 34-41.

Aduayi EA (1985). Making the soil nutritious to plant. Inaugural lecture series. Obafemi Awolowo University press Ltd. Ile Ife, 78: 10-34.

Agbim NN, Adeoye KB (1994). The role of crop residues in soil fertility maintenance and conservation. In G. Lombin, Adeoye KB and Chude VO. Torunana (Eds), Organic fertilizer in the Nigerian Fertilizer seminar 1991, FMANR, Abuja, pp. 21-24.

Akinrinade EA, Iroh I, Obigbesan GO, Hilger T, Romheld V, Neumann G (2006). Response of cowpea varieties to phosphorus supply on an acidic alumi-haplic-Acrisol from Brazil. Nig. J. Soil Sci., 16: 115-120.

Akpegi GI, Ayuba SA, Ali A (2006). Effect of rice husk on the yield of tomato in Benue State of Nigeria. Proceedings of the 30th Annual Conference of the Soil Society of Nigeria 5th – 9th December, 2006. Held at the University of Agriculture, Markurdi, pp. 221-226.

AOAC (1970). Official method of analysis. 11th Ed. Association Analytical Chemists Washington D.C.

Brady NC (1990). The nature and properties of soils. (10th Ed.). Macmillan publ. Co., N.Y., pp. 7, 8,10, 16.

Bray RH, Kurtz LT (1945). Determination of total organic available forms of phosphorus In soils. Soil Sci., 59: 39-45.

Bremner JM, Keeney DR (1966). Determination of isotope ratio analysis of different Forms of nitrogen on soil: 3. Exchangeable ammonium nitrate and nitrite by Extraction distillation methods. Soil Sci. Soc. Am. Proc., 30: 577-582.

Busari MA, Salako FK, Sobulo RA, Adetunji MT, Bello NJ (2005). Variation in soil pH and maize yield as affected by the application of poultry manure and lime. Proceedings of the 29th Annual Conference of the Soil Science Society of Nigeria. Held from December 6-10 2004. pp. 139-142.

Enwezor WO, Udo EJ, Sobulo RA (1981). Fertility status and productivity of the 'Acid Sands'. In, Udo, EJ. and Sobulo, RA. 'Acid Sands' of Southern Nigeria. SSSN Special Pub. Monograph No. 1: 56-53.

Jackson ML (1967). Soil chemical analysis. New Delhi: Prentice Hall of India.

Mba CN (2006). Influence of organic wastes on plant growth parameters and nutrient uptake by maize (*Zea mays* L). Nig. J. Soil Sci., 16: 104-108.

Nottidge DO, Ojeniyi SO, Asawalam DO (2006). Effect of levels of wood ash on soil chemical properties in an acid Utisol of Southeastst Nigeria. Nig. J. Soil Sci., 16: 109-114.

Ombo FI (1994). Self sufficiency in local fertilizer production in Nigeria. In. Proc. 3rd Afr. Soil Sci. Conf., Ibadan.

Nwajiuba C, Akinsanmi A (2002). Organic manure use among smallholders in the rainforest of Southeastern Nigeria. Online-http:www.Tropentag.de/2002/proceedings/node 188.html.

Ojeniyi SO (2000). Effect of goat manure on soil nutrients content and okra yield in rainforest area of Nigeria. Appl. Trop. Agric., 5: 20-23.

Oguike PC, Chukwu GO, Njoku NC (2006). Physico-chemical properties of a Haplic Acrisol in Southeastern Nigeria amended with rice mill waste and NPK fertilizer. Proceedings of the 30th Annual Conference of the Soil Science Society of Nigeria, 5th – 9th December, 2006, held at University of Agriculture, Markudi, pp. 38-45.

Ogunlade MO, Adeoye GO, Ipinmoroti RR, Ibiremo OS, Iloyanomon CI (2006). Comparative effect of organic and NPK fertilizers on the growth and nutrient uptake of cocoa seedlings. Nig. J. Soil Sci., 16: 121-125.

Oguntoyinbo FJ, Aduayi EA, Sobulo RA (1991). Effectiveness of some local liming materials in Nigeria as ameliorants of soil acidity. J. Plant Nutr., 19: 999-1016.

Olatunji O, Ayuba SA, Oboh VU (2006). Growth and yield of okra and tomato as affected by pig dung and other organic manures: Issues for economic consideration in Benue state. Proceedings of the 30th Annual Conference of the Soil Science Society of Nigeria, 5th – 9th December, 2006, held at University of Agriculture, Markudi, pp. 91-98.

Sanchez PA, Miller RH (1986). Organic matter and soil fertility management in acid soils of the tropics. Transactions of the XIII Congress of International Soil Science Society, p. 5.

Spaccini R, Piccolo A, Mbagwu JSC, Zena-Teshale A, Igwe CA (2002). Influence of the addition of organic residues on carbohydrates content and structural stability of some highland soils in Ethopia. Soil Use Manage., 18: 404-411.

United States Department of Agriculture (USDA) (1975). Agricultural handbook SCA/VSA. Washinton D.C. USA, 436: 200.

Changing weather and introduction of pigeon pea intercrop in formerly cold pyrethrum zone

Mwangi, J. A.[1]*, Waweru, S. N.[2], Oeba, V. O.[3] and Wepukhulu, S. B.[4]

[1]Kenya Agriculture Research Institute (KARI), Pyrethrum and Botanical Research Center, P. O. Box 100, Molo, Kenya.
[2]Kenya Forestry College Londiani, Ministry of Forestry and Wildlife, P. O. Box 8, Londiani, Kenya.
[3]Kenya Forestry Research Institute (KEFRI), P. O. Box 20412-00200, Nairobi.
[4]Kenya Agriculture Research Institute (KARI), National Horticultural Research Center, P. O. Box 220, Thika, Kenya.

Pyrethrum is a, herbaceous perennial crop grown commercially for extraction of pyrethrins, which is used in pesticides formulation. Through out the years, it has been grown in a pure stand due to its poor performance as intercrop with other crops that do well in pyrethrum growing areas. A trial was carried out on introduction of a short pigeon pea variety at different line spacing within the pyrethrum crop in randomized complete block design replicated four times. Data on dry pea yields was analyzed using ANOVA and fitted in a generalized linear model, with the distribution assumed to be Gamma and a link function as inverse. The results showed that there was a significant difference ($F_{(4, 132)} = 4.76$; p=0.001) among the intercropping levels of pyrethrum and pigeon pea. The control (pure stand of pigeon pea had the lowest yield (0.92 t ha^{-1}) as compared to 1:1 intercropping of pyrethrum and pigeon pea which had the highest pigeon pea yield (1.12 t ha^{-1}). Comparison of pigeon mean yield showed that there were significant differences ($p<0.05$) between control and 1:1, 2:1 and 3:1 intercropping. However, there were no significant differences ($F_{(1, 132)} = 2.68$; p=0.104) between control and 4:1 intercropping of pyrethrum and pigeon pea. Overall no significant differences ($p>0.05$) in pigeon pea yields were found among all intercropping spacing. These findings indicated that pigeon pea might be a good option to inter crop with pyrethrum. Further monitoring and assessment would be carried out to assess both pyrethrum and pigeon pea yield at different seasons of the year. It was observed that means of the rows were significantly different with pigeon pea pure stand giving the lowest. This observation indicated that pigeon pea is a good option to inter crop with. In all spacings the pigeon pea performed well. If interest would be the quantity of pigeon pea then 1:1 spacing would be resulted to.

Key words: Field productivity, soil microbial symbiotic association, soil nutrient availability.

INTRODUCTION

Pyrethrum is an herbaceous perennial crop grown commercially for extraction of pyrethrin, which is used in pesticides formulation. Pyrethrum has been forth foreign exchange earner in Kenya. Kenya has been the highest producer of pyrethrum in the world. More than 90% of pyrethrum is grown by small-scale farmers. When the world market is good, and often is the case, farmers opts to plant his small piece of land purely with pyrethrum. The main pyrethrum growing areas formerly used to be cold highland where maize and beans would do poorly. Due to continuous climatic changes which has been caused mainly by, global warming, the same areas appear capable of growing some selected varieties of pigeon peas also. To increase soil productivity, pigeon pea (*Cajanus cajan*), a leguminous perennial crop was evaluated for its potential as a short crop for intercropping with pyrethrum. It has been uneconomical to inter-crop pyrethrum with other local food crops. When planted together it will be one of the two but not both that survives. Some of the crops, such as maize and potatoes, that would be planted with pyrethrum are tall and would lead to shading over pyrethrum. Any amount of shading to, this crop that would rarely reach a height of 1 m usually has adverse effect on its growth.

This effect often can lead to total loss of the whole pyrethrum crop. The shorter crops like garden peas and

*Corresponding author. E-mail: mwangiali@yahoo.com.

beans do not survive the nutrient and water competition by pyrethrum. Pyrethum is generally a heavy feeder with most roots growing between, 0 to 20 cm deep. The small scale farmers have limited piece of land to plant both pyrethrum as cash crop and other subsistence crops. Pigeon pea opted for has an advantage in that it is a short variety, and at the same time it is deep rooted. Pigeon pea does well over pH of 4.5 to 8.5, as compared to pyrethrum growing areas in the country are range between, pH of 4.5 to 6.5.

The pigeon pea is known to have high nutritional value when used as food crop, and besides it can fetch ready market locally. It fixes nitrogen through symbiotic association with nitrogen fixing bacteria. From the In contrast from analysis done at in the pyrethrum trials at the research center by the first author, it has been observed that, pyrethrum has a high requirement of nitrogen. Nitrogen application is not recommended and hence this intercropping would possibly enhance availability of the much needed nitrogen. Mychorrhyzal association is known to exist in the plant rhizosphere, (shibata and Yano, 2003; Yano et al., 1998). Certain non-labile forms of P such Ca-phytate can be dissolved by its roots xudates. The association also can improve availability of zinc in pigeon pea (Wellings et al., 1991). From observations of pyrethrum nutrient trials by the first author it was noted that uptake of zinc can go up to 1kg per ha. At one time a field formally occupied by tea, was noted to have very poor performance by pyrethrum and the low zinc in the field was queried. These can lead to higher availability of certain nutrient as phosphorus to the plant. The mycorhyza can be associated with stability in soil structure (Miller and Kling, 2000). Another possible benefit is the control of the common soil pest in pyrethrum, the nematode. Garrido et al. (2008) indicates the potential of pigeon pea and clotalaria in the control of nematodes. One of the interests in this initial trial was to observe the behavior of pigeon peas at Molo altitude of about 2400 m above the sea level. Also observed was performance of the pigeon peas when planted within the rows of pyrethrum. This would act as a foundation to further work to be done later in pyrethrum growing areas.

MATERIALS AND METHODS

Pyrethrum was established through almost uniform splits. In the trial, plots of pyrethrum were planted in sizes of 15 plants by 11 rows. Inter-row spacing was 60 cm and intra-row spacing of 30 cm, where these are the recommended spacing. Clonal material used was L/75/477. Triple Super phosphate was used at the rate of 220 Kg per ha. This is a recommended rate which is based on the uptake rather than the performance. Normal agronomic practice such as, weeding and gapping, was maintained. The crop was left to establish before the pigeon pea intercrop was introduced. This was about 3 months.

For pure crops pigeon pea should be sown 2.5–5 cm deep in rows 40–120 cm by 30–60 cm. When sown as a mixture, it should be sown in widely spaced rows ranging from 1.2–2.1 m depending on the associated crop. About 3–4 seeds may be planted in each hill, and later thinned to 2 plants per hill (James, 1983). On the other hand pigeon pea had an in intra-row spacing of 30 cm. Pigeon pea variety, ICPL 8709, was used. This is a short variety and in the inter-planting, was planted within the normal spacing of pyrethrum rows. The pyrethrum/pigeon pea rows intercrop was laid out in four levels: pure pigeon pea stand, 1:1, 2:1, 3:1, and 4:1. The treatments were replicated four times in randomized complete block design. Pigeon peas were picked when they were dry. After exposing them to sun drying until completely dry, data on the peas' weights was taken from each row of the plants. The yield from the net plot area was converted into t per ha and analyzed using ANOVA and a generalized linear model with a distribution as gamma and link function as inverse. Orthogonal contrasts were used to compare significant differences among treatments. The statistical significant differences were declared at 1 and 5%, unless stated otherwise. All data were entered in micro soft excel 2003 and analyzed using General statistics version 10 (Genstat V10) statistical software.

RESULTS AND DISCUSSION

The results showed that there were significant differences ($F_{(4, 132)} = 4.76$; p=0.001) among the intercropping levels of pyrethrum and pigeon pea. Pure stand of pigeon pea had the lowest yield (0.92 t ha^{-1}) as compared to 1:1 intercropping with pyrethrum, which had the highest pigeon pea yield (1.12 t ha^{-1}; Table 1). Comparison of pigeon mean yield showed that there were significant differences ($p<0.05$) between control and 1:1, 2:1 and 3:1 intercropping. However, there were no significant differences ($F_{(1, 132)} = 2.68$; p=0.104) between control and 4:1 intercropping of pyrethrum and pigeon pea (Table 2). Overall no significant differences ($p>0.05$) in pigeon pea yields were found among all intercropping spacing. Each of the pigeon pea rows had 15 hills, resulting to 55,000 hills per ha. The significant difference in pigeon pea mean yield among treatments was mainly as a result of control. The non-significant statistical differences among intercropping ratios of pigeon pea and pyrethrum showed that any of those ratio spacing was equally likely to produce the same pigeon pea yield. This further demonstrated that the symbiotic association between pyrethrum and pigeon pea with little negative effects on pigeon pea yield. In addition this implied that with successful completion of monitoring and assessment of the trials at different growing seasons compounded with climate data, small scale farmers would be able find a suitable inter spacing of pigeon pea with pyrethrum. This result corroborates well with James (1983) who found that the yields of dried seed were ranging between 0.6 to 2.5 t/ha. Varshney et al. (2009) also found that the average crop productivity of pigeon pea was 0.78, in India.

Conclusion

This initial work indicates that incorporating pigeon pea into pyrethrum field is viable and capable to provide both food and financial gains to the farmer. Pigeon pea has a ready market locally due to its nutritional value, and its

Table 1. Mean yield (t ha^{-1}) of pigeon pea under different interspacing ratios with pyrethrum at Molo.

Treatment	Mean yield (t ha^{-1})	N
Pure stand of pigeon pea	0.9	44
Pyrethrum and pigeon pea in 1:1	1.1	44
Pyrethrum and pigeon pea in 2:1	1.1	24
Pyrethrum and pigeon pea in 3:1	1.1	16
Pyrethrum and pigeon pea in 4:1	1.0	12
s.e.d	0.0763	

Table 2. Mean comparisons of pigeon pea under different interspacing ratios with pyrethrum at Molo.

Source of variation	d.f.	m.s.	p-value
Replication	3	0.12382	
Treatment	4	0.26095	0.001
Control versus 1:1	1	0.93207	<.001
Control versus 2:1	1	0.47335	0.004
Control versus 3:1	1	0.26668	0.029
Control versus 4:1	1	0.14697	0.104
1:1 versus 2:1	1	0.01516	0.6
2:1 versus 3:1	1	0.00545	0.753
3:1 versus 4:1	1	0.0046	0.773
Residual	132	0.05486	
Total	139		

domestic uses. There has been occasional extended draught, during when the farmer can leap from his field through the pigeon peas, this when pyrethrum is dormant and unproductive. More work needs to be done, to observe variations in different pyrethrum growing zones. Also mutual gains as observed in 1:1 plots need to be further observed and quantified. Within pyrethrum cultivars there are certain differences between clones and varieties. There will be a need to observe variation of this response between pyrethrum clones and varieties. There is information on potential of pigeon pea to the control of nematodes. Since nematode is a major soil pest in pyrethrum, this interaction will be beneficial to a pyrethrum farmer. Long term effect on intercropping needs to be observed, with a view that all ratios gave an increase in productivity though at different quantities. Application of nitrogenous and phosphorus fertilizers, in planting pyrethrum, adversely affects establishment of pyrethrum. This has an impact in the later life of the field. Due to the presence of mycorhyzal association and nitrogen fixing bacteria in the pigeon pea rhzosphere, the inter-crop stands impart these beneficial effects to the performance of pyrethrum crop.

REFERENCES

James AD (1983). *Cajanus cajan* (L.) Millsp. Handbook of Energy Crops. Unpublished.

Garrido MS, Soares ACF, Coimbra JL, SousaCS (2008). Management of crotalaria and pigeon pea for control of yam nematode diseases. Summa Phytopathologica. Print version ISSN 0100-5405. Summa. phytopathol. vol. 34 no. 3 Botucatu July/Sept. 2008. doi: 10.1590/S0100-4052008000300003.

Miller RM, Kling M (2000). The importance of integration and scale in the arbuscular mycorrhizal symbiosis. Springer Netherlands. November. Plant Soil, 226(2): 295-309.

Shibata R, Yano K (2003). Phosphorus acquisition from non-labile sources in peanut and pigeonpea with mycorrhizal interaction. Agriculture, ecosystems and environment. Appl. Soil Ecol., 24(2): 133-141.

Wellings NP, Wearing AH, Thompson JP (1991). Vesicular-arbuscular mycorrhizae (VAM) improve phosphorus and zinc nutrition and growth of pigeonpea in a Vertisol. Aust. J. Agric. Res., 42(5): 835–845.

Varshney RK, Penmetsa RV, Dutta S, Kulwal PL, Saxena RK, Datta S, Sharma TR, Rosen B, Carrasquilla-Garcia N, Farmer AD, Dubey A, Saxena KB, Gao J, Fakrudin B, Singh MN, Singh BP, Wanjari KB, Yuan M, Srivastava RK, Kilian A, Upadhyaya HD, Mallikarjuna N Town CD, Bruening GE, He G, May GD, McCombie R, Jackson SA, Singh NK, Cook DR (2009). Pigeonpea genomics initiative (PGI): an international effort to improve crop productivity of pigeonpea (*Cajanus cajan* L.). Springer. 17-Sep-2009. http://hdl.handle.net/123456789/1158.

Yano K, Yamauchi A, Iijima M, Kono Y (1998). Arbuscular mycorrhizal formation in undisturbed soil counteracts compacted soil stress for pigeon pea. School of Agricultural Sciences, Nagoya University, Chikusa Nagoya 464-8601 Japan.

Sources and distribution of polycyclic aromatic hydrocarbons in post flooded soil near Afam power station, South East Niger Delta, Nigeria

Orok E. Oyo-Ita*, Inyang O. Oyo-Ita and Samuel U. Ugim

Department of Pure and Applied Chemistry, University of Calabar, Cross River State, Nigeria.

Post flooded forest soil profile samples obtained near Afam power station, contiguous to Imo River were analysed for quantitative determination of polycyclic aromatic hydrocarbons (PAHs). Relatively moderate bulk density (1.17 g/cm^3), permeability (1.79 cm^3/s) and macro-porosity (60%) values indicate that fluid flow through the soil profile would be unimpeded. PAHs content for the top soil horizon appeared similar to that of one of the point-source samples suggesting possible transport via flooding event on 25th April, 2008. Variation in PAHs content for the lower soil horizons may reflect a record of past PAHs deposition as a consequence of differences in Imo river current trajectory. The decreasing trends in the distribution of ∑thiophenes, ∑alkylPAHs and total PAHs (excluding perylene levels) may be associated with the sorptive character of soil organic matter (SOM) which limits transfer of PAHs deep into the soil. However, the relatively high level of perylene at the bottom soil horizon increases the likelihood of the compound to reach the shallow aquifer and contaminate the ground water. Utilization of six selected molecular ratios such as Fla/(Fla + Pyr) (0.20-0.70), Ant/(Ant + Phe) (0.22-0.41), BaA/(BaA + Chrys) (0.54-0.70), LMW/HMW (0.30-1.47), MPhe/Phe (0.72-1.59) and 1,7/(1,7 + 2,6-DMP) (0.57-0.90) enabled PAH source apportionment to be made. The results indicate a mixed source scenario with predominance of petroleum hydrocarbons input found for the top soil while the lower soil horizons were predominated by PAHs of combustion origin.

Key words: Molecular ratios, natural gas, petroleum condensate, soil, Imo River.

INTRODUCTION

Polycyclic aromatic hydrocarbons (PAHs) are ubiquitous organic pollutants because of their chemical stability and the multiplicity of the emission sources (Oros and Simoneit, 2000; Dendooven et al., 2006; Ekpo et al., 2011). The United State Environmental Protection Agency (USEPA, 1994) has listed 16 PAHs as priority pollutants in wastewater and 24 others in soils and sediments (Christensen and Arora, 2006).

Degradation of the soil environment as a result of human activities is the main effect of several pollutions. Unfortunately, the efforts aimed at limiting the harmful effects have not been quite successful. Flooding is a natural process that still occurs, in spite of expensive protection measures put to minimise its substantial ecological and economic effects (Oleszczuk et al., 2003). Quite often, considerable amounts of PAHs found in ground waters may influence the human food chain or may be taken directly in drinking water (Barbash et al., 1999).

Surface runoff from roads and fields are frequently considered as source of significant amounts of PAHs in soils (Fernandez et al., 2000; Krein and Schorer, 2002; Christensen and Bzdusek, 2005). Flooding is a type of runoff, which causes the transfer of pollutants present in various chemical forms from water to the area covered by the flood. Pollutants remaining in soils after flooding can pollute ground waters as a result of their secondary transfer deep into soil profiles and pose a danger to the human health (Oleszczuk et al., 2003). Thus, it is very important to find out all the consequences of flooding so

*Corresponding author. E-mail: orokoyoita@yahoo.com.

as to be able to minimise them.

In the past two decades, a number of papers shed light on the sources of hazardous PAHs in temperate environments. In contrast, little is known about the sources of PAHs in tropical environments (Wilcke et al., 2003). As persistent organic pollutants released in the tropics may be globally distributed, knowledge of PAH sources in the tropics is a prerequisite for understanding the global dynamics of PAH distribution (Wilcke et al., 2000). Hence, this present investigation from a tropical region should contribute to the knowledge of global PAH distribution.

Although a high level of anthropogenic PAHs to tropical environments is evident, the importance of a biological source input was suggested by Wilcke et al. (2003) on the basis of higher concentrations of certain volatile PAHs (for example, perylene, naphthalene, etc.) in tropics than in temperate environments. Recently, the occurrence of perylene in significant amounts with diagenetically produced PAHs in a tropical environment (Cross River estuary, South East (S.E) Nigeria) bordered by mangrove forest stands was linked to terrestrial organic carbon origin. In contrast to temperate environment, natural PAHs represent the highest percentage (average 76%) of the total PAHs (Ekpo et al., 2011). Finding out the sources of persistent organic pollutants (for example, PAHs) is important in order to legislate the control of influx of these pollutants into the ground water through the soil horizons.

Briefly, the main objectives of this study were to:

i. Identify and quantify all the PAHs in the soil profile covered by flood.
ii. Assess the role of geochemistry on the distribution of PAHs in the soil profile.
iii. Identify the sources of the PAHs using multi-proxy-approach.
iv. Examining the possibility of PAHs transport from potential point-source after the flooding episode on 25th April, 2008.
v. Establish a chronology of PAHs deposition during the past flooding events.

EXPERIMENTAL METHODS

Materials and quality assurance

The following analytical grade solvents, dichloromethane (DCM), methanol (MeOH), hexane, toluene, as well as molecular sieves, potassium hydroxide (KOH) and bistrimethylsilyltrifloroacetamide (BSTF) were purchased from Merck (Darmstad, Germany). Diazald® was provided by Sigma Aldrich (Germany). Perdeuterated polycyclic aromatic hydrocarbons used as internal standards and surrogates: [$^2H_2$]-benzo(a)anthracene, [$^2H_{10}$]-fluoranthene, [$^2H_{12}$]-benzo(ghi)perylene, [$^2H_{10}$]-pyrene and [$^2H_{12}$]-perylene were purchased from Cambridge isotope laboratories (Andoyer, MA, USA) and [$^2H_{10}$]-anthracene from Dr. Ehrenstofer GmbH (Augsburg, Germany).

The vessels used for analysis were pre-cleaned with ethyl acetate followed by acetone and heated at 300°C overnight. We ran one analytical blank with every batch of 15 to 20 samples to check background contamination during the extraction and purification steps. To remove background contamination, we corrected PAH concentrations by subtracting the mean of the analytical blanks.

Sample collection and preparation

The study site belongs to the low-lying coastal deltaic plains of South Eastern Nigeria, about 200 m westward from Imo River and is mostly flooded during intense rainfall (Figure 1). The terrain is virtually flat or gently undulating, sloping generally towards the Atlantic (Ezeayim and Okereke, 1996).

Temperature in the area ranges from 25 to 33°C. The coastal rainfall patterns described by Offiong et al. (2008) reveal this region has rainfall throughout the year with the highest rainfall between the months of April and September.

A combination of heavy rains and good sunshine along with adequate soil nutrients has generated thick vegetation cover in the study area. The vegetation is made up of mangrove swamp forest, tall evergreen trees including pines with prolific undergrowth of entangled shrubs. The main occupations of the native inhabitants are farming (wood burning for agricultural and domestic purposes is the common practice) and fishing. However, there are some industries aligned near the Imo river bank including the Afam power generating station. Geologically, the study site belongs to the south east part of the cretaceous and tertiary sedimentary basin in southern Niger Delta (Ezeayim and Okereke, 1996).

Hydrogeology of the area aforementioned is directly related to the structural setting of the Benin Formation which constitutes part of the water-producing layer for ground water supplies. The layer consists of sands and gravels that are enhanced by high hydraulic conductivity and transmitivity of the water-producing layer (aquifer) such that the water table is invariably at shallow depth level.

To get a deep insight into the sources, transport of PAHs and to assess its impact on ground water through the soil column as well as establish a chronology of PAHs deposition via flooding episodes, two potential point-source top soil samples and one forest soil profile sample were obtained. The potential point-source samples, PS_1 and PS_2 were collected near natural gas combustion turbine and petroleum condensate spill respectively within the Afam power station premises while the soil profile sample was obtained about 1 km southward from Afam power station. Both samples were collected within 24 h after the intense rainfall event on 25th April, 2008 (wet season) as this site keeps being inundated by outflow from the Imo river water through its low-lying bank, especially during wet season. For purposes of comparison as well as examining if diagenetic formation of perylene occurs, another one top soil sample (FS) was taken (about 1.5 km northward from Afam power Station, near the fish settlement) from a non-flooded area.

A 100 cm depth soil profile was dug using an auder (post-hole) sampler after clearing the sampling site of surface debris and removing the first 3 to 6 inches surface soil. Periodically, the accumulated soil of described horizons (5 sections) were removed from the withdrawn auger and thoroughly mixed to obtain homogenized sample representative after discarding about 1 inch of the top soil sample. Samples were placed in pre-cleaned screw-capped bottles and kept in refrigerator at -4°C until laboratory time. Bulk soil sample was collected using a trowel for particle size analysis, determination of hydraulic conductivity, bulk density and micro-porosity. The bulk soil samples were sieved to pass through a 2 µm mesh. Infiltration rate was measured using double-ring infiltrometer method and transmitivity were computed according to Ogban and Ekerette (2001). A total of 8 soil samples were used for the study, 5 of which were sections from the core while the remaining 3 were surface soil samples.

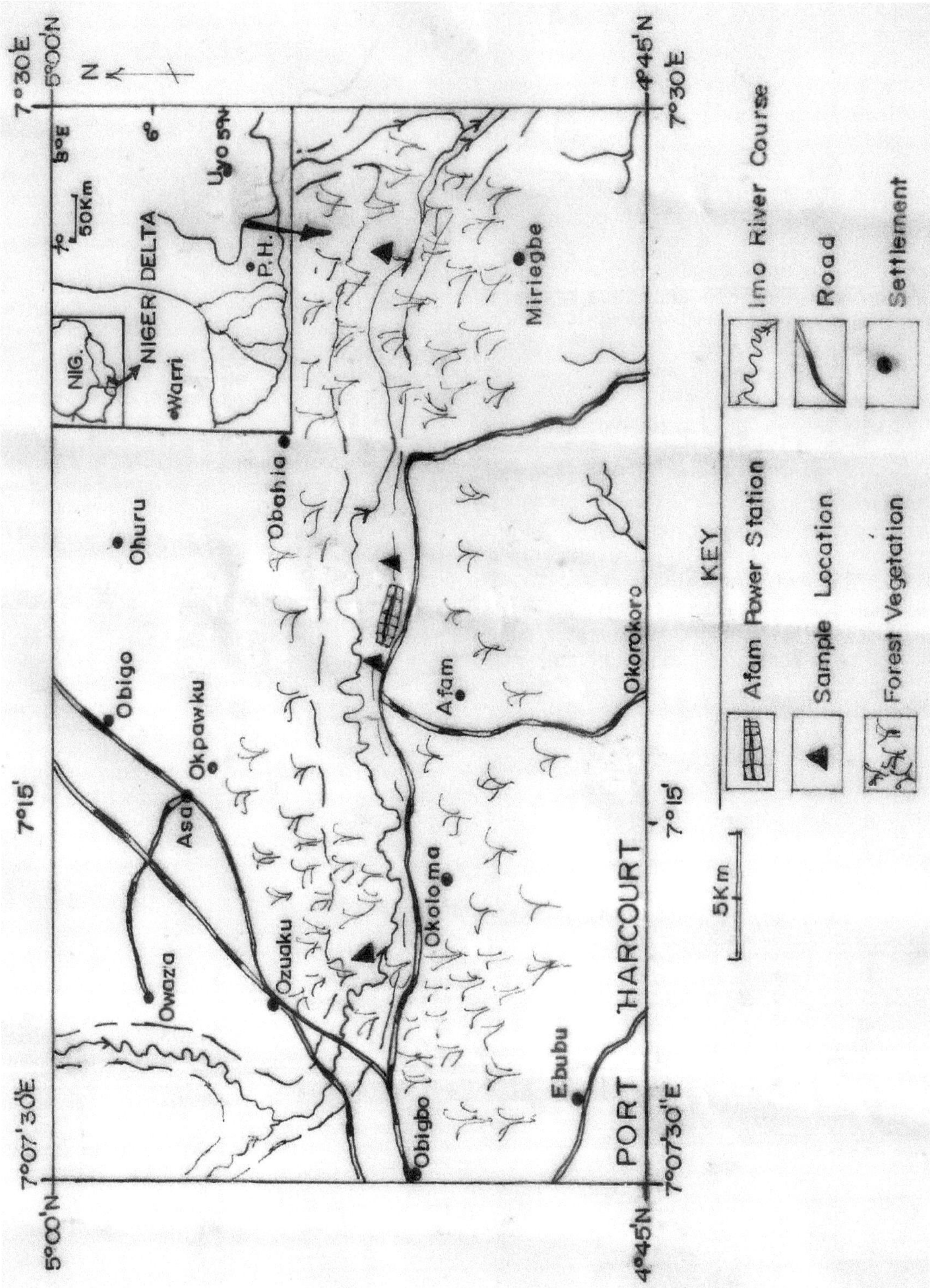

Figure 1. Map of the study area showing sampling locations.

Soil physicochemical properties

Particle size analysis was performed using Bouyoucos hydrometer. Hydraulics conductivity, bulk density and micro-porosity were determined according to the conventional method. Permeability was computed using the equation:

$$K_p = \frac{K_h \Pi}{g \ell g}$$

Where K_p = permeability (cm^3/s); K_h = hydraulic conductivity, (cmh^{-1}); Π = fluid viscosity, ($gcm^{-1}\ s^{-1}$); ℓ = fluid density, (gcm^{-3}); g= acceleration due to gravity, (cms^{-2}).

The permeability classes were determined according to Ogban and Ekerette (2001). Soil pH was measured following equilibration of the glass electrode meter in a 1: 2 soil (20 g) to water (40 g) suspension after calibration with buffer solutions of pH 7.0 and 10. Organic carbon was determined by the dichromate wet oxidation method (Nelson and Sommers, 1996).

Extraction and PAHs analysis

Extraction of 1 g sample in a test tube was performed sequentially by sonification with DCM/MeOH (2:1) and DCM in triplicate (Grimalt et al., 2004). The extracts were hydrolysed overnight at room temperature in a 10% w/v solution of KOH in MeOH. Extraction with hexane yielded a fraction enriched in neutral compounds, after lowering the pH to 2, the acid fraction was extracted with acetone. Hydroxy containing neutral compounds were derivatized to trimethylsilyl ethers with BSTFA and the fatty acids with diazomethane distilled from diazald according to standard procedure (Albaiges et al., 2006).

GC-MS analysis

Identification and quantification of PAHs were carried out by GC-MS (Trace GC-MS, THERMO instruments, Manchester, UK). A 60 m capillary column coated with 5% phenyl-95% methyl-polysiloxane (0.25 mm id and 0.25 µm film thickness; HP-5, Hewlett-Packard, CA, USA) was used. The carrier gas was helium at a flow of 2.1 ml/min. Injection port and transfer line temperatures were 300°C. Samples in toluene were injected in splitless mode. The following oven temperature program was used: the program started at 90°C and held for a minute, raised to 150°C and held for another 1 min, then at incremental rates of 4°C held for 1 min to 320°C and a final holding time of 30 min. The quadrupole mass spectrometer was operated in El mode (70eV), with an ion source temperature of 200°C and scanning from 50 to 700 m/z in one second. Selected ion monitoring (SIM) of the PAH and the deuterated internal standards molecular ions were carried out using retention time windows.

A calibration curve (detector response versus amounts injected) were performed for each compound to be quantified. The lineal range of the detector was estimated from the curve generated by plotting detector signal versus amount injected. All measurements were performed in the lineal ranges for each target analyte. In few cases, the samples were re-diluted and re-injected for fitting within the lineal range of the instrument. The quantitative data were corrected for surrogate recovery; recoveries ranged from 70 to 110 % (average 85%). Procedural blanks were lower than 5 ng/g, the limit of detection (LOD) in the full scan mode ranged from 0.004 to 0.013 ng/g dw. The lower limit of determination of PAHs was estimated from the smallest peak (signal-to-noise ratio > 3:1 in the chromatograms) that could be integrated. We validated the accuracy of PAHs quantification by analysing with the certified reference materials "CRM-104" (Resource Technology Corporation, USA), obtaining accuracy from 75 to 107% compared with the certified value (Grimalt et al., 2004).

RESULTS AND DISCUSSION

Bulk properties

The soils bulk properties including particle size, bulk density, micro-porosity, permeability and pH as well as extractable organic matter (EOM) and total organic carbon (TOC) contents are given in Table 1. Bulk density, micro-porosity and permeability did not vary with soil profile depth and reflected the predominance of the sand fraction particle in the size distribution. The observed relatively moderate bulk density and micro-porosity values when compared with soil profile study reported by Oleszczuk et al. (2003) in Poland are attributed to the fact that sand particles were not completely in close contact because of minor bridging organic material (TOC ~ 1.38 to 2.35%; Table 1).

Morphologically, the soils were loose when moist indicating that root penetration would not be hindered. As a consequence, PAHs in flooding rain water could easily leach deep into the soil horizons. Generally, permeability was moderately rapid in the soil horizon studied which indicates that fluid flow would be un-hindered (Table 1).

PAHs content and distribution in flooded and non-flooded forest soils

Concentrations expressed in ng/g dry weight (dw) in the study are representative as they were estimated from homogenized soil samples at each interval of soil profile. Among the 16 USEPA priority listed PAHs, 10 were found in the entire forest soil profile inundated by flood (Figure 2a to e) and the concentrations estimated (2.01 to 5.20 ng/g dw) were in sum lower than those found for the potential point-source samples PS_1 (11.33 ng/g dw) and PS_2 (47.07 ng/g dw) (Figure 3a, b) and higher than those for the non-flooded forest soil (4.43 ng/g dw). The parent PAH sum determined for the top forest soil was low (5.20 ng/g dw) and was about 40 order of magnitude higher than those for the forest top soil horizon in Poland and other European areas with limited anthropopressure. These PAHs include acenapthylene (Acy), fluorene (F), phenanthrene (Phe), anthracene (Ant), fluoranthene (Fla), pyrene (Pyr), benzo(a)anthracene (BaA), chrysene (Chrys), indene(1,2,3-cd)pyrene (IcdP) and dibenzo(a,h)anthracene (DBA). Other non-priority listed PAHs identified were perylene (Pery), retene (Ret), benzo(e)pyrene (BaP), dibenzothiophene (DBT), and isomeric naphto(1,2-b) thiophenes.

Other various alkyl PAHs also identified at low

Table 1. Soil horizons, geochemical characteristics and bulk properties.

Sample code	Horizon (cm)	Coordinate	pH	Micro porosity (%)	Permeability (cm^3/s)	Particles density (g/cm^3)	Bulk density (g/cm^3)	Colour	Particle size distribution (%)			TOC (%)	EOM (mg/kg)
									Sand	Silt	Clay		
AS1	A_p (0-15)	04°51'N 07°16'E	5.52	60.10	1.79	2.55	1.17	Dark grey	33.80	46.00	30.20	2.35	1020.00
AS2	A_2 (15-45)	04°51'N 07°16'E	5.54	60.10	1.79	2.55	1.17	Dark grey	33.80	46.00	30.20	2.08	700.00
AS3	BA_1 (45-55)	04°51'N 07°16'E	5.60	60.10	1.79	2.55	1.17	Brownish grey	33.80	46.00	30.20	1.87	450.00
AS4	BA_2 (55-80)	04°51'N 07°16'E	5.62	60.10	1.79	2.55	1.17	Brownish grey	33.80	46.00	30.20	1.40	100.00
AS5	BA_3 (80-100)	04°51'N 07°16'E	5.71	60.10	1.71	2.65	1.19	Brown	33.80	46.00	30.20	1.38	290.00
FS	A_p (0-15)	04°59'N 07°15'E	6.20	-	-	-	-	Dark grey	43.50	40.00	16.50	2.65	1860.00
PS_1	A_p (0-15)	04°51'N 07°18'E	5.10	-	-	-	-	Dark grey	68.90	6.000	15.10	2.50	2150.00
PS_2	A_p (0-15)	04°50'N 07°18'E	5.10	-	-	-	-	Dark grey	58.90	22.00	19.20	2.68	1150.00
Standard deviation			0.35	31.10	0.92	1.33	0.61		13.86	15.12	6.96	0.52	735.02
Standard error (+/-)			0.12	11.00	0.32	0.47	0.21		4.90	5.34	2.46	0.19	259.87

concentrations in the soil profile were isomeric alkylphenanthrenes [including 3-methylphenanthrene (3-MPbe), 2-methylphenanthrene (2-MPhe), 9/4-methylphenanthrene (9/4-MPhe), 1-methylphenanthrene (1-MPhe), 1-ethylphenanthrene (1-EPhe), 3,5-dimethylphenanthrene (3,5-DMPhe), 2,6-dimethylphenanthrene (2,6-DMPhe), 2,7-dimethylphenanthrene (2,7-DMPhe), 1,3-dimethylphenanthrene (1,3-DMP), 1,6-dimethylphenanthrene (1,6-DMPhe), 1,7-dimethylphenanthrene (1,7-DMPhe), 2,3-dimethylphenanthrene (2,3-DMPhe), 4,9-dimethylphenanthrene (4,9-DMPhe), 1,8-dimethylphenanthrene (1,8-DMPhe) and 9,10-dimethylphenanthrene (9,10-DMPhe)]. Low molecular weights PAHs were predominant, for instance, the percentage content of phenanthrene was 20% with respect to the total parent priority listed PAHs determined. A clear differentiation in PAH content was observed. Unlike the top horizons, higher content levels of high molecular weight PAHs were found in the bottom horizon. The concentrations of 3-ring priority listed PAHs varied between 0.01 and 1.30 ng/g dw, with their maximum value recorded for phenanthrene at the top soil horizon [A_P – (0 to 15 cm)]. For the 4-ring PAHs, concentrations were in the range 0.14 to 0.95 ng/g dw, maximizing in flouranthene at the A_2 horizon (15 to 45 cm) while 0.01 to 0.17 ng/g dw were recorded for the 5-ring PAHs, maximizing in dibenzo(a,h)anthracene (DBA) at the bottom horizon [BA_3 – (80 to 100 cm)](Figure 2a to e). The higher contribution of PAHs with high molecular weight (for example, DBA and IcdP) at the bottom horizon can be explained by the limitation of biodegradation processes towards heavy PAHs compared to the light counterparts and the higher leaching tendency of light PAHs (for example, flourene, acenaphthylene and acenaphthene) through the soil profile to much deeper depth than their heavy counterparts (Oleszcuk et al., 2003).

To establish a chronology of PAHs deposition during past flooding events, we examine the distribution of individual PAH compositions down the soil horizons. The top soil horizon (A_P ~ 0 to 15 cm) had PAH composition similar to that found for one of the potential point-source top soil samples (PS2) (except the detection of low level of acenapthylene and high level of retene), indicating that the top soil horizon was mostly influenced by petroleum related PAHs transported in flooded water through the potential point source sites following the intense rainfall event on 25th April, 2008. Despite the low solubility of PAHs in water, high molecular weight PAHs such as benzo(a)anthracene are transferred in the form adsorbed on particles, carried away by water (in general, only very fine grained particles < 63 μm are transported by river current) (Dorzd et al., 2002; Oleszcuk et al., 2003; Zhihuan et al., 2006). The detection of retene at high levels in the top soil horizon may be indicative of pine wood combustion source to the soil profile study site (Zhihuan et al., 2006). Retene concentration profiles reveal a decreasing trend down core except at BA_2 horizon (55 to 80 cm) (Figure 4a). The retene level at this horizon may reflect higher retene production, a consequence of past increased utilization/burning of pine wood for agricultural and domestic purposes.

Similar PAH composition and distribution were observed for the lower soil horizons (A_2 - BA_3 ~ 15 to 100 cm) except in the case of

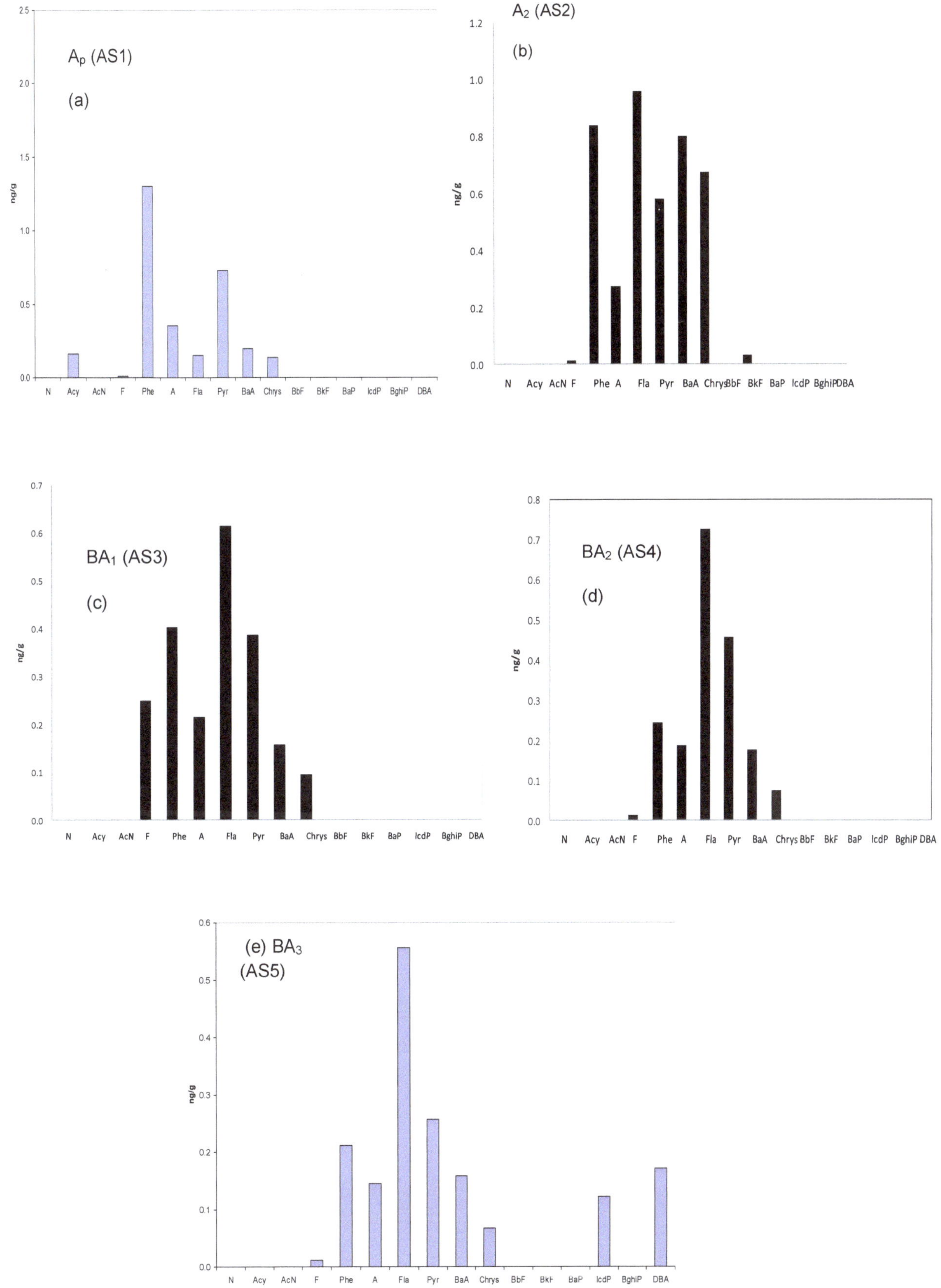

Figure 2. Distribution of individual PAH compositions down the forest soil profile; a) A_p (AS1); b) A_2 (AS2); c) BA_1 (AS3); d) BA_2 (AS4); e) BA_3 (AS5).

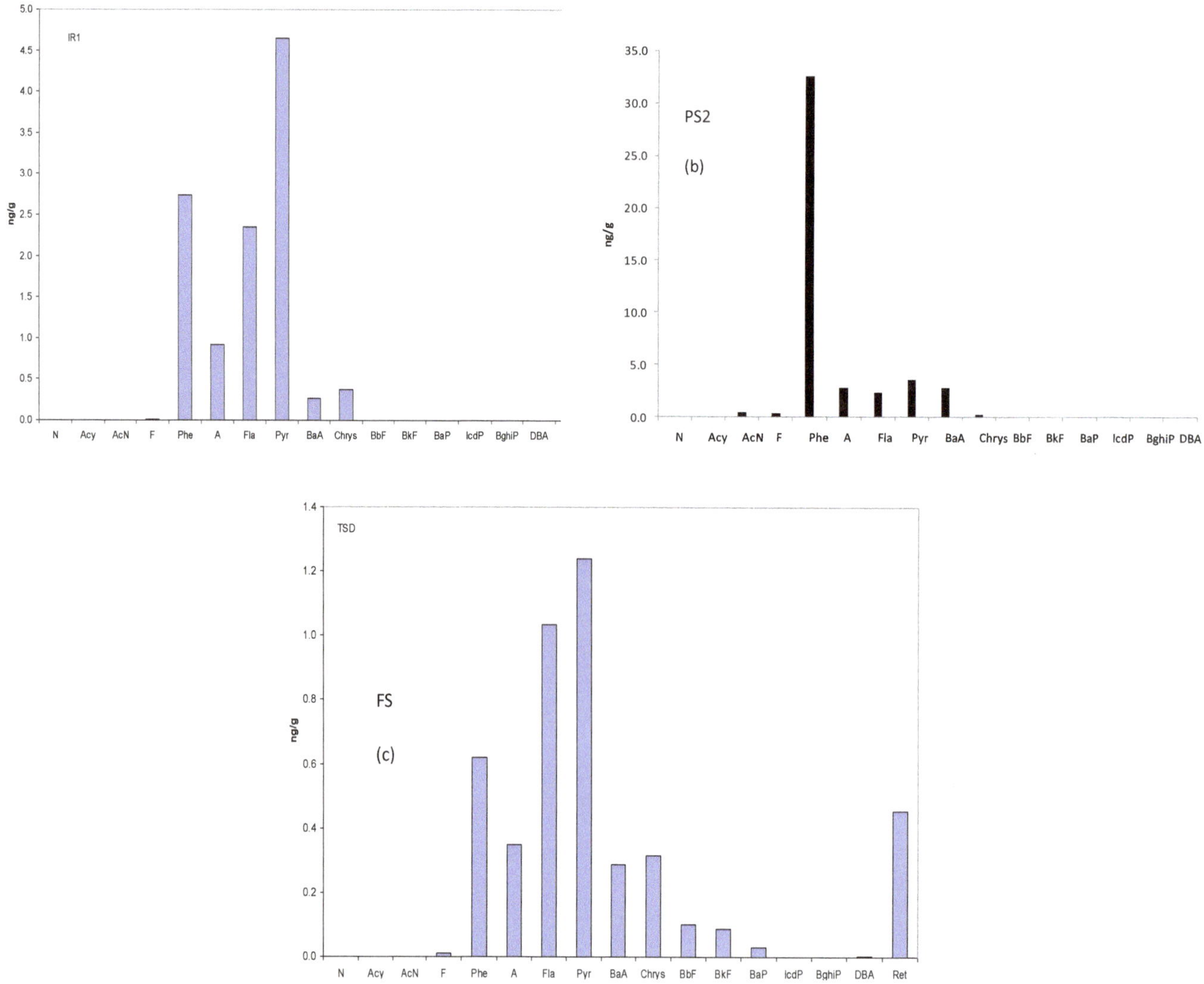

Figure 3. PAH contents for (a) potential point-source sample collected near gas turbine, (b) potential point-source sample collected near petroleum condensate spill and (c) non-flooded top soil sample collected near fish settlement.

benzo(k)flouranthene at A_2 horizon as well as indeno(123-cd)pyrene and dibenzo(a,h)anthracene at the bottom horizon (BA_3). These mixed source scenario at A_2 soil horizon may be associated with additional input of vehicular exhaust emission linked to boat traffic, carried away by the Imo River current during past flooding events (Harrison et al., 1996; Simcik et al., 1999; Yunker et al., 2002; Zhang et al., 2004). The detection of dibenzo(a,h)anthracene and indeno(123-cd)pyrene at the bottom horizon (BA_3) also suggests additional input from gasoline exhaust emission and diesel spill besides natural gas combustion /petroleum condensate spill inputs (Rogge et al., 1998; Zhihuan et al., 2006). Variation in PAHs content for the lower soil horizons may reflect a record of past PAHs deposition via flooding caused by differences in Imo river current trajectory.

The levels and types of PAHs in flooded top soil (A_p) and non-flooded top soils (FS) varied and depended upon the nature of anthropogenic activities in these areas (Figures 2a and 3c). First, there were clear differences in the PAH compositions for these two sites with higher levels recorded for the flooded top soil. Comparing data in Figures 2a and 3c, it can be seen that 5-ring PAHs (for example, benzo(b)flouranthene and benzo(a)pyrene) were detected in appreciable levels for the non-flooded soil only. Furthermore, there were clear differences in the framework of ∑thiophenes, ∑alkylphenanthrenes and total PAHs (sum of parents and alkyl PAHs) contents for the flooded soil profile down core (Figure 4b, c and d). The decreasing trends down core observed are a natural phenomenon reported by many researchers (Olesczuk et al., 2003) arising mainly from sorptive character of soil organic matter in relation to organic pollutants. Organic matter is a strong sorbent which limits transfer of PAHs deep into the soil. This influence had already been observed in soils with organic carbon contents above

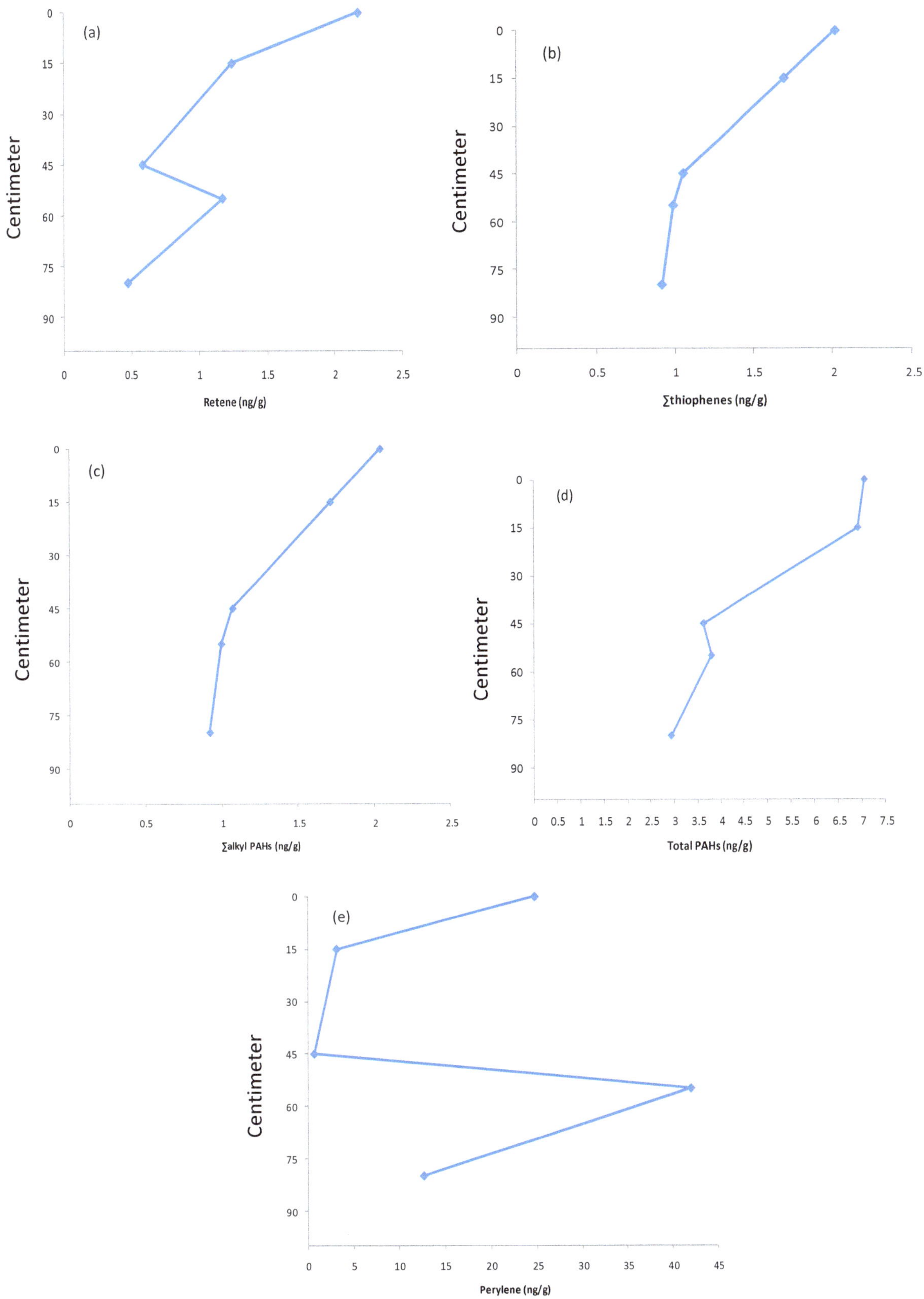

Figure 4. Distribution of (a) retene, (b) ∑ thiophenes, (c) ∑ alkylphenanthrenes, (d) total PAHs and (e) perylene in flooded soil profile.

0.1% as in our case study (Table 1). These result profiles indicate that the levels of PAHs leaching through the soil profiles would be insignificant at a depth close to the shallow aquifer if extrapolated and their possible health effect would be negligible.

Generally, naphthalene and its alkylated analogues were not detected in the entire soil profile. As low molecular weight PAH, these compounds are more volatile and remain in the atmosphere for a longer time, and are thus more susceptible to long distance transport compared to high molecular weights PAHs. In North America, climatic condition was shown to modify the PAH pattern in topsoil (Wilcke et al., 2003). The contribution of naphthalene according to these authors to the sum of the PAH concentrations in a climatically different zone (tropical) decreased in increasing mean annual temperature because of enhanced volatilisation and degradation with increasing temperature. This phenomenon may partly be responsible for the non-detection of naphthalene and its alkylated analogues as well as the low levels of acenapthylene besides leaching as previously mentioned above in the entire soil profile studied (Baran and Oleszczuk, 2002).

Among the non-priority listed PAHs identified, perylene showed the highest concentrations (0.73 to 41.99 ng/g dw) for the soil profile under study. Perylene concentrations were not included in the calculations of the total PAHs down core because perylene is reported in many environments to have biological/diagenetic origin (Wilcke et al., 2003) and its concentrations were considered outliers especially for the point-source sample PS_1 (1601.72 ng/g dw). In a near shore area of Prince Williams sound, Alaska, for example, perylene levels exceeding several hundred ppm were attributed to abundant supply of terrigenous organic debris in combination with a restricted circulation regime (Page et al., 1996). Besides this natural biogenic production, perylene could also be produced by fossil fuel (for example, natural gas) combustion (Wang et al., 1999). According to these authors and others (wilcke et al., 2000), high level of perylene in temperate soils have been consistently linked with frequent water logging arising from frequent flooding. However, in the present study, perylene concentrations for the flooded top soil profile and well aerated non-flooded top soil (near fish settlement) were not consistent with this hypothesis as the flooded top soil horizon had lower perylene concentration (24.69 ng/g dw) than the well aerated non-flooded forest top soil (64.65 ng/g dw). These results indicate that perylene concentrations in the soil profile were not of biogenic origin but rather exclusively derived from fossil fuel (for example, natural gas) combustion origin. This is supported by the extremely high perylene concentrations detected in one of the potential point- source samples (PS_1) obtained near natural gas combustion turbine. The distribution of perylene down core reveals an irregular pattern. This irregularity may reflect a record of PAHs production pattern from natural gas combustion process taking place within Afam power station occasioned by past flooding events. Therefore, perylene could be utilized as marker for natural gas combustion pollution for the region. It appears there is an extent to which soil organic matter could limit sipping of organic pollutants down the soil profile. At high concentrations, the sorptive capacity of soil organic matter (SOM) dwindles, thus increasing the likelihood of perylene to reach the shallow aquifer.

PAH source identification in soils

A large number of parent and alkyl PAH has been used for interpreting PAH distribution and assessing their sources in soils and sediments (Yunker et al., 2002; Wilcke et al., 2003; Yunker and McDonald, 2003; Aichner et al., 2007; Bechtel et al., 2007; Ekpo et al., 2011). Based on our study, limited set of six PAH ratios were selected using parent, alkyl and PAH molecular masses that exhibited the best potential for interpreting PAHs' geochemistry in soils near Afam power station.

To assess combustion versus petroleum inputs, the ratio of Fla to Fla +Pyr was used. This ratio has demonstrated to provide a better diagnostic index of PAHs formation processes than ratios of other commonly quantified parent PAHs (Yunker et al., 2002). Fla/(Fla + Pyr) ratios smaller than 0.40 usually indicate petroleum input such as crude oil, diesel, etc., between 0.40 and 0.50 indicate liquid fossil fuel combustion such as vehicular exhaust emission. Ratios over 0.50 are attributed to grass, wood or coal combustion (Yunker and Mcdonald, 2003). As the soil profile sample including the point-source samples were collected within 24 h after flooding episode accompanying the intense rainfall event that occurred on 25th April, 2008, only the top soil horizon (A_p) was observed to exhibit a Fla/(Fla + Pyr) ratio < 0.40 similar to that of one of the point-source samples (PS_2). This similar characteristic petroleum source input suggests transport of PAHs via flood water (Table 2). Other lower horizons (A_2-BA_3) had Fla/(Fla + Pyr) ratios > 0.50 characteristic of combustion source input. The non-flooded forest soil sample had Fla/(Fla + Pyr) ratio of 0.5, typical of liquid fossil fuel combustion input associated with utilization of gasoline engine boat for commercial fishery activities by the inhabitants of the fish settlement (Table 2).

Methylphenanthrene to phenanthrene (MPhe/Phe) ratio has been widely used by many researchers as a diagnostic index for petroleum hydrocarbons contamination in different compartments of the environment (Zakaria et al., 2002; Walker et al., 2005). This ratio is defined as a sum of 3-methylphenanthrene(3-MP) + 2-methylphenanthrene (2-MP) + 9/4-methylphenanthrene (9/4-MP) + 1-methylphenanthrene (1-MP) relative to phenanthrene concentrations. The basis for this index is on the observation that alkylated homologues are typically present

Table 2. PAH source indices in forest soil profile near Afam power station.

Parameter	AS1 (A_p)	AS2 (A_2)	AS3 (BA_1)	AS4 (BA_2)	AS5 (BA_3)	FS (A_p)	PS_1 (A_p)	PS_2 (A_p)
Fla/(Fla + Pyr)	0.20	0.60	0.60	0.60	0.70	0.50	0.30	0.40
BaA/(BaA + Chrys)	0.58	0.54	0.62	0.70	0.69	0.47	0.42	0.93
Ant/(Ant + Phe)	0.22	0.25	0.31	0.41	0.40	0.36	0.25	0.07
1,7/(1,7+2,6DMP)	0.80	0.60	0.57	0.90	0.88	0.95	0.61	0.60
LPAH/HPAH	1.47	0.30	0.73	0.30	0.31	0.31	0.48	4.12
MPhe/Phe	1.59	1.57	0.96	0.86	0.93	0.72	0.35	2.24

in higher proportions relative to their parent PAHs (for example, methylphenanthrene > phenanthrene) in petrogenic source. In contrast, higher temperature combustion processes form PAHs through condensation reactions and/or ring fusion limiting the presence of alkyl groups (Walker et al., 2005). Consequently, MPhe/Phe ratios of > 1 is indicative of petrogenic source input while ratios < 1 implies input from pyrogenic origin. In our study, the MPhe/Phe ratios for the soil profile were in the range 0.72 to 1.59 with the lowest value found for the bottom soil horizon whereas the top soil sample exhibited the highest MPhe/Phe value. The results show a decreasing trend of petroleum hydrocarbon input down core with a corresponding gradual increase in pyrogenic source input.

Fla/(Fla+Pyr) ratio is supplemented by Ant/(Ant+Phe) and BaA/(BaA+Chrys) ratios. Ant/(Ant+Phe) ratios < 0.10 are usually taken as indicative of petroleum input while ratios > 0.10 indicate combustion source input, although ratios > 0.10 have been reported for shale oil and coal (Radke et al., 1991). In the soil profile under study, the Ant/(Ant+Phe) ratio for the top soil horizon (A_p) had the least value (0.22) compared to those for the lower horizons (A_2-BA_3) (> 0.22), supporting predominant petroleum source input to the top soil while the lower horizons were predominated by PAHs of combustion origin (Table 2).

However, a boundary plot of Ant/(Ant+Phe) versus Fla/(Fla+Pyr) indicating apportionment of soil profile samples to source shows that top soil sample was mainly influenced by petroleum while the lower horizons had a mixed source origin dominated by combustion source (Figure 5a). These results are consistent with the distribution pattern down the soil profile.

Similarly, according to Yunker et al. (2002), values of BaA/(BaA+Chrys) ratio < 0.20 imply petroleum source input, between 0.20 and 0.35 indicate either petroleum or combustion input whereas ratios > 0.35 imply combustion. In the present study, the lower horizons had BaA/(BaA+Chrys) values that are consistent with above mentioned indices for combustion input except for the top soil horizon. As many possible contamination sources coexist in the soil profile, the observed complication in molecular ratios may be linked to weathering processes associated with microbial or photo- degradation effect (Doong and Lin, 2004) as well as leaching through the soil profile.

A boundary plot of BaA/(BaA+Chrys) versus Fla/(Fla+Pyr) shows apportionment of soil profile samples to source (Figure 5b) where the top soil horizon sample was mainly influenced by petroleum and the lower horizons by a mixed source, dominated by combustion source input. Fla/(Fla+Pyr), Ant/(Ant+Phe) and BaA/(BaA+Chrys) ratios for the non-flooded forest top soil sample (obtained near fish settlement) show values (0.50, 0.36 and 0.47 respectively) that are typical of liquid fossil fuel combustion source, a consequence of utilization of gasoline engine boat for commercial fishery activity.

Ratio calculations are traditionally restricted to PAHs within a given molecular mass to minimise differences arising from volatility, water solubility, adsorption, etc. (McVeety and Hites, 1988). Despite variations in water solubility among PAH isomers in most cases, mass ratios seem to reflect faithfully the characteristic source of PAHs, likely because only strongly bound refractory phase compounds, protected from dissolution or degradation, remain after transport (McVeety and Hites, 1988).

As an abundance of high molecular weight PAH (for example, 4-, 5- and 6- ring) is typically characteristic of pyrogenic origin whereas low molecular weight PAHs (for example, 2- and 3- ring species) are more abundant in petroleum source, ratios of low molecular weight (LMW)/high molecular weight (HMW) were calculated to further distinguish petroleum from combustion source inputs. LMW/HMW ratios > 1 indicate petroleum source whereas values < 1 imply combustion (Walker et al., 2005). For the soil profile under study, only the forest top soil horizon exhibited LMW/HMW ratio > 1 supporting the dominant petroleum source input. Other lower horizons including the non-flooded forest top soil had LMW/HMW ratios < 1, supporting major combustion source input (Table 2).

Although several studies have shown that some low molecular weight PAHs are biosusceptible and could be degraded more rapidly than their high molecular weight counterparts, this ratio has been proven useful for source discrimation. The molecular ratio 1,7/(1,7 + 2,6-DMP) has been specifically used to characterise input from wood burning (Yunker and Mcdonald, 2003). 1,7/(1,7 + 2,6-DMP) > 0.70 indicate wood combustion, those less than 0.45 indicate vehicle exhaust emission while ratios in between may imply other combustion sources or the

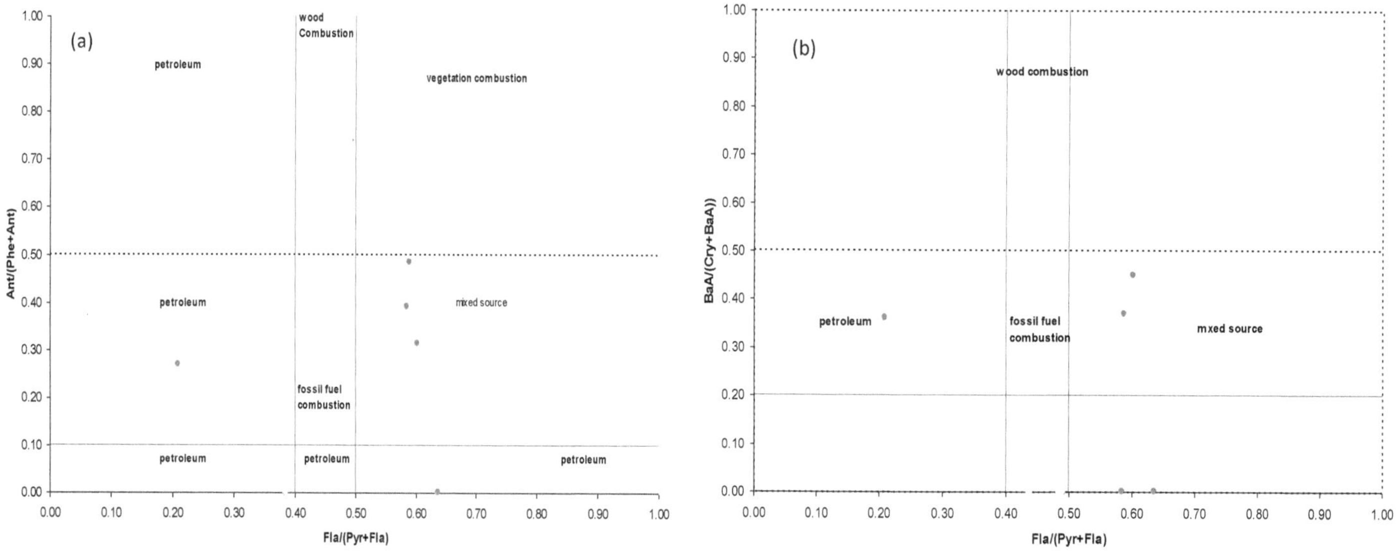

Figure 5. Boundary plots between (a) Ant/(Ant + Phe) and Fla/(Fla + Pyr) and (b) BaA/(BaA + Chrys and Fla/(Fla + Pyr) for the forest soil profile showing apportionments of samples to source.

presence of shale oil or coal. In the soil profile under study, the top soil horizons (A_p and BA_2) exhibited 1,7/(1,7 + 2,6-DMP) ratios typical of input from wood burning while others show evidence of other combustion sources (Table 2).

Conclusion

Flooding had a slight impact on the total PAH concentrations of the forest soil profile sample obtained near Afam power station contiguous to Imo River. The flooded forest top soil horizon was characterised by a higher contribution of PAHs of 3- and 4-ring types compared to content of the non-flooded forest top soil with additional 5- and 6-ring species, indicating differences in the nature of anthropogenic activity in these two sites. The similarity in PAHs content in flooded top soil with that of one of the potential point-source samples suggest transport of PAHs via flood water following the intense rainfall event on 25th April, 2008.

Due to the observed decreasing trends in the concentrations of ∑ thiophenes, ∑alkylphenanthrene and total PAHs (excluding perylene concentrations), no excessive levels which could reach the shallow water table and threaten human health was observed. However, the relatively high perylene concentration at the bottom horizon increases the likelihood of the compound to reach the shallow aquifer.

Variation in PAH contents observed for the lower soil horizons may reflect a record of past PAHs deposition via flooding, a consequence of differences in Imo river current trajectory. Utilization of six selected PAHs molecular ratios enabled source apportionment to be made. The results indicate a mixed source scenario with predominance of petroleum source input for the top soil horizon while the lower soil horizons were predominated with PAHs of combustion origin.

ACKNOWLEDGEMENTS

The authors are thankful to the management and laboratory staff of Fugro Nig. Ltd. Port Harcourt, Nigeria, for the assistance in the comprehensive analysis of our fractions by GC-MS and our postgraduate student, Samuel Ugim who assisted in the field work and are also grateful to the laboratory staff of Pure and Applied Chemistry Department of the University of Calabar for the assistant in the preliminary investigation of our samples.

REFERENCES

Aichner B, Glaser B, Zech W (2007). Polycyclic aromatic hydrocarbons and polychlorinated biphenyls in urban soils from Kathmandu, Nepal. Org. Geochem., 38: 700-715.

Albaiges J, Frano MA, Vinas L, Bairas R, Salas N, Bayona JM, Soriano JA, Armas D, Gonzalez JJL (2006). Spatial distribution and ecotoxicity of petroleum hydrocarbons in sediments from the Galicia continental shelf (NW Spain). Mar. Pollut. Bull., 53: 260 – 271.

Baran S, Oleszczuk P (2002). Chromatographic determination of polycyclic aromatic hydrocarbons in soil, sewage sludge and sewagesludge-amended soil. Pollution. J. Environ. Stud., 11: 609–615.

Barbash JE, Thelin GP, Kolpin DW, Gilliom RJ (1999). Distribution of major herbicides in ground water of the United States. Water Res. Investig. Report, 98: 42- 45.

Bechtel A, Gawlick H, Gratzer R, Tomaselli M, Puttmann W (2007). Molecular indicators of palaeosalinity and depositional environment of small scale basins within carbonate platforms: The Late Triassic Hauptdolomite Wiestalstausee section near Hallein (Northern Calcareous Alps, Austria). Org. Geochem., 38: 92–111.

Christensen ER, Bzdusek PA (2005). PAH in sediments of the Black River and the Astitabula River, Ohio: source apportionment by factor analysis. Water Res. 59: 511–524.
Christensen ER, Arora S (2006). Source apportionment of PAHs in sediments using factor analysis by time records: application to Lake Michigan, USA. Water Res., 23: 1–8.
Dendooven G, Amos-Vald A, Batancur A (2006). Bioremediation of polycyclic aromatic hydrocarbons contaminated saline-alkaline soils and sediments of former Lake Texaco. Chemosphere, 62: 1749 – 1760.
Doong R, Lin Y (2004). Characterization and distribution of polycyclic aromatic hydrocarbon contaminants in surface sediments and water from Gao-pin River, Taiwan. Water Res. 38: 1733–1744.
Dorzd J, Licznar M, Licznar SE, Weber J (2002). Soil science with mineralogy and petrography, elements. Wyo AR Wroclaw, Poland.
Ekpo BO, Oyo-Ita OE, Oros DH, Simoneit BRT (2011). Distribution and sources of polycyclic aromatic hydrocarbons in surface sediments from Cross River estuary, S. E. Niger Delta, Nigeria. Environ. Assess. Monit. DOI: 10.1007/s10661-011-2019-5
Ezeayim V, Okereke CA (1996). The effect of petroleum activities in Nigerian soil environment. Glob. J. Pure Appl. Sci., 11: 285–297.
Fernandez P, Vilanova R, Martinez C, Appleby P, Grimalt JO (2000). The historical record of atmospheric pyrolytic pollution in Europe. Environ. Sci. Technol., 34: 1906–1913.
Grimalt JO, Van-Drooge BL, Ribes A, Vilanova RM, Fernandex P, Appleby P (2004). Persistent Organochlorine compounds in soils and sediments of European High Altitude Mountain Lakes. Chemosphere, 54: 1549–1561.
Harrison RM, Smith DJT, Luhana L (1996). Source apportionment of atmospheric polycyclic aromatic hydrocarbons collected from an urban location in Birmingham, UK. Env. Sci. Technol., 30: 825-832.
Jenkins BM, Turn SQ, Williams RB (1996). Emission factors for polycyclic aromatic hydrocarbons from biomass burning. Environ. Sci. Technol., 30: 2462–2469.
Krein A, Schorer M (2002). Road runoff pollution by polycyclic aromatic hydrocarbons and its contribution to river sediments. Water Resour., 34: 4100–4115.
McVeety BD, Hites RA (1988). Atmospheric deposition of polycyclic aromatic hydrocarbons to water surfaces: A mass balance approach. Atmos. Environ., 22: 511–536.
Meyers PA, Ishiwatari P (1998). The early diagenesis of organic matter in lacustrine sediments. Org. Geochem., 22: 185–200.
Nelson DW, Sommers LE (1996). Total carbon, organic carbon, and organic matter. In: Methods of Soil Analysis, Part 2, 2nd ed., Page AL, Ed. American Society of Agronomy. Incorporated Madison, Wisconsin. Agronomy, 9: 961-1010.
Offiong RA, Atu JE, Nja GN, Anieyou UA (2008). Problems and prospect of poor drainage system of urban sustenability, Calabar, Nigeria. Glob. J. Soc. Sci., 7: 121–127.
Ogban PI, Ekerette IO (2001). Physical and chemical properties of the coastal plain sand soils of South-Eastern Nigeria. Niger. J. Soil. Res., 2: 6-14.
Oleszczuk P, Baran S, Baranowska E (2003). Degradation of soil environment in the post-flooding area: content of polycyclic aromatic hydrocarbons (PAHs) and s-triasin herbicides. Environ. Sci. Health, 38(6): 799–812.
Oros DR, Simoneit BRT (2000). Identification and emission rates of molecular tracers in coal smoke particulate matter. Fuel, 79: 615–636.
Page DS, Boehm PD, Douglas GS, Bence AE, Burns PJ, Mankiewicz PJ (1996). The natural petroleum hydrocarbon background in subtidal sediments Prince William Sound, Alaska, USA. Environ. Toxicol. Chem., 15(8): 1266-1281.
Radke M, Willsch J, Rapp JB (1991). A statistical approach to the interpretation of aliphatic hydrocarbons distribution in marine sediments. Chem. Geol., 93: 163–177.
Rogge WF, Hilderman LM, Mazurek MA, Cas GR, Simoneit BRT (1998). PAHs molecular marker characteristics of specific source emissions. Environ. Sci. Technol., 27: 1902–1904.
US Environmental Protection Agency USEPA (1994). Test Methods for Evaluating Solid Waste, Physical/Chemical Methods SW846, Revision 2. Office of Solid Waste and Emergency Response, Washington, DC.
Walker SE, Dickhut RM, Chisholm-Brause C, Sylva S, Reddy CM (2005). Molecular and Isotopic Identification of PAH sources in a highly industrialized urban estuary. Org. Geochem., 36: 619-632.
Wang MB, McDonald RM, Vingarzan R, Mitchell RH, Govette D, Sylvestre S (1999). PAHs in the Fraser River Basin: a critical appraisal of PAH ratios as indicators of PAH sources and composition. Org. Geochem., 33: 489–515.
Wilcke W, Amelung W, Martins C, Gracia MC, Zech W (2000). Biological sources of polycyclic aromatic hydrocarbons in the Amazonian rainforest. Soil. Sci. Plant. Nutr., 163: 27–30.
Wilcke W, Amelung W, Krauss M, Martins C, Bandeir RA, Gracia M (2003). Polycyclic aromatic hydrocarbons (PAHs) patterns in climatically different ecological zones of Brazil. Org. Geochem., 34: 1405–1418.
Yunker MB, McDonald RW, Goyette D, Paton DW, Fowler BR, Sullivan D, Boyd J (2001). Natural and anthropogenic inputs of hydrocarbons to the straits of Georgia. Sci. Total. Environ., 225: 181–209.
Yunker MB, Backus SM, Pannatier EG, Jeffries DS, Macdonald RW (2002). Sources and significance of alkane and PAH hydrocarbons in Canadian Arctic Rivers. Estuar. Coastal Shelf Sci., 55: 131.
Yunker MB, McDonald RW (2003). Alkanes and PAH depositional history, sources and fluxes in sediments from the Frazer River Basin and Strait of Georgia, Canada. Org. Geochem., 34: 1429–1454.
Zakaria MP, Takada H, Tsutsumi S, Ohno K, Yamada J, Kouno E (2002). Distribution of PAHs in rivers and estuaries in Malaysia: a widespread input of petrogenic PAHs. Environ. Sci. Technol., 36: 1907 -1918.
Zhihuan Z, Ring M, Bixtong Y (2006). Pollution source identification of polycyclic aromatic hydrocarbons of soils in Tianjin area, China. Chemosphere, 64: 525–534.

Integrating indigenous knowledge and soil science approaches to detailed soil survey in Kaduna State, Nigeria

Raji, B. A.[1]*, Malgwi, W. B.[1], Berding, F. R.[2] and Chude, V. O.[2]

[1]Department of Soil Science, Ahmadu Bello University, Zaria, Nigeria.
[2]National Special Program for Food Security (NSPFS), FAO Nigeria Office, Abuja, Nigeria.

Farmers' knowledge about their soils and their management constitutes a complex wisdom system, which if integrated with modern soil science, could provide the necessary synergy for sustainable agricultural development. A field work was undertaken in parts of Kaduna State to ascertain the indigenous knowledge of farmers and compared it with scientific classification of the same soils. The farmers used non-taxonomic classification with two levels of classification, topography or position in the landscape and inherent fertility as the first classifiers, while soil colour, surface characteristics and texture were the second. Overall, the indigenous classification used morphological and physical properties of the surface horizon as their main diagnostic attributes. In the Jagindi site, farmers were able to classify the soil units into *Tudu Jar-Kasa* and *Tudu Yunbu* for the soil units JD1 and JD2 respectively while the seasonally flooded unit; JD3 was classified as *Fadama*, an indigenous soil name that has been found acceptable in scientific literature. Correlation would be easier with the World Reference Base (WRB) system, which also place emphasis on morphological properties observable in the field as shown by the chromic subgroup in WRB and *Jar* (red) in the indigenous classification of soil unit JD1. The indigenous classification scheme in the study areas, though simple, was adequate in most cases in grouping soils into classes that could be managed using similar management practices, re-emphasizing the use-oriented nature of indigenous classification. For sustainable development in the study areas and to improve communication between the scientists, the extension agents and the farmers, it is suggested that local soil name be integrated into the soil map legend.

Key words: Conventional soil survey, *Fadama*, indigenous knowledge, Nigeria.

INTRODUCTION

The soil resource of any country is its most valuable natural resource which requires careful management for sustainable development. Soil information is required for soil-related agrotechnology transfer (Braimoh, 2002), the basis for the planning and execution of sustainable agricultural landuse and development and other non-agricultural projects (Braimoh, 2002). Given the central role of soil resources information to sustainable development, soil survey at various scales has been carried out in many countries of the world, including Nigeria. However, according to Tabor and Hutchinson (1994), most soil resource survey information in developing countries are not effectively utilized, because the audience they are intended to serve perceives them as irrelevant and totally 'foreign' to them (Ettema, 1994). If the resulting soil maps ever reach the farmers, they often contain scientific terms which are inherently complex and alien to them. They also contain recommendations on sole crops which are often not of interest to farmers who are interested in mixed farming

*Corresponding author. E-mail: rajibash2009@gmail.com.

(Osunade, 1994). Another reason of interest is that while farmers' focus is primarily use-oriented and concern with the surface horizons that are most relevant for agricultural production, the scientist is more focus on the deeper soil horizons; subject to pedogenesis and soil forming factors (Kundirir et al., 1997; Corbeels et al., 2000).

As imported scientific interpretation of soils in developing countries has failed to bring about the desired sustainable development, it has become more and more apparent that the indigenous knowledge of soils (ethnopedology) could offer many insights about sustainable management of soils in these countries (Hecht, 1990; Osunade, 1994; Corbeels et al., 2000). In a worldwide review of ethnopedological research (EPR), Barrera-Bassols and Zinck (2000) reported an average of 33 studies per year worldwide with Africa having the highest number of EPR. The increasing interest in indigenous soil knowledge (ISK) is fully justified. For one thing, detailed soil survey is often faster and cheaper using indigenous soil classification than conventional soil survey techniques (Tabor, 1990). In addition, communication between farmers, scientists and extensionists will be greatly improved if local soil nomenclature is incorporated into soil survey reports and maps (Dialla, 1993; Niemeijer, 1995; Talawar and Rhoades, 1998). However, in practice, there are some complications. The most important ones being the heterogeneity of local soil knowledge even within a single village (Tabor, 1990) and, the problem of correlation between local soil types with scientific groupings (Niemeijer, 1995). The question is whether scientific and indigenous knowledge could be linked to improve the success of cooperation in sustainable agricultural development?

Study area

To test the hypothesis two locations were selected within Kaduna State. These were National Special Program for Food Security (NSPFS) sites at Likoro and Jagindi. The Likoro site is located on the Nigerian topographic map sheet 102 (Zaria) SE which is defined by latitudes 11° 09' 40" and 11° 10' 08" N and longitudes 7° 48' 57" and 7° 50' 57"E. The Jagindi site is located on the Jama'a sheet No. 188 NW and is defined by latitudes 9° 23' 43" and 9° 24' 33" N and longitudes 8°12' 14" and 8° 13' 06" E. The inhabitants of the two sites were mostly Hausa – Fulani except that the Jagindi site was located within an area inhibited by the Southern Kaduna indigenes. Even then the bulk of the farmers at the two locations remain Hausa / Fulani, while Hausa is the language of communication.

The project site at Jagindi has a humid Southern Guinea Savanna climate while the Likoro site has a tropical Northern Guinea Savanna climate, characterized by a long dry season and a shorter wet season. The rains usually start in April/May with the intensity and amount rising gradually to a peak in August and ends first week of October. The average mean monthly temperature over most of the survey area was 24 to 38°C. The harmattan months, November to January, are usually cool, with monthly means in the range of 21 to 24°C (Kowal and Knabe, 1972). Within the study sites, rainfed cultivation of annual crops like sorghum, millet, cowpea and soya beans take place on the upland while small scale irrigation of maize, vegetables, and rice commonly take place during the dry season on the *Fadama* soil units.

The study sites lie within the Kaduna Plains. Generally, the landforms of the project sites were one of a gently undulating to flat plains with only slight contrasting features except for the soil Unit JD 2 which was highly dissected with broad U-shaped valleys. Much of the site in Likoro lies between 600 and 700 m above sea level rising to between 700 and 800 m in the Jagindi site. The geology of the Kaduna plains, that is, the study sites; Likoro and Jagindi, was of the Nigerian Basement Complex as described by McCurry (1976) and Maclead et al. (1971). In general, the geology of the sites was underlain by rocks of the lower Palaeozoic to Precambrian Basement complex. In the Likoro site, the Basement complex rocks were dominated by gneisses with scattered occurrence of older granite. Rock outcrops were few and low lying and were found mainly within the older granites. The geology of the Jagindi site is described by Maclead et al. (1971) as being dominated by newer basalt emplaced in the quaternary.

The objective of this paper is to test the integrated use of indigenous and scientific knowledge in a detailed soil survey of two agricultural sites in Kaduna State of Nigeria.

MATERIALS AND METHODS

Field studies

The conventional method of soil survey involving the rigid grid procedure with boundary transects checking was adapted for the survey. For both sites, traverses and auger points were at 100 m interval. During the actual soil survey work, two farmers were selected from among the villagers to partake in the field work. The age of the farmers within the study area is between 25 and 35 years old. Structured questionnaires were prepared (Table 1). At each auger points, one of the farmers was first asked to answer the structured questionnaires concerning the soils, while the other was asked to confirm or give alternative answers. Thereafter, the scientists proceeded with augering and recorded their observations. The attention of the farmers was drawn to soil variability with depth.

Augering was made to a depth of 125 cm or to an impenetrable layer whichever is deeper. Soil descriptions were made at 25 cm depth interval or according to horizontal sequence where possible. At each observation point, the local relief, soil erosion or deposition hazard, rock outcrops, surface characteristics, vegetation and land use were recorded. The full range of soil morphological characteristics was recorded. The auger descriptions were classified and plotted along the traverses on a base map, along with the farmers' local soil name and soil boundary lines were drawn to delineate soil mapping unit polygons. Bulk soil samples were taken from the soil profiles, put in plastic bags and labeled.

Table 1. Structured questionnaire asked the farmers.

What is the local name of this soil?
Why is the name given to the soil?
What is the source of the name?
How do you recognize the soil?
What properties or indicators would you use in recognizing the soil?
For farming which one do you prefer?
Why do you prefer that one over the others?
Which of the soils yield better?
How much labour do you require to farm the soil
What kind of crops would do better on each soil type?
How much energy is required to till the land?
Which of the soil type would require too much fertilizer?
How did you know all the different soils?

Laboratory studies

Soil samples collected from the soil profiles pits were air-dried, ground and sieved to remove materials larger than 2 mm. The amount of gravel in each of the soil samples was weighed and the percentage gravel recorded. Soil pH was determined in 1: 2.5 soil/water suspensions using a glass electrode pH meter. Organic carbon was determined by the wet combustion method of Walkley-Black (Nelson and Summer, 1982). Particle-size distribution was determined by the hydrometer method using sodium hexametaphosphate (Calgon) as the dispersant (Gee and Bauder, 1986). Exchangeable bases (Ca, Mg, K, and Na) were brought into solution by repeated extraction procedure with neutral 1 M NH_4OAc (pH 7) solution (IITA, 1979); Ca^{2+} and Mg^{2+} in solution were read on an atomic absorption spectrophotometer; while K^+ and Na^+ were read on the flame photometer. Cation exchange capacity was determined by the 1 N NH_4OAc saturation method while the effective cation exchange capacity (ECEC) was calculated as the sum of the exchangeable bases and exchange acidity. Exchangeable acidity was by the 1 N KCl method. Available phosphorus was extracted by the Bray-1 method (Bray and Kurtz, 1945) and P in solution determined colorimetrically using the ascorbic acid method (Murphy and Riley, 1962). Electrical conductivity (EC) was measured in a 1:2.5 saturation extract (Richards, 1954).

RESULTS AND DISCUSSION

Scientists' classification of the soils

A total of three soil mapping units were delineated at Jagindi while four mapping units were identified at the Likoro site. Delineation by the scientists was based on both morphological and chemical properties of the soil and had as diagnostic attributes, properties related to pedogenesis. The farmers map delineation was however based mainly on soil surface characteristics. While the indigenous classification had two levels, the scientist had four levels of classification (Table 2) with similarity at the highest level between the two classifications. Even at the highest level of classification, while the farmers had principally two landscape types (*Fadama* and Upland), the scientists could have up to six landscape types (Table 2). Details of the soil properties of each soil units for only the Jagindi site are discussed below while cross references are made to the Likoro sites.

Jagindi site

Soil unit JD1

Soil unit JD1 was deep, well drained soils occupying nearly level to moderately sloping (4 to 7% slope) crest to middle slope positions of the upland. The soils colour range from brown (10YR 4/3) to dark brown (10YR 3/3) sandy loam over strong brown (7.5YR 5/6) to red (2.5YR 4/6) sandy clay loam to clay loam subsoil. The soils were weakly structured in the upper horizons to moderately structured subangular blocky in the subsoil. The clay content increased sharply from the Ap horizon from about 7% to over 40% in the subsoil indicating possible argilluviation or increased weathering in the subsoil (Table 3). On the field, the whole of the unit was observed to be under cultivation with limited soil erosion.

Soil unit JD2

The soils were deep and imperfectly drained. They were found in many topographic positions ranging from crest, upper slope and middle slope positions in a highly dissected upland. Soils of this unit were sandy loam

Table 2. Criteria for classification by scientists and indigenous farmers.

Diagnostic attribute	Indigenous	Level	scientist	Diagnostic attribute
Position in landscape	Landscape (*Tudu* and *Fadama*)	1	Landscape	Similar relief types, example, valley, mountain, plateau, hill, etc
Soil colour, texture, drainage, soil surface characteristics	Surface Soil morphology	2	Relief	Combination of topography, example, terrace, glacis etc
-	-	3	Lithology	Geology or hard rock, example, limestone, alluvial etc
-	-	4	Landform	Basic geoforms, example, levee, basin etc.

Table 3. Physico-chemical properties of the soil mapping units at Jagindi site, Kaduna State.

Pedon JDIP1			Soil unit JDI							Exchangeable Bases ($cmol\ kg^{-1}$)						
Horizon	Depth (cm)	Sand	Silt	Clay	Textural class	pH	ECe (dSm^{-1})	Organic carbon (%)	Available phosphorus ($mg\ kg^{-1}$)	Ca	Mg	K	Na	Total Exch. CEC ($cmol\ kg^{-1}$)	Bases ($cmol\ kg^{-1}$)	Acidity($cmol\ kg^{-1}$)
Ap	0-20	81	12	7	LS	5.4	0.06	0.54	5.84	1.70	0.95	0.55	0.08	3.28	0.4	15.8
Bt1	20-37	43	12	45	C	5.5	0.06	0.46	2.28	2.00	1.05	0.46	0.09	3.60	0.2	18.3
Bt2	37-140	41	10	49	C	5.7	0.06	0.23	3.65	1.50	0.70	2.51	0.06	2.77	0.2	7.7
									Soil unit JD2							
Pedon JD2P2																
Ap	0-28	69	18	13	SL	5.2	0.10	0.60	42.95	2.20	0.85	0.33	0.10	3.48	0.6	6.8
AB	28-63	31	26	43	C	4.8	0.07	0.56	10.30	2.70	1.30	0.34	0.10	4.44	1.2	10.5
Bt1	63-107	25	22	53	C	5.7	0.04	0.35	6.22	3.00	1.75	0.33	0.10	5.18	0.2	12.1
Bt2	107-135	41	20	39	CL	5.5	0.08	0.15	6.02	3.10	1.60	0.28	0.12	5.15	0.2	10.8
									Soil unit JD3							
Pedon JD3P1																
Ap	0-34	35	36	29	CL	4.7	0.02	1.52	8.61	2.00	1.05	0.26	0.10	3.41	1.8	11.9
AC	34-71	27	42	31	CL	5.0	0.06	0.93	4.37	2.40	1.20	0.11	0.10	3.81	1.2	12.4
C1	71-119	13	38	49	C	4.2	0.06	1.08	3.31	2.60	1.45	0.13	0.13	4.31	2.2	12.8
C2	119-160	17	34	49	C	5.0	0.02	1.00	2.96	2.90	1.60	0.16	0.15	4.81	1.2	13.7

underlain by sandy clay loam horizons at depth. Fine mica flakes could be found in all horizons while cracks, 1 mm wide extend to a depth of 70 cm. In some cases distinct mottles and gleying occur mostly in the subsoil. This unit was the least cultivated and remnants of the forest reserve could be found along valley of streams. Total porosity values were also less than 40% except in the Ap horizon where 45% was recorded. These total porosity values were below the optimum values for normal crop growth suggesting impeded air and water circulation in the soils. These soils were therefore, structurally weak and prone to severe erosion. On the field, the bulk of the unit was observed to be under secondary forest regrowth and limited crop cultivation.

Soil unit JD3

This mapping unit included deep, imperfectly to well drained, very dark brown (10YR 3/2) sandy loamy to clay loam surface horizons, overlying brown (10YR 4/3) to dark grayish brown 10YR 4/2) sandy loam to sandy clay loam or clay loam subsoil. The soils occupied nearly level to level (0 to 2%) seasonally flooded positions (*Fadama*) along the river and constituted about 10% of the total site area. Although levees and basins could be identified within the present floodplain they could however, not be mapped out. The structure was predominantly weak but tended towards granular structureless with depth. Common fine mica flakes and soft iron nodules occurred especially in the subsoil.

The soils of the three units; JD1, JD2 and JD3 all had strong acid reaction in the Ap horizon to moderately acid reactions in the subsoil. Exchangeable bases, CEC, ECe, base saturation were all low except in profile JD1P1 where moderate values were obtained for CEC and BS. Both cation exchange capacity (CEC) and effective cation exchange capacity (ECEC) of the soil were also low in the Ap horizon but moderate in the subsoil, the distribution was similar to the pattern of clay distribution. Total exchangeable bases increased with soil depth, indicative of possible leaching of bases downward. Organic carbon and the other organic matter based nutrients like total nitrogen and available phosphorus were all low except in the available phosphorus content of the Ap horizon of profile JD2P2 where values as high as 42 mg kg^{-1} was recorded (Table 3). However, cationic micronutrients and boron were moderate in content and theoretically, would be adequate in the first few years of cropping. The very low values of electrical conductivity of less than 0.08 dSm^{-1}, indicate that salinization is presently not a problem. Sodication is also not likely to pose management problems because of the very low ESP and SAR values that are less than 1% and 0.10 in all the horizons. In summary, the soils of unit JD1 were classified as Typic Haplustults and Fluventic Dystrustepts while soils of JD2 were also classified as Typic Haplustults and Ombroaquic Haplustults by the USDA Soil Taxonomy (Soil Survey Staff, 1981) and Chromic Acrisols and Epidystric Cambisols for soil unit JD1while soils of JD2 were mostly Chromic Acrisols and Haplic Acrisols by WRB system (FAO/ISSS/ISRIC, 1998) (Table 4).

However, soils of unit JD3 (the *Fadama* unit) were classified as Mollic Ustifluvents by the USDA Soil Taxonomy (Soil Survey Staff, 1981) and Umbric Gleysols and Epigleyic Fluvisols by WRB system (FAO/ISSS/ISRIC, 1998)

Farmers' classification of the soils

Fadama/Tsaki Kogi/Baki Kogi (soil unit JD3)

The hydromorphic soil units or the soil unit occupying valley floors and low lying flat lands adjacent to rivers were also identified by all the farmers from the two sites. The famers' and the scientist's maps were similar but the main differences were found in the total land area marked as *Fadama* by the farmers. This was because all low lying areas adjacent to the river were considered *Fadama,* which were considered so purely from use-oriented perspective. This further elaborates the fact that farmers' restrict their classification to mostly limiting factors similar to what is obtainable in Fertility Capability Classification (FCC) or in land suitability classification system. There were similarities in the semantics used at the two study areas because of the common language of communication which is Hausa, but there were very slight dialectic differences. Farmers in the two sites named the hydromorphic soils as *Fadama* at the first level of classification but in addition farmers in Jagindi refer to it as *Tsaki Kogi,* while those at Likoro calls it *Baki kogi.* When translated they mean literarily "middle of the river" and "back of the river" respectively. The two mean virtually the same and refer to the same soil materials. The units, according to the farmers were more fertile than other soil units and were lands liable to periodic flooding by river and suitable for the cultivation of rice, sugar cane and vegetables. The unit was also used for the dry season farming because of the nearness to the river. The farmers' perception of the *Fadama* soils was quite similar to the scientists' perception. However, the scientists' could further subdivide the *Fadama* into levee and basins, depending on the scale of the survey, and based on the vertical variability which farmers though were aware, but do not take into account in their classification. During the course of interview, the farmers claimed that *yashi*, *farin kasa* and *bakin kasa* were other soil types that could also be found in some other *Fadama* areas. Literally, *yashi* means sand, *farin* means light coloured while *bakin* refers to dark coloured. In effect the *Fadama* landscape could be sub-divided, at the second level of classification, based on soil colour and texture (Figure 1).

Table 4. Summary of soil classification.

Site	Soil unit	Soil taxonomy	WRB	Indigenous
	LKIP1	Petroferric Haplustult	Chromic Acrisol	*Tudu Jar Kasa*
	LK1P2	Typic Haplustult	Haplic Acrisol	
	LK2P1	Typic Dystrustept	Hyperdystric Cambisols	*Tudu Mara-mara*
	LK2P2	Lithic Dystrustept	Hyperdystric Cambisols	
Likoro	LK3P1	Petroferric Haplustult	Haplic Acrisol	*Tudu Mara-mara*
	LK3P2	Petroferric Haplustult	Chromic Plinthosol	
	LK4P1	Vertic Ustorthent	Umbric Gleysols	*Fadama Baki-Kogi*
	LK4P2	Vertic Ustorthent	Mollic Gleysol	
	JD1P1	Typic Haplustult	Chromic Acrisol	*Tudu Jar-Kasa*
	JD1P2	Fluventic Dystrustept	Chromic Acrisol	
	JD2P1	Typic Haplustult	Haplic Acrisol	*Tudu Yunbu*
	JD2P2	Ombroaquic Haplustult	Chromic Acrisol	
Jagindi				
	JD3P1	Mollic Ustifluvent	Umbric Gleysol	*Fadama Tsaki-Kogi*
	JD3P2	Mollic Ustifluvent	Epigleyic Fluvisol	

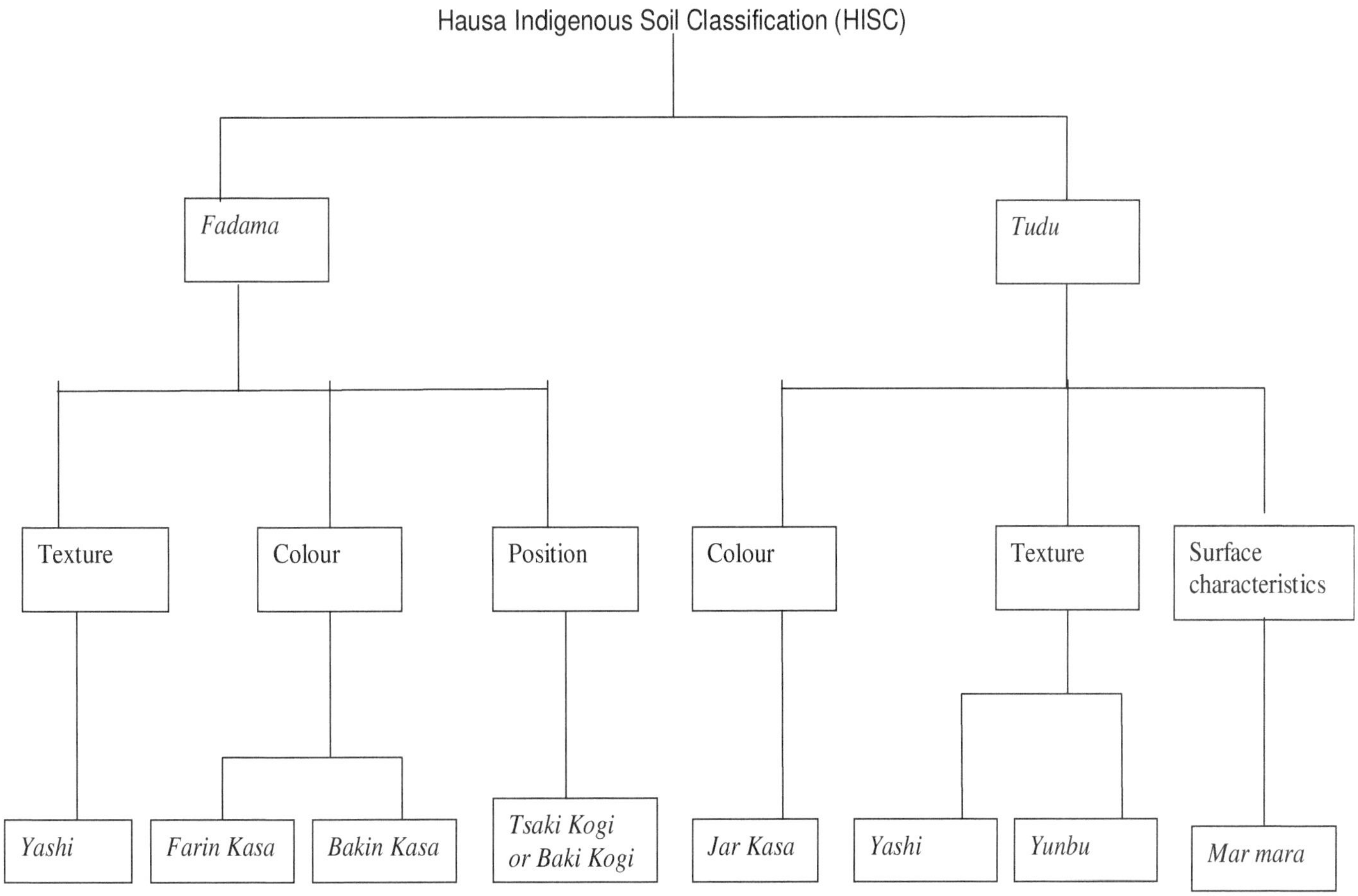

Figure 1. Structure of the Hausa indigenous soil classification (HISC) scheme.

Tudu/Jar Kasa/Jar (soil units JD1 and JD2)

All the upland soils were classified by farmers as *Tudu* which literarily means upland, or well drained soils. According to the farmers, the unit is best suited for crops like sorghum, millet or maize and is not as fertile as the *Fadama* soils. In the Jagindi site, soil unit JD1 which was classified as Chromic Acrisol by the WRB system was classified by the farmers as *Tudu Jar Kasa* while similar soils in the Likoro site was also classified as *Tudu Jar* by the farmers of the site (Figure 1). Chromic refers to the redder hues which were also recognized by the farmers as *Jar* (meaning red) in their classification. In the Jagindi site, soil unit JD2 which was highly dissected, imperfectly drained and had clay loam in the Ap horizon was classified as *Tudu Yunbu* by the farmers. Literarily it means clay or upland clayey soils with poor drainage. In the upland, farmers also claimed that *yashi* and *Mar mara* were two other soil types common within the *Tudu* landscape but which were however not encountered in the present study area. *Mar mara* means rocks or laterite and it refers to soils which have surface gravels, stones or rock outcrop that impede cultivation. In effect the *Tudu* was sub-divided based on texture, soil surface characteristics and soil colour, similar to what was done in the *Fadama*.

Conclusion

In conclusion, it was apparent that the farmers within the study areas have a fair knowledge of their soils in terms of soil colour, texture, surface characteristics, drainage and inherent soil fertility (Taylor-Powell et al., 1991; Guillet, 1992). As a first classifier, the farmers use position in the landscape while as second classifier, soil colour, texture, surface characteristics and drainage were used. Some of these properties have great pedo-transfer functions. Soil colour for instance is associated with organic matter content, soil moisture, drainage and the sesquioxide contents. Their classification scheme, though looks simple (two levels only), but was adequate in most cases to group soils into classes that could be managed using similar management practices. A strong correlation of the Hausa indigenous soil classification could be better achieved with the WRB system, both of which have two levels. It was also obvious that the indigenous classification could not differentiate the soils on the bases of subtle differences, especially in the subsoil, unless translated onto the soil surface. However, the farmers' soil map was very similar in outline to the soil map by the scientists. The major difference is in the extent of the mapping units especially the *Fadama*. For sustainable development in the study areas and to improve communication between the scientists, the extension agents and the farmers, an integration of local soil name into the soil map legend will go a long way in this direction.

REFERENCES

Barrera-Bassols N, Zinck JA (2000). Ethnopedology in a worldwide Perspective: An Annotated Bibliography. ITC, p. 77.

Braimoh AK (2002). Integrating indigenous knowledge and soil science to develop a national soil classification system for Nigeria. Agric. Hum. Values, 19: 75-80.

Bray RH, Kurtz LT (1945). Determination of total, organic and available forms of phosphorus in soils. Soil Sci., 59: 39-45.

Corbeels M, Shiferaw A, Haile M (2000). Farmers' knowledge of soil fertility and local management strategies in Tigray, Ethiopia. Managing African soils No 10. International Institute for Environment and Developmet (IIED) Drylands Programme, London.

Dialla BE (1993). Tha Mossi indigenous soil classification in Burkina Faso. Indigenous Knowl. Dev. Monit., 1(3): 17-18.

Ettema CH (1994). Indigenous soil classifications. What are their structure an functions, and How do they compare with scientific soil classifications? Institute of Ecology, University of Georgia, Athens 30602-2202, http://www.nrel.colostate.edu.

FAO, ISRIC, ISSS (1998). World Reference Base for Soil Resources, Food and Agricultural Organization of United Nations, Rome, p. 112.

Gee GW, Bauder JW (1986). Particle-size analysis. In: A Klute (ed). Methods of soil analysis. Part 1. Physical and mineralogical methods, 2nd ed., Agronomy, 9: 383-411.

Guillet DW (1992). Crop choice and soil management. In D.W. Guillet (ed). Covering ground: Communnal water management and the state in the Peruvian highlands. The University of Michigan Press, Ann Arbor, Michigan, USA, pp. 67-84.

Hecht SB (1990). Indigenous soil management in the Latin American tropics: Neglected knowledge of native peoples. In M.A. Altieri and S.B. Hecht (eds), Agroecology and small farm development, Boca Raton, Florida, CRC Press, pp. 151-158.

IITA (1979). Selected methods for soil and plant analysis. International Institute of Tropical Agriculture Manual Series No. 1, Ibadan, p. 62.

Kowal JM, Knabe DT (1972). An agricultural Atlas of the Northern States of Nigeria with Explanatory Notes, Ahmadu Bello University Press, Zaria, p. 111.

Kundiri AM, Jarvis MG, Bullock P (1997). Traditional soil and land appraisal on *Fadama* lands in northeast Nigeria. Soil Use Manage., 13: 205-208.

Maclead WN, Turner DC, Wright EP (1971). The geology of the Jos Plateau. General Geology, Fed. Government of Nigeria, Lagos, 1: 91-101.

McCurry P (1976). The geology of the Precambrian to Lower Palaeozoic rocks of northern Nigeria- A review. In: Kogbe, C.A. (ed.) Geology of Nigeria, Elizabethan Publishing Co., Ibadan, Nigeria, p 15-38.

Murphy T, Riley JP (1962). A modified single solution method for the determination of phosphate in natural water. Anal. Chim. Acta, 27: 31-36.

Nelson DW, Sommers LE (1982). Total C, organic C and organic matter. In: A.L. Page, R.H. Miller, and D.R. Keeney (ed.). Methods of Soil analysis. Part 2, Chemical and microbiological properties. 2nd ed. Agronomy, 9: 539-579.

Niemeijer D (1995). Indigenous soil classifications: Complications and considerations. Indigenous knowl. Dev. Monit., 3: 20-21.

Osunade MAA (1994). Indigenous climate knowledge and agricultural practices in southwestern Nigeria. Malaysian J. Trop. Geogr., 25(1): 21-28.

Richards LA (1954). Diagnosis and improvement of saline and alkali soils. USDA Handbook 60, US Govt. Printing Office, Washington D.C., p. 503.

Soil Survey Staff (1981). Examination and description of soils in the field, In Soil Survey Manual, USDA Agricultural Handbook 18, United States Government Printing Office, Washington, D.C. p. 503.

Soil Survey Staff (1999). Soil Taxonomy: A basic System of Soil Classification for making and interpreting Soil Surveys. Second edition, USDA Agric. Handbook 436, US Govt. Printing Office, Washington, D.C., p. 863.

Tabor J (1990). Ethnopedology: using indigenous knowledge to classify soils. Aridd lands Newslett., 30: 28-90.

Tabor JA, Hutchinson CF (1994). Using indigenous knowledge, remote sensing and GIS for sustainable development. Indigenous Knowl. Dev. Monit., 2(1): 17-19.

Talawar S, Rhoades RE (1998). Scientifican and local classification and management of soils. Agric. Hum. Values, 15(1): 3-14.
Taylor-Powell E (1991). A. Manu; SC. Geiger; M. Ouattara and ASR. Juo. Integrated management of agriculturl watersheds: land tenure and indigenous knowledge of soil and crop management. Texas Trop. Soils Bull., 30. 91-04.

Permissions

All chapters in this book were first published in JSSEM, by Academic Journals; hereby published with permission under the Creative Commons Attribution License or equivalent. Every chapter published in this book has been scrutinized by our experts. Their significance has been extensively debated. The topics covered herein carry significant findings which will fuel the growth of the discipline. They may even be implemented as practical applications or may be referred to as a beginning point for another development.

The contributors of this book come from diverse backgrounds, making this book a truly international effort. This book will bring forth new frontiers with its revolutionizing research information and detailed analysis of the nascent developments around the world.

We would like to thank all the contributing authors for lending their expertise to make the book truly unique. They have played a crucial role in the development of this book. Without their invaluable contributions this book wouldn't have been possible. They have made vital efforts to compile up to date information on the varied aspects of this subject to make this book a valuable addition to the collection of many professionals and students.

This book was conceptualized with the vision of imparting up-to-date information and advanced data in this field. To ensure the same, a matchless editorial board was set up. Every individual on the board went through rigorous rounds of assessment to prove their worth. After which they invested a large part of their time researching and compiling the most relevant data for our readers.

The editorial board has been involved in producing this book since its inception. They have spent rigorous hours researching and exploring the diverse topics which have resulted in the successful publishing of this book. They have passed on their knowledge of decades through this book. To expedite this challenging task, the publisher supported the team at every step. A small team of assistant editors was also appointed to further simplify the editing procedure and attain best results for the readers.

Apart from the editorial board, the designing team has also invested a significant amount of their time in understanding the subject and creating the most relevant covers. They scrutinized every image to scout for the most suitable representation of the subject and create an appropriate cover for the book.

The publishing team has been an ardent support to the editorial, designing and production team. Their endless efforts to recruit the best for this project, has resulted in the accomplishment of this book. They are a veteran in the field of academics and their pool of knowledge is as vast as their experience in printing. Their expertise and guidance has proved useful at every step. Their uncompromising quality standards have made this book an exceptional effort. Their encouragement from time to time has been an inspiration for everyone.

The publisher and the editorial board hope that this book will prove to be a valuable piece of knowledge for researchers, students, practitioners and scholars across the globe.

List of Contributors

Clifford Tafangenyasha
Department of Geography and Environmental Science, University of Zululand, P, Bag X10003, KwaDlangezwa 3886, South Africa

Amos T. Mthembu
Department of Geography and Environmental Science, University of Zululand, P, Bag X10003, KwaDlangezwa 3886, South Africa

Hector Chikoore
Department of Geography and Environmental Science, University of Zululand, P, Bag X10003, KwaDlangezwa 3886, South Africa

Nothile Ndimande
Department of Geography and Environmental Science, University of Zululand, P, Bag X10003, KwaDlangezwa 3886, South Africa

Sifiso Xulu
Department of Geography and Environmental Science, University of Zululand, P, Bag X10003, KwaDlangezwa 3886, South Africa

Nonkululeko Gwcensa
Department of Geography and Environmental Science, University of Zululand, P, Bag X10003, KwaDlangezwa 3886, South Africa

Faruk Özkutlu
Universiy of Ordu, Faculty of Agriculture, Department of Soil Science, Ordu-Turkey

Yusuf Ziya Doğru
Universiy of Atatürk, Narman Vocational School, Erzurum-Turkey

Nedim Özenç
Giresun Hazelnut Research Institute, Giresun-Turkey

Gizem Yazici
Universiy of Ordu, Faculty of Agriculture, Department of Soil Science, Ordu-Turkey

Metin Turan
Universiy of Atatürk Faculty of Agriculture, Department of Soil Science, Erzurum-Turkey

Fatih Akçay
Universiy of Atatürk, Faculty of Medicine, Department of Biochemistry, Erzurum-Turkey

Kevin Z. Mganga
Department of Range & Wildlife Sciences, South Eastern University College (SEUCO), P. O. BOX 170-90200, Kitui, Kenya

Nashon K. R. Musimba
Department of Range & Wildlife Sciences, South Eastern University College (SEUCO), P. O. BOX 170-90200, Kitui, Kenya

Dickson M. Nyariki
Department of Range & Wildlife Sciences, South Eastern University College (SEUCO), P. O. BOX 170-90200, Kitui, Kenya

Moses M. Nyangito
Department of Land Resources Management & Agricultural Technology, University of Nairobi

Wellington N. Ekaya
RUFORUM, P. O. BOX 7062, Kampala, Uganda

William M. Muiru
Department of Plant & Crop Science, University of Nairobi, P. O. BOX 29053-00625, Nairobi, Kenya

Agnes, W. Mwang'ombe
Department of Plant & Crop Science, University of Nairobi, P. O. BOX 29053-00625, Nairobi, Kenya

S. Mustapha
Crop Production Programme, Abubakar Tafawa Balewa University, Bauchi, Nigeria

H. K. Mamman
Crop Production Programme, Abubakar Tafawa Balewa University, Bauchi, Nigeria

N. A. Abdulhamid
Department of Agricultural Engineering, Faculty of Engineering, University of Maiduguri, Nigeria

Jane Muthoni
Kenya Agricultural Research Institute (KARI), National Potato Research Centre, Tigoni, Kenya

Jackson N. Kabira
Kenya Agricultural Research Institute, National Potato Research Centre, Tigoni, P. O. Box 338 00217, Limuru, Kenya

Jane Muthoni
Kenya Agricultural Research Institute (KARI), National Potato Research Centre, Tigoni, P. O. Box 338 00217, Limuru, Kenya

Jackson N. Kabira
Kenya Agricultural Research Institute (KARI), National Potato Research Centre, Tigoni, P. O. Box 338 00217, Limuru, Kenya

P. O. Youdeowei
Institute of Geosciences and Space Technology, Rivers State University of Science and Technology, P. M. B. 5080, Nkpolu-Oroworukwo, Port Harcourt, Nigeria

H. O. Nwankwoala
Department of Geology, University of Port Harcourt, P. M. B. 5323, Choba, Port Harcourt, Nigeria

A. J. Adeyemo
Department of Crop, Soil and Pest Management. Federal University of Technology, Akure, Nigeria

S. O. Agele
Department of Crop, Soil and Pest Management. Federal University of Technology, Akure, Nigeria

M. Y. Mohamed
Faculty of Agriculture, Alzaiem Al-Azhari University, Khartoum North, Box 1432, Sudan

B. M. Abdo
Ministry of Agriculture and Forestry, Sudan

BENICHA Mohamed
Laboratory of Pesticides residues, UR research on nuclear techniques, environment and quality, Regional Center of Agronomic research, 78, Bd Mohamed Ben Abdellah, 90010 Tangier, Morocco

MRABET Rachid
Regional Center of Agronomic research, 78, Bd Mohamed Ben Abdellah, 90010 Tangier, Morocco

AZMANI Amina
University Abdel Malik Essaadi, Faculty of Science and Technology, Ancienne Route de l'Aéroport, Km 10, Ziaten. BP:416. Tanger - Maroc

Kurt D. Nolte
The University of Arizona, Yuma County Cooperative Extension, Yuma, Arizona (USA) 85364, USA

O. T. Ande
Institute of Agriculture, Research and Training, Obafemi Awolowo University, P. M. B. 5029, Apata, Ibadan, Oyo State, Nigeria

S. Adjei-Nsiah
University of Ghana Agricultural Research Centre, Kade, Institute of Agricultural Research, University of Ghana, Legon, P. O. Box 38, Legon, Ghana

M. Alamgir
Department of Soil Science, University of Chittagong, Chittagong- 4331, Bangladesh
School of Agriculture, Food and Wine, University of Adelaide, SA 5005, Australia

M. G. Kibria
Department of Soil Science, University of Chittagong, Chittagong- 4331, Bangladesh

M. Islam
Department of Soil Science, University of Chittagong, Chittagong- 4331, Bangladesh

F. O. Ekundayo
Department of Microbiology, Federal University of Technology, P. M. B. 704, Akure, Ondo State, Nigeria

Jane Mutune
Department of Agricultural Economics, University of Nairobi, Kenya

John Mburu
Department of Agricultural Economics, University of Nairobi, Kenya

Rose Nyikal
Department of Agricultural Economics, University of Nairobi, Kenya

Geoffrey Kironchi
Land Resource Management and Agricultural Technology, University of Nairobi, Kenya

A. K. Ibrahim
Federal College of Horticulture, Dadin Kowa, P. M. B 108 Gombe, Gombe State, Nigeria

A. Usman
Federal College of Horticulture, Dadin Kowa, P. M. B 108 Gombe, Gombe State, Nigeria

B. Abubakar
Federal College of Horticulture, Dadin Kowa, P. M. B 108 Gombe, Gombe State, Nigeria

U. H Aminu
Federal College of Horticulture, Dadin Kowa, P. M. B 108 Gombe, Gombe State, Nigeria

J. R. Orimoloye
Rubber Research Institute of Nigeria, P. M. B. 1049, Benin City, Edo State, Nigeria

I. K. Ugwa
Rubber Research Institute of Nigeria, P. M. B. 1049, Benin City, Edo State, Nigeria

S. O. Idoko
Rubber Research Institute of Nigeria, P. M. B. 1049, Benin City, Edo State, Nigeria

Hailemariam Kassa
1Tigray Agricultural Research Institute, Humera Agricultural Research Centre, Humera, P. O. Box 230 Tigray, Ethiopia

Kindeya Gebrehiwet
Department of Land Resource Management and Environmental Protection, Faculty of Dryland Agriculture, Mekelle University, P. O. Box 231, Ethiopia

Charles Yamoah
Faculty of Renewable Natural Resources, University of Development Studies, P. O. Box TL 1350 Tamale, Ghana

E. M. Muya
National Agricultural Research Laboratories, P.O. Box 14733, Nairobi, Kenya

S. Obanyi
National Agricultural Research Laboratories, P.O. Box 14733, Nairobi, Kenya

M. Ngutu
Kenya Agricultural Research Institute (KARI) P.O. Box 147, Marsabit, Kenya

I. V. Sijali
National Agricultural Research Laboratories, P.O. Box 14733, Nairobi, Kenya

M. Okoti
Kenya Agricultural Research Institute (KARI) P.O. Box 147, Marsabit, Kenya

P. M. Maingi
National Agricultural Research Laboratories, P.O. Box 14733, Nairobi, Kenya

H. Bulle
Kenya Agricultural Research Institute (KARI) P.O. Box 147, Marsabit, Kenya

Ahmed Al-wadaey
Department of Agronomy and Horticulture, University of Nebraska, Lincoln NE 68583-0915, USA ICARDA, P. O. Box 5466, Aleppo, Syrian Arab Republic

S. Charles Wortmann
Department of Agronomy and Horticulture, University of Nebraska, 58C Filley Hall, Lincoln NE 68583-0951, 402 4722902, USA

A. Charles Shapiro
Department of Agronomy and Horticulture, University of Nebraska, 279 Plant Science, Lincoln NE 68583-0951, 402 5843803, USA

G. Thomas Franti
Department of Biosystems Engineering, University of Nebraska, 242 Chase Hall, Lincoln NE 68583-0726, 402 4729872, USA

E. Dean Eisenhauer
Department of Biosystems Engineering, University of Nebraska, 238 Chase Hall, Lincoln NE 68583-0726, 402 4721637, USA

Farooq Ahmad
Department of Geography, University of the Punjab, Lahore, Pakistan

K. Osei
Crops Research Institute, Box 3785, Kumasi, Ghana

J. O. Fening
Soil Research Institute, Academy Post Office, PMB, Kwadaso Kunmsi, Ghana

S. R. Gowen
University of Reading, School of Agriculture, Policy and Development, Box 237, Earley Gate, RG6 6AR, Reading, UK

A. Jama
University of Reading, School of Agriculture, Policy and Development, Box 237, Earley Gate, RG6 6AR, Reading, UK

Clifford Tafangenyasha
Department of Geography and Environmental Science, University of Zululand, P, Bag X10003, KwaDlangezwa 3886, South Africa

Amos T. Mthembu
Department of Geography and Environmental Science, University of Zululand, P, Bag X10003, KwaDlangezwa 3886, South Africa

Hector Chikoore
Department of Geography and Environmental Science, University of Zululand, P, Bag X10003, KwaDlangezwa 3886, South Africa

Nothile Ndimande
Department of Geography and Environmental Science, University of Zululand, P, Bag X10003, KwaDlangezwa 3886, South Africa

Sifiso Xulu
Department of Geography and Environmental Science, University of Zululand, P, Bag X10003, KwaDlangezwa 3886, South Africa

Nonkululeko Gcwensa
Department of Geography and Environmental Science, University of Zululand, P, Bag X10003, KwaDlangezwa 3886, South Africa

E. M. Muya
National Agricultural Laboratories, P. O. Box 14733, Nairobi, Kenya

K. Gitau
National Agricultural Laboratories, P. O. Box 14733, Nairobi, Kenya

S. Obanyi
National Agricultural Laboratories, P. O. Box 14733, Nairobi, Kenya

M. Ngutu
KARI-Marsabit, P. O. Box 147, Marsabit, Nairobi, Kenya

S. Kuria
KARI-Marsabit, P. O. Box 147, Marsabit, Nairobi, Kenya

H. Sharfi
KARI-Marsabit, P. O. Box 147, Marsabit, Nairobi, Kenya

D. G. Dub
KARI-Marsabit, P. O. Box 147, Marsabit, Nairobi, Kenya

H. Bulle
KARI-Marsabit, P. O. Box 147, Marsabit, Nairobi, Kenya

D. M. Miano
Kenya Agricultural Research Institute, P. O. Box 57811, Nairobi, Kenya

J. Muriethi
Kenya Agricultural Research Institute, P. O. Box 57811, Nairobi, Kenya

G. Keya
Kenya Agricultural Research Institute, P. O. Box 57811, Nairobi, Kenya

I. V. Sijali
National Agricultural Laboratories, P. O. Box 14733, Nairobi, Kenya

S. Mamo
KARI-Marsabit, P. O. Box 147, Marsabit, Nairobi, Kenya

P. M. Maingi
National Agricultural Laboratories, P. O. Box 14733, Nairobi, Kenya

M. Okoti
KARI-Marsabit, P. O. Box 147, Marsabit, Nairobi, Kenya

Eng. M. J. Omedi
Ministry of Water and Irrigation, P. O. Box 29703, Nairobi, Kenya

A. Esilaba
Kenya Agricultural Research Institute, P. O. Box 57811, Nairobi, Kenya

A. M. Ibrahim
Soils and Water Department, Faculty of Agriculture, El-Fayoum University, Fayoum, Postal Code 63514, Egypt

N. Persaud
Crop and Soil Environmental Sciences Department, Virginia Polytechnic Institute and State University, Blacksburg, Virginia 24061-0404, USA

R. W. Zobel
USDA/ARS Appalachian Farming Systems Research Center, 1224 Airport Road, Beaver, West Virginia, USA

A. Hass
West Virginia State University, Gus R. Douglass Institute, P. O. Box 1000, Institute, WV 25112-1000,USA

Obidimma C. Ezezika
Yale School of Forestry and Environmental Studies, 195 Prospect Street, New Haven, CT 06511 USA
McLaughlin-Rotman Centre for Global Health 101 College Street, Suite 406, Toronto, Ontario, Canada M5G 1L7

Olorunfemi Adetona
Yale School of Forestry and Environmental Studies, 195 Prospect Street, New Haven, CT 06511 USA
Department of Environmental Health Science, College of Public Health, University of Georgia, Athens, GA 30602, USA

F. O. Adekayode
Department of Crop, Soil and Pest Management, Federal University of Technology, Akure, Ondo State, Nigeria

M. R. Olojugba
Department of Soil Science and Management, Joseph Ayo Babalola University, Ikeji Arakeji, Osun State, Nigeria

F. O. Adekayode
Department of Crop, Soil and Pest Management, Federal University of Technology, Akure, Nigeria

M. O. Ogunkoya
Federal College of Agriculture, Akure, Ondo State, Nigeria

O. N. Adeniyan
Institute of Agricultural Research and Training, Obafemi Awolowo University (OAU), P. M. B. 5029 Ibadan, Nigeria

A. O. Ojo
Institute of Agricultural Research and Training, Obafemi Awolowo University (OAU), P. M. B. 5029 Ibadan, Nigeria

O. A. Akinbode
Institute of Agricultural Research and Training, Obafemi Awolowo University (OAU), P. M. B. 5029 Ibadan, Nigeria

J. A. Adediran
Institute of Agricultural Research and Training, Obafemi Awolowo University (OAU), P. M. B. 5029 Ibadan, Nigeria

J. A. Mwangi
Kenya Agriculture Research Institute (KARI), Pyrethrum and Botanical Research Center, P. O. Box 100, Molo, Kenya

S. N. Waweru
Kenya Forestry College Londiani, Ministry of Forestry and Wildlife, P. O. Box 8, Londiani, Kenya

V. O. Oeba
Kenya Forestry Research Institute (KEFRI), P. O. Box 20412-00200, Nairobi

S. B. Wepukhulu
Kenya Agriculture Research Institute (KARI), National Horticultural Research Center, P. O. Box 220, Thika, Kenya

Orok E. Oyo-Ita
Department of Pure and Applied Chemistry, University of Calabar, Cross River State, Nigeria

O. Inyang
Department of Pure and Applied Chemistry, University of Calabar, Cross River State, Nigeria

Oyo-Ita
Department of Pure and Applied Chemistry, University of Calabar, Cross River State, Nigeria

Samuel U. Ugim
Department of Pure and Applied Chemistry, University of Calabar, Cross River State, Nigeria

B. A. Raji
Department of Soil Science, Ahmadu Bello University, Zaria, Nigeria

W. B. Malgwi
Department of Soil Science, Ahmadu Bello University, Zaria, Nigeria

F. R. Berding
National Special Program for Food Security (NSPFS), FAO Nigeria Office, Abuja, Nigeria

V. O. Chude
National Special Program for Food Security (NSPFS), FAO Nigeria Office, Abuja, Nigeria